McGraw-Hill

Dictionary of
Mathematics

Second
Edition

McGraw-Hill

New York   Chicago   San Francisco   Lisbon   London   Madrid
Mexico City   Milan   New Delhi   San Juan   Seoul   Singapore
Sydney   Toronto

The McGraw·Hill Companies

1 2 3 4 5 6 7 8 9 0   DOC/DOC   0 9 8 7 6 5 4 3

ISBN 0-07-141049-X

This book is printed on recycled, acid-free paper containing a minimum of 50% recycled, de-inked fiber.

This book was set in Century ITC Book, Book Italic, Bold, Bold Italic, Novarese Book, Book Italic, Bold, and Bold Italic by the Clarinda Company, Clarinda, Iowa. It was printed and bound by RR Donnelley, The Lakeside Press.

McGraw-Hill books are available at special quantity discounts to use as premiums and sales promotions, or for use in corporate training programs. For more information, please write to the Director of Special Sales, McGraw-Hill Professional, Two Penn Plaza, New York, NY 10121-2298. Or contact your local bookstore.

Library of Congress Cataloging-in-Publication Data

McGraw-Hill dictionary of mathematics — 2nd. ed.
        p.   cm.
    "All text in this dictionary was published previously in the McGraw-Hill dictionary of scientific and technical terms, sixth edition, c2003..."
— T.p. verso.
    ISBN 0-07-141049-X (alk, paper)
    1. Mathematics—Dictionaries.   I. Title: Dictionary of mathematics.
II. McGraw-Hill dictionary of scientific and technical terms. 6th ed.

QA5.M425   2003
510'.3—dc21   2002033168

Contents

Preface

The *McGraw-Hill Dictionary of Mathematics* provides a compendium of more than 5000 terms that are central to mathematics and statistics but may also be encountered in virtually any field of science and engineering. The coverage in this Second Edition includes branches of mathematics taught at the secondary school, college, and university levels, such as algebra, geometry, analytic geometry, trigonometry, calculus, and vector analysis, group theory, and topology, as well as statistics.

All of the definitions are drawn from the *McGraw-Hill Dictionary of Scientific and Technical Terms*, Sixth Edition (2003). The pronunciation of each term is provided along with synonyms, acronyms, and abbreviations where appropriate. A guide to the use of the Dictionary appears on pages vii-viii, explaining the alphabetical organization of terms, the format of the book, cross referencing, and how synonyms, variant spellings, and similar information are handled. The Pronunciation Key is provided on page ix. The Appendix provides conversion tables for commonly used scientific units, extensive listings of mathematical notation along with definitions, and useful tables of mathematical data.

It is the editors' hope that the Second Edition of the *McGraw-Hill Dictionary of Mathematics* will serve the needs of scientists, engineers, students, teachers, librarians, and writers for high-quality information, and that it will contribute to scientific literacy and communication.

Mark D. Licker
Publisher

Staff

Mark D. Licker, Publisher—Science

Elizabeth Geller, Managing Editor
Jonathan Weil, Senior Staff Editor
David Blumel, Staff Editor
Alyssa Rappaport, Staff Editor
Charles Wagner, Digital Content Manager
Renee Taylor, Editorial Assistant

Roger Kasunic, Vice President—Editing, Design, and Production

Joe Faulk, Editing Manager
Frank Kotowski, Jr., Senior Editing Supervisor

Ron Lane, Art Director

Thomas G. Kowalczyk, Production Manager
Pamela A. Pelton, Senior Production Supervisor

Henry F. Beechhold, Pronunciation Editor
Professor Emeritus of English
Former Chairman, Linguistics Program
The College of New Jersey
Trenton, New Jersey

How to Use the Dictionary

ALPHABETIZATION. The terms in the McGraw-Hill Dictionary of Mathematics, Second Edition, are alphabetized on a letter-by-letter basis; word spacing, hyphen, comma, solidus, and apostrophe in a term are ignored in the sequencing. For example, an ordering of terms would be:

Abelian group	**binary system**
Abel's problem	**binary-to-decimal conversion**
Abel theorem	**binomial**

FORMAT. The basic format for a defining entry provides the term in boldface, and the single definition in lightface:

term Definition.

A term may be followed by multiple definitions, each introduced by a boldface number:

term **1.** Definition. **2.** Definition. **3.** Definition.

A simple cross-reference entry appears as:

term *See* another term.

A cross reference may also appear in combination with definitions:

term **1.** Definition. **2.** *See* another term.

CROSS REFERENCING. A cross-reference entry directs the user to the defining entry. For example, the user looking up "abac" finds:

abac *See* nomograph.

The user then turns to the "N" terms for the definition. Cross references are also made from variant spellings, acronyms, abbreviations, and symbols.

AD *See* average deviation.
cot *See* cotangent.
geodetic triangle *See* spheroidal triangle.

ALSO KNOWN AS . . . , etc. A definition may conclude with a mention of a synonym of the term, a variant spelling, an abbreviation for the term, or other such information, introduced by "Also known as . . . ," "Also spelled . . . ," "Abbreviated . . . ," "Symbolized . . . ," "Derived from . . . ." When a term has more than one definition, the positioning of any of these phrases conveys the extent of applicability. For example:

term   **1.** Definition. Also known as synonym. **2.** Definition. Symbolized T.

In the above arrangement, "Also known as . . ." applies only to the first definition; "Symbolized . . ." applies only to the second definition.

term   Also known as synonym. **1.** Definition. **2.** Definition.

In the above arrangement, "Also known as . . ." applies to both definitions.

Pronunciation Key

Vowels

a	as in b**a**t, th**a**t
ā	as in b**ai**t, cr**a**te
ä	as in b**o**ther, f**a**ther
e	as in b**e**t, n**e**t
ē	as in b**ee**t, tr**ea**t
i	as in b**i**t, sk**i**t
ī	as in b**i**te, l**igh**t
ō	as in b**oa**t, n**o**te
ȯ	as in b**ough**t, t**au**t
u̇	as in b**oo**k, p**u**ll
ü	as in b**oo**t, p**oo**l
ə	as in b**u**t, sof**a**
au̇	as in cr**ow**d, p**ow**er
ȯi	as in b**oi**l, sp**oi**l
yə	as in form**u**la, spectac**u**lar
yü	as in f**ue**l, m**u**le

Semivowels/Semiconsonants

w	as in **w**ind, t**w**in
y	as in **y**et, on**i**on

Stress (Accent)

'	precedes syllable with primary stress
͵	precedes syllable with secondary stress
⁞	precedes syllable with variable or indeterminate primary/ secondary stress

Consonants

b	as in **b**i**b**, dri**bb**le
ch	as in **ch**arge, stre**tch**
d	as in **d**og, ba**d**
f	as in **f**ix, sa**f**e
g	as in **g**ood, si**g**nal
h	as in **h**and, be**h**ind
j	as in **j**oint, di**g**it
k	as in **c**ast, bri**ck**
k̲	as in Ba**ch** (used rarely)
l	as in **l**oud, be**ll**
m	as in **m**ild, su**mm**er
n	as in **n**ew, de**n**t
n̲	indicates nasalization of preceding vowel
ŋ	as in ri**ng**, si**ng**le
p	as in **p**ier, sli**p**
r	as in **r**ed, sca**r**
s	as in **s**ign, po**s**t
sh	as in **s**ugar, **sh**oe
t	as in **t**imid, ca**t**
th	as in **th**in, brea**th**
t̲h̲	as in **th**en, brea**the**
v	as in **v**eil, wea**v**e
z	as in **z**oo, crui**s**e
zh	as in bei**g**e, trea**s**ure

Syllabication

·	Indicates syllable boundary when following syllable is unstressed

A

abac *See* nomograph. { ə'bak }

abacus An instrument for performing arithmetical calculations manually by sliding markers on rods or in grooves. { 'ab·ə₁kəs }

Abelian domain *See* Abelian field. { ə'bēl·yən dō'mān }

Abelian extension A Galois extension whose Galois group is Abelian. { ə'bēl·yən ik 'sten·chən }

Abelian field A set of elements a, b, c, . . . forming Abelian groups with addition and multiplication as group operations where $a(b + c) = ab + ac$. Also known as Abelian domain; domain. { ə'bēl·yən 'fēld }

Abelian group A group whose binary operation is commutative; that is, $ab = ba$ for each a and b in the group. Also known as commutative group. { ə'bēl·yən 'grüp }

Abelian operation *See* commutative operation. { ə'bēl·yən ₁äp·ə·ə'rā·shən }

Abelian ring *See* commutative ring. { ə'bēl·yən 'riŋ }

Abelian theorems A class of theorems which assert that if a sequence or function behaves regularly, then some average of the sequence or function behaves regularly; examples include the Abel theorem (second definition) and the statement that if a sequence converges to s, then its Cesaro summation exists and is equal to s. { ə'bēl·yən 'thir·əmz }

Abel's inequality An inequality which states that the absolute value of the sum of n terms, each in the form ab, where the b's are positive numbers, is not greater than the product of the largest b with the largest absolute value of a partial sum of the a's. { 'ä·bəlz ₁in·ē'kwäl·i·dē }

Abel's integral equation The equation

$$f(x) = \int_a^x u(z)(x - z)^{-a} dz \ (0 < a < 1, x \geq a)$$

where $f(x)$ is a known function and $u(z)$ is the function to be determined; when $a = 1/2$, this equation has application to Abel's problem. { 'ä·bəlz 'in·tə·grəl i'kwā·zhən }

Abel's problem The problem which asks what path a particle will follow if it moves under the influence of gravity alone and its altitude-time function is to follow a specific law. { 'ä·bəlz 'präb·ləm }

Abel's summation method A method of attributing a sum to an infinite series whose nth term is a_n by taking the limit on the left at $x = 1$ of the sum of the series whose nth term is $a_n x^n$ { 'ä·bəlz sə'mā·shən ₁meth·əd }

Abel theorem 1. A theorem stating that if a power series in z converges for $z = a$, it converges absolutely for $|z| < |a|$. **2.** A theorem stating that if a power series in z converges to $f(z)$ for $|z| < 1$ and to a for $z = 1$, then the limit of $f(z)$ as z approaches 1 equals a. **3.** A theorem stating that if the three series with nth term a_n, b_n, and $c_n = a_0 b_n + a_1 b_{n-1} + \cdots + a_n b_0$, respectively, converge, then the third series equals the product of the first two series. { 'ä·bəl 'thir·əm }

abscissa One of the coordinates of a two-dimensional coordinate system, usually the horizontal coordinate, denoted by x. { ab'sis·ə }

absolute convergence That property of an infinite series (or infinite product) of real or complex numbers if the series (product) of absolute values converges; absolute convergence implies convergence. { 'ab·sə‚lüt kən'vərj·əns }

absolute coordinates Coordinates given with reference to a fixed point of origin. { 'ab·sə‚lüt kō'órd·ən·əts }

absolute deviation The difference, without regard to sign, between a variate value and a given value. { 'ab·sə‚lüt dēv·ē'ā·shən }

absolute error In an approximate number, the numerical difference between the number and a number considered exact. { 'ab·sə‚lüt 'er·ər }

absolute inequality See unconditional inequality. { 'ab·sə‚lüt ‚in·ē'kwäl·ə·dē }

absolutely continuous function A function defined on a closed interval with the property that for any positive number ϵ there is another positive number η such that, for any finite set of nonoverlapping intervals, (a_1,b_1), (a_2,b_2), . . . , (a_n,b_n), whose lengths have a sum less than η, the sum over the intervals of the absolute values of the differences in the values of the function at the ends of the intervals is less than ϵ. { ‚ab·sə‚lüt·lē kən‚tin·yə·wəs 'fəŋk·shən }

absolutely continuous measure A sigma finite measure m on a sigma algebra is absolutely continuous with respect to another sigma finite measure n on the same sigma algebra if every element of the sigma algebra whose measure n is zero also has measure m equal to zero. { ab·sə‚lüt·lē kən‚tin·yə·wəs 'mezh·ər }

absolute magnitude The absolute value of a number or quantity. { 'ab·sə‚lüt 'mag·nə·tüd }

absolute mean deviation The arithmetic mean of the absolute values of the deviations of a variable from its expected value. { 'ab·sə‚lüt ‚mēn dē·vē'ā·shən }

absolute moment The nth absolute moment of a distribution $f(x)$ about a point x_0 is the expected value of the nth power of the absolute value of $x - x_0$ { ‚ab·sə‚lüt mō·mənt }

absolute number A number represented by numerals rather than by letters. { 'ab·sə‚lüt 'nəm·bər }

absolute retract A topological space, A, such that, if B is a closed subset of another topological space, C, and if A is homeomorphic to B, then B is a retract of C. { ‚ab·sə‚lüt ri'trakt }

absolute term See constant term. { 'ab·sə‚lüt 'tərm }

absorbing state A special case of recurrent state in a Markov process in which the transition probability, P_{ii}, equals 1; a process will never leave an absorbing state once it enters. { əb'sórb·iŋ ‚stāt }

absorbing subset A subset, A, of a vector space such that, for any point, x, there exists a number, b, greater than zero such that ax is a member of A whenever the absolute value of a is greater than zero and less than b. { əb‚sórb·iŋ 'səb‚set }

absorption property For set theory or for a Boolean algebra, the property that the union of a set, A, with the intersection of A and any set is equal to A, or the property that the intersection of A with the union of A and any set is also equal to A. { əb'sórp·shən ‚präp·ərd·ē }

absorptive laws Either of two laws satisfied by the operations, usually denoted ∪ and ∩, on a Boolean algebra, namely $a \cup (a \cap b) = a$ and $a \cap (a \cup b) = a$, where a and b are any two elements of the algebra; if the elements of the algebra are sets, then ∪ and ∩ represent union and intersection of sets. { əb'sórp·tiv ‚lóz }

abstract algebra The study of mathematical systems consisting of a set of elements, one or more binary operations by which two elements may be combined to yield a third, and several rules (axioms) for the interaction of the elements and the operations; includes group theory, ring theory, and number theory. { 'abz·trakt 'al·jə·brə }

abundant number A positive integer that is greater than the sum of all its divisors, including unity. Also known as redundant number. { ə'bən·dənt 'nəm·bər }

accessibility condition The condition that any state of a finite Markov chain can be reached from any other state. { ak‚ses·ə'bil·əd·ē kən‚dish·ən }

accretive operator A linear operator T defined on a subspace D of a Hilbert space which satisfies the following condition: the real part of the inner product of Tu with u is nonnegative for all u belonging to D. { ə¦krēd·iv 'äp·ə,rād·ər }

accumulation factor The quantity $(1 + r)$ in the formula for compound interest, where r is the rate of interest; measures the rate at which the principal grows. { ə·kyü·myə'lā·shən 'fak·tər }

accumulation point *See* cluster point. { ə·kyü·myə'lā·shən ,pȯint }

accumulative error *See* cumulative error. { ə'kyü·myə,lād·iv 'er·ər }

acnode *See* isolated point. { 'ak·nōd }

acute angle An angle of less than 90°. { ə'kyüt 'aŋ·gəl }

acute triangle A triangle each of whose angles is less than 90°. { ə'kyüt 'trī,aŋ·gəl }

acyclic 1. A transformation on a set to itself for which no nonzero power leaves an element fixed. **2.** A chain complex all of whose homology groups are trivial. { ā'sik·lik }

acyclic digraph A directed graph with no directed cycles. { ā¦sīk·lik 'dī,graf }

acyclic graph A graph with no cycles. Also known as forest. { ā¦sīk·lik 'graf }

AD *See* average deviation.

Adams-Bashforth process A method of numerically integrating a differential equation of the form $(dy/dx) = f(x,y)$ that uses one of Gregory's interpolation formulas to expand f. { 'a·dəmz 'bash,fȯrth ,prä·səs }

adaptive integration A numerical technique for obtaining the definite integral of a function whose smoothness, or lack thereof, is unknown, to a desired degree of accuracy, while doing only as much work as necessary on each subinterval of the interval in question. { ə'dap·tiv ,int·ə'grā·shən }

add To perform addition. { ad }

addend One of a collection of numbers to be added. { 'a,dend }

addition 1. An operation by which two elements of a set are combined to yield a third; denoted $+$; usually reserved for the operation in an Abelian group or the group operation in a ring or vector space. **2.** The combining of complex quantities in which the individual real parts and the individual imaginary parts are separately added. **3.** The combining of vectors in a prescribed way; for example, by algebraically adding corresponding components of vectors or by forming the third side of the triangle whose other sides each represent a vector. Also known as composition. { ə'di·shən }

addition formula An equation expressing a function of the sum of two quantities in terms of functions of the quantities themselves. { ə'dish·ən ,fȯr·myə·lə }

addition sign The symbol $+$, used to indicate addition. Also known as plus sign. { ə'di·shən ,sīn }

additive Pertaining to addition. That property of a process in which increments of the dependent variable are independent for nonoverlapping intervals of the independent variable. { 'ad·əd·iv }

additive function Any function f that preserves addition; that is, $f(x + y) = f(x) + f(y)$. { 'ad·əd·iv 'fəŋ·shən }

additive identity In a mathematical system with an operation of addition denoted $+$, an element 0 such that $0 + e = e + 0 = e$ for any element e in the system. { 'ad·ə·div ī'den·ə·dē }

additive inverse In a mathematical system with an operation of addition denoted $+$, an additive inverse of an element e is an element $-e$ such that $e + (-e) = (-e) + e = 0$, where 0 is the additive identity. { 'ad·ə·div 'in,vərs }

additive set function A set function with the properties that (1) the union of any two sets in the range of the function is also in this range and (2) the value of the function at a finite union of disjoint sets in the range of the set function is equal to the sum of the values at each set in the union. Also known as finitely additive set function. { ¦ad·əd·iv ¦set ,fəŋk·shən }

adherent point For a set in a topological space, a point that is either a member of the set or an accumulation point of the set. { ad¦hir·ənt 'pȯint }

adjacency matrix 1. For a graph with n vertices, the $n \times n$ matrix $A = a_{ij}$, where the nondiagonal entry a_{ij} is the number of edges joining vertex i and vertex j, and the diagonal entry a_{ii} is twice the number of loops at vertex i. **2.** For a diagraph with no loops and not more than one are joining any two vertices, an $n \times n$ matrix $A = [a_{ij}]$, in which $a_{ij} = 1$ if there is an are directed from vertex i to vertex j, and otherwise $a_{ij} = 0$. { ə'jäs·ən·sē ˌmā·triks }

adjacency structure A listing, for each vertex of a graph, of all the other vertices adjacent to it. { ə'jäs·ən·sē ˌstrək·chər }

adjacent angle One of a pair of angles with a common side formed by two intersecting straight lines. { ə'jäs·ənt 'aŋ·gəl }

adjacent side For a given vertex of a polygon, one of the sides of the polygon that terminates at the vertex. { ə'jäs·ənt 'sīd }

adjoined number A number z that is added to a number field F to form a new field consisting of all numbers that can be derived from z and the numbers in F by the operations of addition, subtraction, multiplication, and division. { əˈjȯind 'nəm·bər }

adjoint of a matrix *See* adjugate; Hermitian conjugate. { 'aj͵ȯint əv ə 'mā·triks }

adjoint operator An operator B such that the inner products (Ax,y) and (x,By) are equal for a given operator A and for all elements x and y of a Hilbert space. Also known as associate operator; Hermitian conjugate operator. { 'aj͵ȯint 'äp·ə͵räd·ər }

adjoint vector space The complete normed vector space constituted by a class of bounded, linear, homogeneous scalar functions defined on a normed vector space. { 'aj͵ȯint 'vek·tər ͵spās }

adjugate For a matrix A, the matrix obtained by replacing each element of A with the cofactor of the transposed element. Also known as adjoint of a matrix. { 'aj·ə͵gät }

affine connection A structure on an n-dimensional space that, for any pair of neighboring points P and Q, specifies a rule whereby a definite vector at Q is associated with each vector at P; the two vectors are said to be parallel. { ə'fīn kə'nek·shən }

affine geometry The study of geometry using the methods of linear algebra. { ə'fīn jē'äm·ə·trē }

affine Hjelmslev plane A generalization of an affine plane in which more than one line may pass through two distinct points. Also known as Hjelmslev plane. { əˈfīn 'hyelm͵slev ͵plān }

affine plane In projective geometry, a plane in which (1) every two points lie on exactly one line, (2) if p and L are a given point and line such that p is not on L, then there exists exactly one line that passes through p and does not intersect L, and (3) there exist three noncollinear points. { ə'fīn ͵plān }

affine space An n-dimensional vector space which has an affine connection defined on it. { ə'fīn ͵spās }

affine transformation A function on a linear space to itself, which is the sum of a linear transformation and a fixed vector. { ə'fīn ͵tranz·fər'mā·shən }

Airy differential equation The differential equation $(d^2f/dz^2) - zf = 0$, where z is the independent variable and f is the value of the function; used in studying the diffraction of light near caustic surface. { ͵er·ē ͵dif·ə͵ren·chəl i'kwā·zhən }

Airy function Either of the solutions of the Airy differential equation. { ͵er·ē ˈfəŋk·shən }

aleph null The cardinal number of any set which can be put in one-to-one correspondence with the set of positive integers. Also known as aleph zero. { ͵ä͵lef ˈnəl }

aleph one The smallest cardinal number that is larger than aleph zero. { ͵äl͵ef 'wən }

aleph zero *See* aleph null. { ˈäl͵ef 'zir·ō }

Alexander's subbase theorem The theorem that a topological space is compact if and only if its topology has a subbase with the property that any set that is contained in the union of a collection of members of the subbase is contained in the union of a finite number of members of this collection. { ͵al·ig'zan·dərz ˈsəb͵bās ͵thir·əm }

Alexandroff compactification *See* one-point compactification. { al·ik'san͵drȯf kəm͵pak·tə·fə'kā·shən }

algebra 1. A method of solving practical problems by using symbols, usually letters, for unknown quantities. **2.** The study of the formal manipulations of equations involving symbols and numbers. **3.** An abstract mathematical system consisting of a vector space together with a multiplication by which two vectors may be combined to yield a third, and some axioms relating this multiplication to vector addition and scalar multiplication. Also known as hypercomplex system. { 'al·jə·brə }

algebraic addition The addition of algebraic quantities in the sense that adding a negative quantity is the same as subtracting a positive one. { ¦al·jə¦brā·ik ə'dish·ən }

algebraically closed field 1. A field F such that every polynomial of degree equal to or greater than 1 with coefficients in F has a root in F. **2.** A field F is said to be algebraically closed in an extension field K if any root in K of a polynominal with coefficients in F also lies in F. Also known as algebraically complete field. { ¦al·jə¦brā·ik·lē ¦klōzd 'fēld }

algebraically complete field *See* algebraically closed field. { ‚al·jə‚brā·ik·lē kəm‚plēt 'fēld }

algebraically independent A subset S of a commutative ring B is said to be algebraically independent over a subring A of B (or the elements of S are said to be algebraically independent over A) if, whenever a polynominal in elements of S, with coefficients in A, is equal to 0, then all the coefficients in the polynomial equal 0. { ¦al·jə¦brā·ik·lē ‚in·də'pen·dənt }

algebraic closure of a field An algebraic extension field which has no algebraic extensions but itself. { ¦al·jə¦brā·ik 'klō·zhər əv ə 'fēld }

algebraic curve 1. The set of points in the plane satisfying a polynomial equation in two variables. **2.** More generally, the set of points in n-space satisfying a polynomial equation in n variables. { ¦al·jə¦brā·ik 'kərv }

algebraic deviation The difference between a variate and a given value, which is counted positive if the variate is greater than the given value, and negative if less. { ¦al·jə¦brā·ik ‚dē·vē'ā·shən }

algebraic equation An equation in which zero is set equal to an algebraic expression. { ¦al·jə¦brā·ik i'kwā·zhən }

algebraic expression An expression which is obtained by performing a finite number of the following operations on symbols representing numbers: addition, subtraction, multiplication, division, raising to a power. { ¦al·jə¦brā·ik ik'spresh·ən }

algebraic extension of a field A field which contains both the given field and all roots of polynomials with coefficients in the given field. { ¦al·jə¦brā·ik ik'sten·shən əv ə 'fēld }

algebraic function A function whose value is obtained by performing only the following operations to its argument: addition, subtraction, multiplication, division, raising to a rational power. { ¦al·jə¦brā·ik 'fəŋk·shən }

algebraic geometry The study of geometric properties of figures using methods of abstract algebra. { ¦al·jə¦brā·ik jē'äm·ə·trē }

algebraic hypersurface For an n-dimensional Euclidean space with coordinates x_1, x_2, ..., x_n, the set of points that satisfy an equation of the form $f(x_1, x_2, ..., x_n) = 0$, where f is a polynomial in the coordinates. { ‚al·jə‚brā·ik 'hī·pər‚sər·fəs }

algebraic identity A relation which holds true for all possible values of the literal symbols occurring in it; for example, $(x + y)(x - y) = x^2 - y^2$. { ¦al·jə¦brā·ik i'den·ə·tē }

algebraic integer The root of a polynomial whose coefficients are integers and whose leading coefficient is equal to 1. { ¦al·jə¦brā·ik 'in·tə·jər }

algebraic invariant A polynomial in coefficients of a quadratic or higher form in a collection of variables whose value is unchanged by a specified class of linear transformations of the variables. { ¦al·jə¦brā·ik in'ver·ē·ənt }

algebraic K theory The study of the mathematical structure resulting from associating with each ring A the group $K(A)$, the Grothendieck group of A. { ¦al·jə¦brā·ik 'kā ‚thē·ə·rē }

algebraic language The conventional method of writing the symbols, parentheses, and other signs of formulas and mathematical expressions. { ¦al·jə¦brā·ik 'laŋ·gwij }

algebraic number

algebraic number Any root of a polynomial with rational coefficients. { ¦al·jə¦brā·ik 'nəm·bər }

algebraic number field A finite extension field of the field of rational numbers. { ¦al·jə¦brā·ik 'nəm·bər ‚fēld }

algebraic number theory The study of properties of real numbers, especially integers, using the methods of abstract algebra. { ¦al·jə¦brā·ik 'nəm·bər ‚thē·ə·rē }

algebraic object Either an algebraic structure, such as a group, ring, or field, or an element of such an algebraic structure. { ¦al·jə¦brā·ik 'äb‚jekt }

algebraic operation Any of the operations of addition, subtraction, multiplication, division, raising to a power, or extraction of roots. { ¦al·jə·¦brā·ik ‚äp·ə'rā·shən }

algebraic set A set made up of all zeros of some specified set of polynomials in n variables with coefficients in a specified field F, in a specified extension field of F. { ¦al·jə¦brā·ik 'set }

algebraic subtraction The subtraction of signed numbers, equivalent to reversing the sign of the subtrahend and adding it to the minuend. { ‚al·jə‚brā·ik səb'trak·shən }

algebraic sum 1. The result of the addition of two or more quantities, with the addition of a negative quantity equivalent to subtraction of the corresponding positive quantity. **2.** For two fuzzy sets A and B, with membership functions m_A and m_B, that fuzzy set whose membership function m_{A+B} satisfies the equation $m_{A+B}(x) = m_A(x) + m_B(x) - [m_A(x) \cdot m^B(x)]$ for every element x. { ¦al·jə¦brā·ik 'səm }

algebraic surface A subset S of a complex n-space which consists of the set of complex solutions of a system of polynomial equations in n variables such that S is a complex two-manifold in the neighborhood of most of its points. { ¦al·jə¦brā·ik 'sər·fəs }

algebraic symbol A letter that represents a number or a symbol indicating an algebraic operation. { ¦al·jə¦brā·ik 'sim·bəl }

algebraic term In an expression, a term that contains only numbers and algebraic symbols. { ¦al·jə¦brā·ik 'tərm }

algebraic topology The study of topological properties of figures using the methods of abstract algebra; includes homotopy theory, homology theory, and cohomology theory. { ¦al·jə¦brā·ik tə'päl·ə·jē }

algebraic variety A set of points in a vector space that satisfy each of a set of polynomial equations with coefficients in the underlying field of the vector space. { ‚al·jə‚brā·ik və'rī·əd·ē }

algebra of subsets An algebra of subsets of a set S is a family of subsets of S that contains the null set, the complement (relative to S) of each of its members, and the union of any two of its members. { ¦al·jə·brə əv 'səb‚sets }

algebra with identity An algebra which has an element, not equal to 0 and denoted by 1, such that, for any element x in the algebra, $x1 = 1x = x$. { ¦al·jə·brə with i'den·ə·tē }

algorithm A set of well-defined rules for the solution of a problem in a finite number of steps. { 'al·gə‚rith·əm }

alias Either of two effects in a factorial experiment which cannot be differentiated from each other on the basis of the experiment. { 'ā·lē·əs }

aliasing Introduction of error into the computed amplitudes of the lower frequencies in a Fourier analysis of a function carried out using discrete time samplings whose interval does not allow the proper analysis of the higher frequencies present in the analyzed function. { 'āl·yəs·iŋ }

alignment chart *See* nomograph. { ə'līn·mənt ‚chärt }

aliquant A divisor that does not divide a quantity into equal parts. { 'al·ə‚kwänt }

aliquot A divisor that divides a quantity into equal parts with no remainder. { 'al·ə‚kwät }

allometry A relation between two variables x and y that can be written in the form $y = ax^n$, where a and n are constants. { ə'läm·ə·trē }

almost every A proposition concerning the points of a measure space is said to be true at almost every point, or to be true almost everywhere, if it is true for every

6

point in the space, with the exception at most of a set of points which form a measurable set of measure zero. { 'ȯl‚mōst 'ev·rē }

almost-perfect number An integer that is 1 greater than the sum of all its factors other than itself. { 'ȯl‚mōst 'pər·fik 'nəm·bər }

almost-periodic function A continuous function $f(x)$ such that for any positive number ϵ there is a number M so that for any real number x, any interval of length M contains a nonzero number t such that $|f(x + t) - f(x)| < \epsilon$. { 'ȯl‚mōst ‚pir·ē'äd·ik 'fəŋk·shən }

alpha rule See renaming rule. { 'al·fə ‚rül }

alternate angles A pair of nonadjacent angles that a transversal forms with each of two lines; they lie on opposite sides of the transversal, and are both interior, or both exterior, to the two lines. { 'ȯl·tər·nət 'aŋ·gəlz }

alternating form A bilinear form f which changes sign under interchange of its independent variables; that is, $f(x,y) = -f(y, x)$ for all values of the independent variables x and y. { 'ȯl·tər‚nād·iŋ 'fȯrm }

alternating function A function in which the interchange of two independent variables causes the dependent variable to change sign. { 'ȯl·tər·nād·iŋ 'fəŋk·shən }

alternating group A group made up of all the even permutations of n objects. { 'ȯl·tər·nād·iŋ 'grüp }

alternating series Any series of real numbers in which consecutive terms have opposite signs. { 'ȯl·tər·nād·iŋ 'sir·ēz }

alternation See disjunction. { ‚ȯl·tər'nā·shən }

alternative algebra A nonassociative algebra in which any two elements generate an associative algebra. { ȯl'tər·nəd·iv 'al·jə·brə }

alternative hypothesis Value of the parameter of a population other than the value hypothesized or believed to be true by the investigator. { ȯl'tər·nət·iv hī'päth·ə·səs }

altitude Abbreviated alt. The perpendicular distance from the base to the top (a vertex or parallel line) of a geometric figure such as a triangle or parallelogram. { 'al·tə‚tüd }

ambiguous case 1. For the solution of a plane triangle, the case in which two sides and the angle opposite one of them is given, and there are two distinct solutions. **2.** For the solution of a spherical triangle, the case in which two sides and the angle opposite one of them is given, or two angles and the side opposite one of them is given, and there are two distinct solutions. { am'big·yə·wəs 'kās }

amicable numbers Two numbers such that the exact divisors of each number (except the number itself) add up to the other number. { 'am·ə·kə·bəl 'nəm·bərz }

amplitude The angle between a vector representing a specified complex number on an Argand diagram and the positive real axis. Also known as argument. { 'am·plə‚tüd }

anallagmatic curve A curve that is its own inverse curve with respect to some circle. { ə‚nal·ig‚mad·ik 'kərv }

analysis The branch of mathematics most explicitly concerned with the limit process or the concept of convergence; includes the theories of differentiation, integration and measure, infinite series, and analytic functions. Also known as mathematical analysis. { ə'nal·ə·səs }

analysis of variance A method for partitioning the total variance in experimental data into components assignable to specific sources. { ə‚nal·ə·səs əv 'ver·ē·əns }

analytic continuation The process of extending an analytic function to a domain larger than the one on which it was originally defined. { ‚an·əl'id·ik kən·tin·yü'ā·shən }

analytic curve A curve whose parametric equations are real analytic functions of the same real variable. { ‚an·əl'id·ik 'kərv }

analytic function A function which can be represented by a convergent Taylor series. Also known as holomorphic function. { ‚an·əl'id·ik 'fuŋk·shən }

analytic geometry The study of geometric figures and curves using a coordinate system and the methods of algebra. Also known as Cartesian geometry. { ‚an·əl'id·ik jē'äm·ə·trē }

analytic hierarchy A systematic procedure for representing the elements of any problem which breaks down the problem into its smaller constituents and then calls for only simple pairwise comparison judgments to develop priorities at each level. { ‚an·əl'id·ik 'hī·ər‚är·kē }

analytic number theory The study of problems concerning the discrete domain of integers by means of the mathematics of continuity. { ‚an·əl'id·ik 'nəm·bər ‚thē·ə·rē }

analytic set A subset of a separable, complete metric space that is a continuous image of a Borel set in this metric space. { ‚an·ə¦lid·ik 'set }

analytic structure A covering of a locally Euclidean topological space by open sets, each of which is homeomorphic to an open set in Euclidean space, such that the coordinate transformation (in both directions) between the overlap of any two of these sets is given by analytic functions. { an·əl'id·ik 'strək·chər }

analytic trigonometry The study of the properties and relations of the trigonometric functions. { ‚an·əl'id·ik ‚trig·ə'näm·ə·trē }

anchor point Either of the two end points of a Bézier curve. { 'aŋ·kər ‚póint }

AND function An operation in logical algebra on statements P, Q, R, such that the operation is true if all the statements P, Q, R, . . . are true, and the operation is false if at least one statement is false. { 'and ‚fuŋk·shən }

angle The geometric figure, arithmetic quantity, or algebraic signed quantity determined by two rays emanating from a common point or by two planes emanating from a common line. { 'aŋ·gəl }

angle bisection The division of an angle by a line or plane into two equal angles. { 'aŋ·gəl bī'sek·shən }

angle of contingence For two points on a plane curve, the angle between the tangents to the curve at those points. { 'aŋ·gəl əv kən'tin·jəns }

angle of geodesic contingence For two points on a curve on a surface, the angle of intersection of the geodesics tangent to the curve at those points. { 'aŋ·gəl əv ‚jē·ə¦des·ik kən'tin·jəns }

angular distance 1. For two points, the angle between the lines from a point of observation to the points. 2. The angular difference between two directions, numerically equal to the angle between two lines extending in the given directions. 3. The arc of the great circle joining two points, expressed in angular units. { 'an·gyə·lər 'dis·təns }

angular radius For a circle drawn on a sphere, the smaller of the angular distances from one of the two poles of the circle to any point on the circle. { 'aŋ·gyə·lər 'rād·ē·əs }

annihilator For a set S, the class of all functions of specified type whose value is zero at each point of S. { ə'nī·ə‚lād·ər }

annular solid A solid generated by rotating a closed plane curve about a line which lies in the plane of the curve and does not intersect the curve. { 'an·yə·lər 'säl·əd }

annulus The ringlike figure that lies between two concentric circles. { 'an·yə·ləs }

annulus conjecture For dimension n, the assertion that if f and g are locally flat embeddings of the $(n - 1)$ sphere, S^{n-1}, in real n space, R^n, with $f(S^{n-1})$ in the bounded component of $R^n - g(S^{n-1})$, then the closed region in R^n bounded by $f(S^{n-1})$ and $g(S^{n-1})$ is homeomorphic to the direct product of S^{n-1} and the closed interval [0,1]; it is established for $n \neq 4$. { 'an·yə·ləs kən'jek·chər }

antecedent 1. The numerator of a ratio. 2. The first of the two statements in an implication. 3. For an integer, n, that is greater than 1, the preceding integer, $n - 1$. { 'an·tə‚sēd·ənt }

antiautomorphism An antiisomorphism of a ring, field, or integral domain with itself. { ‚an·tē‚òd·ə'mòr‚fiz·əm }

antichain 1. A subset of a partially ordered set in which no pair is a comparable pair. 2. *See* Sperner set. { 'an·tē‚chān }

anticlastic Having the property of a surface or portion of a surface whose two principal curvatures at each point have opposite signs, so that one normal section is concave and the other convex. { ¦an·tē¦klas·tik }

anticommutative operation A method of combining two objects, $a \cdot b$, such that $a \cdot b = -b \cdot a$ { ‚an·tē‚käm·yə‚tād·iv ‚äp·ə'rā·shən }

anticommutator The anticommutator of two operators, A and B, is the operator $AB + BA$. { ‚an·tē'käm·yə‚tād·ər }

anticommute Two operators anticommute if their anticommutator is equal to zero. { ‚an·tē·kə'myüt }

anticosecant See arc cosecant. { ‚an·tē·kō'sē‚kant }

anticosine See arc cosine. { ‚an·te'kō‚sīn }

anticotangent See arc contangent. { ‚an·tē·kō'tan·jənt }

antiderivative See indefinite integral. { ¦an·tē·di¦riv·əd·iv }

anti-isomorphism A one-to-one correspondence between two rings, fields, or integral domains such that, if x' corresponds to x and y' corresponds to y, then $x' + y$ corresponds to $x + y$, but $y'x'$ corresponds to xy. { ‚an·tē‚ī·sə'mȯr‚fiz·əm }

antilog See antilogarithm. { 'an·ti‚läg }

antilogarithm For a number x, a second number whose logarithm equals x. Abbreviated antilog. Also known as inverse logarithm. { ¦an·ti'läg·ə‚rith·əm }

antiparallel Property of two nonzero vectors in a vector space over the real numbers such that one vector equals the product of the other vector and a negative number. { ¦an·tē'par·ə‚lel }

antiparallel lines Two lines that make equal angles in opposite order with two specified lines. { ‚an·tē‚par·ə‚lel 'līnz }

antipodal points The points at opposite ends of a diameter of a sphere. { an¦tip·əd·əl 'pȯins }

antisecant See arc secant. { ‚an·tē'sē‚kant }

antisine See arc sine. { ‚an·tē'sīn }

antisymmetric determinant The determinant of an antisymmetric matrix. Also known as skew-symmetric determinant. { ‚an·tē·sə‚me·trik di'tər·mə·nənt }

antisymmetric dyadic A dyadic equal to the negative of its conjugate. { ¦an·tē·si¦me·trik dī'ad·ik }

antisymmetric matrix A matrix which is equal to the negative of its transpose. Also known as skew matrix; skew-symmetric matrix. { ¦an·tē·si¦me·trik 'mā·triks }

antisymmetric relation A relation, which may be denoted $\in$, among the elements of a set such that if $a \in b$ and $b \in a$ then $a = b$. { ‚ant·i·si¦me·trik ri'lā·shən }

antisymmetric tensor A tensor in which interchanging two indices of an element changes the sign of the element. { ¦an·tē·si¦me·trik 'ten·sər }

antitangent See arc tangent. { ‚an·tē'tan·jənt }

antithetic variable One of two random variables having high negative correlation, used in the antithetic variate method of estimating the mean of a series of observations. { ¦an·tē¦thed·ik 'ver·ē·ə·bəl }

apex 1. The vertex of a triangle opposite to the side which is regarded as the base. **2.** The vertex of a cone or pyramid. { 'ā‚peks }

Apollonius' problem The problem of constructing a circle that is tangent to three given circles. { ‚ap·ə¦lōn·ē·əs ¦präb·ləm }

a posteriori probability See empirical probability. { ¦ā ‚pä‚stir·ē'ȯr‚ē ‚präb·ə'bil·əd·ē }

apothem The perpendicular distance from the center of a regular polygon to one of its sides. Also known as short radius. { 'ap·ə‚them }

applicable surfaces Surfaces such that there is a length-preserving map of one onto the other. { ¦ap·lə·kə·bəl 'sər·fəs·əz }

approximate 1. To obtain a result that is not exact but is near enough to the correct result for some specified purpose. **2.** To obtain a series of results approaching the correct result. { ə'präk·sə‚māt }

approximate reasoning The process by which a possibly imprecise conclusion is deduced from a collection of imprecise premises. { ə¦präks·ə·mət 'rēz·ən·iŋ }

approximation 1. A result that is not exact but is near enough to the correct result for some specified purpose. **2.** A procedure for obtaining such a result. { ə¦präk·sə¦mā·shən }

approximation property The property of a Banach space, B, in which compact sets are approximately finite-dimensional in the sense that, for any compact set, K, continuous linear transformations, L, from K to finite-dimensional subspaces of B can be found with arbitrarily small upper bounds on the norm of $L(x) - x$ for all points x in K.   { ə‚präk·sə'mā·shən ‚präp·ərd·ē }

a priori Pertaining to deductive reasoning from assumed axioms or supposedly self-evident principles, supposedly without reference to experience.   { ¦ā prē¦ȯr·ē }

a priori probability *See* mathematical probability.   { ¦ā prē¦ȯr·ē ‚präb·ə'bil·əd·ē }

arabic numerals The numerals 0, 1, 2, 3, 4, 5, 6, 7, 8, and 9.   Also known as Hindu-Arabic numerals.   { 'ar·ə·bik 'nüm·rəlz }

arbilos A plane figure bounded by a semicircle and two smaller semicircles which lie inside the larger semicircle, have diameters along the diameter of the larger semicircle, and are tangent to the larger semicircle and to each other.   Also known as shoemaker's knife.   { 'är·bī‚lōs }

arc 1. A continuous piece of the circumference of a circle.   Also known as circular arc.   **2.** *See* edge.   { ärk }

arc cosecant Also known as anticosecant; inverse cosecant.   **1.** For a number x, any angle whose cosecant equals x.   **2.** For a number x, the angle between $-\pi/2$ radians and $\pi/2$ radians whose cosecant equals x; it is the value at x of the inverse of the restriction of the cosecant function to the interval between $-\pi/2$ and $\pi/2$.   { 'ärk kō'sē‚kant }

arc cosine Also known as anticosine; inverse cosine.   **1.** For a number x, any angle whose cosine equals x.   **2.** For a number x, the angle between 0 radians and π radians whose cosine equals x; it is the value at x of the inverse of the restriction of the cosine function to the interval between 0 and π.   { 'ärk 'kō‚sīn }

arc cotangent Also known as anticotangent; inverse cotangent.   **1.** For a number x, any angle whose cotangent equals x.   **2.** For a number x, the angle between 0 radians and π radians whose cotangent equals x; it is the value at x of the inverse of the restriction of the cotangent function to the interval between 0 and π.   { 'ärk kō'tan·jənt }

arc-disjoint paths In a graph, two paths with common end points that have no arcs in common.   { ‚ärk'dis‚jȯint pa<u>th</u>z }

Archimedean ordered field A field with a linear order that satisfies the axiom of Archimedes.   { ‚är·kə¦mē·dē·ən ¦ȯrd·ərd 'fēld }

Archimedean solid One of 13 possible solids whose faces are all regular polygons, though not necessarily all of the same type, and whose polyhedral angles are all equal.   Also known as semiregular solid.   { ¦är·kə¦mēd·ē·ən 'säl·əd }

Archimedean spiral A plane curve whose equation in polar coordinates (r, θ) is $r^m = a^m\theta$, where a and m are constants.   { är·kə¦mēd·ē·ən 'spī·rəl }

Archimedes' axiom *See* axiom of Archimedes.   { ¦är·kə¦mēd‚ēz 'ak·sē·əm }

Archimedes' problem The problem of dividing a hemisphere into two parts of equal volume with a plane parallel to the base of the hemisphere; it cannot be solved by Euclidean methods.   { ¦är·kə¦mēd‚ēz 'präb·ləm }

Archimedes' spiral *See* spiral of Archimedes.   { ¦är·kə¦mēd'ēz 'spī·rəl }

arc-hyperbolic cosecant For a number, x, not equal to zero, the number whose hyperbolic cosecant equals x; it is the value at x of the inverse of the hyperbolic cosecant function.   Also known as inverse hyperbolic cosecant.   { ¦ärk ‚hī·pər‚bäl·ik kō'sē‚kant }

arc-hyperbolic cosine Also known as inverse hyperbolic cosine.   **1.** For a number, x, equal to or greater than 1, either of the two numbers whose hyperbolic cosine equals x.   **2.** For a number, x, equal to or greater than 1, the positive number whose hyperbolic cosine equals x; it is the value at x of the restriction of the inverse of the hyperbolic cosine function to the positive numbers.   { ‚ärk ‚hī·pər‚bäl·ik 'kō‚sīn }

arc-hyperbolic cotangent For a number, x, with absolute value greater than 1, the number whose hyperbolic cotangent equals x; it is the value at x of the inverse of the hyperbolic cotangent function.   Also known as inverse hyperbolic cotangent.   { ‚ärk ‚hī·pər‚bäl·ik kō'tan·jənt }

arc-hyperbolic secant Also known as inverse hyperbolic secant.   **1.** For a number, x, equal to or greater than 0 and equal to or less than 1, either of the two numbers whose hyperbolic secant equals x.   **2.** For a number, x, equal to or greater than 0, and equal to or less than 1, the positive number whose hyperbolic cosecant equals x; it is the value at x of the restriction of the hyperbolic secant function to the positive numbers.   { ˌärk ˌhī·pər‚bäl·ik 'sē‚kant }

arc-hyperbolic sine For a number, x, the number whose hyperbolic sine equals x; it is the value at x of the inverse of the hyperbolic sine function.   Also known as inverse hyperbolic sine.   { ˌärk ˌhī·pər‚bäl·ik 'sīn }

arc-hyperbolic tangent For a number, x, with absolute value less than 1, the number whose hyperbolic tangent equals x; it is the value at x of the inverse of the hyperbolic tangent function.   Also known as inverse hyperbolic tangent.   { ˌärk ˌhī·pər‚bäl·ik 'tan·jənt }

arcmin *See* minute.

arc secant Also known as antisecant; inverse secant.   **1.** For a number x, any angle whose secant equals x.   **2.** For a number x, the angle between 0 radians and π radians whose secant equals x; it is the value at x of the inverse of the restriction of the secant function to the interval between 0 and π.   { ¦ärk 'sē‚kant }

arc sine Also known as antisine; inverse sine.   **1.** For a number x, any angle whose sine equals x.   **2.** For a number x, the angle between $-\pi/2$ radians and $\pi/2$ radians whose sine equals x; it is the value at x of the inverse of the restriction of the sine function to the interval between $-\pi/2$ and $\pi/2$.   { ¦ärk ¦sīn }

arc sine transformation A technique used to convert data made up of frequencies or proportions into a form that can be analyzed by analysis of variance or by regression analysis.   { ¦ärk ¦sīn ‚tranz·fər'mā·shən }

arc tangent Also known as antitangent; inverse tangent.   **1.** For a number x, any angle whose tangent equals x.   **2.** For a number x, the angle between $-\pi/2$ radians and $\pi/2$ radians whose tangent equals x; it is the value at x of the inverse of the restriction of the tangent function to the interval between $-\pi/2$ and $\pi/2$.   { ¦ärk 'tan·jənt }

arcwise-connected set A set in which each pair of points can be joined by a simple arc whose points are all in the set.   Also known as path-connected set; pathwise-connected set.   { 'ärk‚wīz kə‚nek·təd 'set }

area A measure of the size of a two-dimensional surface, or of a region on such a surface.   { 'er·ē·ə }

area sampling A method in which the area to be sampled is subdivided into smaller blocks which are selected at random and then subsampled or fully surveyed; method is used when a complete frame of reference is not available.   { ¦er·ē·ə ¦samp·liŋ }

Argand diagram A two-dimensional Cartesian coordinate system for representing the complex numbers, the number $x + iy$ being represented by the point whose coordinates are x and y.   { 'är‚gän 'di·ə‚gram }

Arguesian plane *See* Desarguesian plane.   { är¦gesh·ən 'plān }

argument *See* amplitude; independent variable.   { 'är·gyə·mənt }

arithlog paper Graph paper marked with a semilogarithmic coordinate system.   { ə'rith‚läg ‚pā·pər }

arithmetic Addition, subtraction, multiplication, and division, usually of integers, rational numbers, real numbers, or complex numbers.   { ə'rith·mə‚tik }

arithmetical addition The addition of positive numbers or of the absolute values of signed numbers.   { ¦a·rith¦med·ə·kəl ə'dish·ən }

arithmetic average *See* arithmetic mean.   { ¦a·rith¦med·ik 'av·rij }

arithmetic-geometric mean For two positive numbers a_1 and b_1, the common limit of the sequences $\{a_n\}$ and $\{b_n\}$ defined recursively by the equations $a_{n+1} = \frac{1}{2}(a_n + b_n)$ and $b_{n+1} = (a_n b_n)^{1/2}$.   { ¦a·rith¦med·ik jē·ə¦me·trik 'mēn }

arithmetic mean The average of a collection of numbers obtained by dividing the sum of the numbers by the quantity of numbers.   Also known as arithmetic average; average (av).   { ¦a·rith¦med·ik 'mēn }

arithmetic progression A sequence of numbers for which there is a constant d such that the difference between any two successive terms is equal to d. Also known as arithmetic sequence. { ¦a·rith¦med·ik prə'gresh·ən }

arithmetic sequence *See* arithmetic progression. { ¦a·rith¦med·ik 'sē·kwəns }

arithmetic series A series whose terms form an arithmetic progression. { ¦a·rith¦med·ik ,sir,ēz }

arithmetic sum 1. The result of the addition of two or more positive quantities. 2. The result of the addition of the absolute values of two or more quantities. { ¦a·rith¦med·ik 'səm }

arithmetization 1. The study of various branches of higher mathematics by methods that make use of only the basic concepts and operations of arithmetic. 2. Representation of the elements of a finite or denumerable set by nonnegative integers. Also known as Gödel numbering. { ə,rith·məd·ə'zā·shən }

arm A side of an angle. { ärm }

array The arrangement of a sequence of items in statistics according to their values, such as from largest to smallest. { ə'rā }

Artinian ring A ring is Artinian on left ideals (or right ideals) if every descending sequence of left ideals (or right ideals) has only a finite number of distinct members. { ar¦tin·ē·ən 'riŋ }

ascending chain condition The condition on a ring that every ascending sequence of left ideals (or right ideals) has only a finite number of distinct members. { ə,sen·diŋ 'chān kən,dish·ən }

ascending sequence 1. A sequence of elements of a partially ordered set such that each member of the sequence is equal to or less than the following one. 2. In particular, a sequence of sets such that each member of the sequence is a subset of the following one. { ə,sen·diŋ 'sē·kwəns }

ascending series 1. A series each of whose terms is greater than the preceding term. 2. *See* power series. { ə'send·iŋ 'sir·ēz }

Ascoli's theorem The theorem that a set of uniformly bounded, equicontinuous, real-valued functions on a closed set of a real Euclidean n-dimensional space contains a sequence of functions which converges uniformly on compact subsets. { as'kō,lēz ,thir·əm }

associate curve *See* Bertrand curve. { ə'sō·sē·ət ,kərv }

associated prime ideal A prime ideal I in a commutative ring R is said to be associated with a module M over R if there exists an element x in M such that I is the annihilator of x. { ə'sō·sē,ād·əd 'prīm ,ī·dēl }

associated radii of convergence For a power series in n variables, $z_1, . . ,z_n$, any set of numbers, $r_1, . . . , r_n$, such that the series converges when $|z_i| < r_i$, $i = 1, . . . ,$ n, and diverges when $|z_i| > r_i$, $i = 1, .. ,n$. { ə'sō·sē,ād·əd ¦rād·dē,ī əv kən'vər·jəns }

associated tensor A tensor obtained by taking the inner product of a given tensor with the metric tensor, or by performing a series of such operations. { ə'sō·sē,ād·əd 'ten·sər }

associate matrix *See* Hermitian conjugate. { ə'sō·sē·ət 'mā·triks }

associate operator *See* adjoint operator. { ə'sō·sē·ət 'äp·ə,rād·ər }

associates Two elements x and y in a commutative ring with identity such that $x = ay$, where a is a unit. Also known as equivalent elements. { ə'sō·sē·ətz }

associative algebra An algebra in which the vector multiplication obeys the associative law. { ə'sō·sē,ād·iv 'al·jə·brə }

associative law For a binary operation that is designated ∘, the relationship expressed by $a ∘ (b ∘ c)=(a ∘ b) ∘ c$. { ə'sō·sē,ād·iv 'lò }

astroid A hypocycloid for which the diameter of the fixed circle is four times the diameter of the rolling circle. { 'a,stròid }

asymptote 1. A line approached by a curve in the limit as the curve approaches infinity. 2. The limit of the tangents to a curve as the point of contact approaches infinity. { 'as·əm,tōt }

asymptotic curve A curve on a surface whose osculating plane at each point is the same as the tangent plane to the surface. { ā,sim'täd·ik 'kərv }

asymptotic directions For a hyperbolic point on a surface, the two directions in which the normal curvature vanishes; equivalently, the directions of the asymptotic curves passing through the point. { ‚ā·sim'täd·ik də'rek·shənz }

asymptotic efficiency The efficiency of an estimator within the limiting value as the size of the sample increases. { ‚ā·sim'täd·ik ə'fish·ən·sē }

asymptotic expansion A series of the form $a_0 + (a_1/x) + (a_2/x^2) + \cdots + (a_n/x_n) + \cdots$ is an asymptotic expansion of the function $f(x)$ if there exists a number N such that for all $n > N$ the quantity $x_n[f(x) - S_n(x)]$ approaches zero as x approaches infinity, where $S_n(x)$ is the sum of the first n terms in the series. Also known as asymptotic series. { ā‚sim'täd·ik ik'span·shən }

asymptotic formula A statement of equality between two functions which is not a true equality but which means the ratio of the two functions approaches 1 as the variable approaches some value, usually infinity. { ā‚sim'täd·ik 'fòr·myə·lə }

asymptotic series See asymptotic expansion. { ā‚sim'täd·ik 'sir·ēz }

asymptotic stability The property of a vector differential equation which satisfies the conditions that (1) whenever the magnitude of the initial condition is sufficiently small, small perturbations in the initial condition produce small perturbations in the solution; and (2) there is a domain of attraction such that whenever the initial condition belongs to this domain the solution approaches zero at large times. { ā‚sim'täd·ik stə'bil·əd·ē }

atlas An atlas for a manifold is a collection of coordinate patches that covers the manifold. { 'at·ləs }

atom An element, A, of a measure algebra, other than the zero element, which has the property that any element which is equal to or less than A is either equal to A or equal to the zero element. { 'ad·əm }

augend A quantity to which another quantity is added. { 'ò‚jənd }

augmented matrix The matrix of the coefficients, together with the constant terms, in a system of linear equations. { 'òg·men·təd 'mā·triks }

autocorrelation In a time series, the relationship between values of a variable taken at certain times in the series and values of a variable taken at other, usually earlier times. { ¦òd·ō‚kär·ə'lā·shən }

autocorrelation function For a specified function $f(t)$, the average value of the product $f(t)f(t - \tau)$, where τ is a time-delay parameter; more precisely, the limit as T approaches infinity of $1/(2T)$ times the integral from $-T$ to T of $F(t)f(t - \tau) \, dt$. { ¦òd·ō‚kär·ə'lā·shən ‚fuŋk·shən }

automata theory A theory concerned with models used to simulate objects and processes such as computers, digital circuits, nervous systems, cellular growth and reproduction. { ò'täm·əd·ə 'thē·ə·rē }

automorphism An isomorphism of an algebraic structure with itself. { ¦òd·ō'mòr‚fiz·əm }

autoregressive series A function of the form $f(t) = a_1 f(t - 1) + a_2 f(t - 2) + \cdots + a_m f(t - m) + k$, where k is any constant. { ¦òd·ō·ri¦gres·iv 'sir·ēz }

auxiliary equation The equation that is obtained from a given linear differential equation by replacing with zero the term that involves only the independent variable. Also known as reduced equation. { òg¦zil·yə·re i'kwā·zhən }

av See arithmetic mean.

average See arithmetic mean. { 'av·rij }

average curvature For a given arc of a plane curve, the ratio of the change in inclination of the tangent to the curve, over the arc, to the arc length. { ¦av·rij 'kərv·ə·chər }

average deviation In statistics, the average or arithmetic mean of the deviation, taken without regard to sign, from some fixed value, usually the arithmetic mean of the data. Abbreviated AD. Also known as mean deviation. { 'av·rij ‚dē·vē'ā·shən }

axial symmetry Property of a geometric configuration which is unchanged when rotated about a given line. { 'ak·sē·əl 'sim·ə·trē }

axiom Any of the assumptions upon which a mathematical theory (such as geometry, ring theory, and the real numbers) is based. Also known as postulate. { 'ak·sē·əm }

axiom of Archimedes The postulate that if x is any real number, there exists an integer n such that n is greater than x. Also known as Archimedes' axiom. { ¦ak·sē·əm əv ˌärk·ə'mē͟dēz }

axiom of choice The axiom that for any family A of sets there is a function that assigns to each set S of the family A a member of S. { ¦ak·sē·əm əv 'chȯis }

axis 1. In a coordinate system, the line determining one of the coordinates, obtained by setting all other coordinates to zero. **2.** A line of symmetry for a geometric figure. **3.** For a cone whose base has a center, a line passing through this center and the vertex of the cone. { 'ak·səs }

axis of abscissas The horizontal or x axis of a two-dimensional Cartesian coordinate system, parallel to which abscissas are measured. { 'ak·səs əv ab'sis·əz }

axis of ordinates The vertical or y axis of a two-dimensional Cartesian coordinate system, parallel to which ordinates are measured. { 'ak·səs əv 'ȯrd·nəts }

B

backward difference One of a series of quantities obtained from a function whose values are known at a series of equally spaced points by repeatedly applying the backward difference operator to these values; used in interpolation and numerical calculation and integration of functions. { ¦bak·wərd 'dif·rəns }

backward difference operator A difference operator, denoted ∇, defined by the equation $\nabla f(x) = f(x) - f(x - h)$, where h is a constant denoting the difference between successive points of interpolation or calculation. { ¦bak·wərd ¦dif·rəns 'äp·ə‚rād·ər }

Baire function The smallest class of functions on a topological space which contains the continuous functions and is closed under pointwise limits. { 'ber ‚fəŋk·shən }

Baire measure A measure defined on the class of all Baire sets such that the measure of any closed, compact set is finite. { 'ber ‚mezh·ər }

Baire's category theorem The theorem that a complete metric space is of second category; equivalently, the intersection of any sequence of open dense sets in a complete metric space is dense. { ¦berz 'kad·ə‚gór·ē ‚thir·əm }

Baire set A member of the smallest sigma algebra containing all closed, compact subsets of a topological space. { 'ber ‚set }

Baire space A topological space in which every countable intersection of dense, open subsets is dense in the space. { 'ber ‚spās }

balanced digit system A number system in which the allowable digits in each position range in value from $-n$ to n, where n is some positive integer, and $n + 1$ is greater than one-half the base. { 'bal·ənst 'dij·ət ‚sis·təm }

balanced incomplete block design For positive integers b, v, r, k, and λ, an arrangement of v elements into b subsets or blocks so that each block contains exactly k distinct elements, each element occurs in r blocks, and every combination of two elements occurs together in exactly λ blocks. Also known as (b,v,r,k,λ)-design. { ¦bal·ənst ‚iŋ·kəm‚plēt 'bläk di‚zīn }

balanced range of error A range of error in which the maximum and minimum possible errors are opposite in sign and equal in magnitude. { 'bal·ənst ¦rānj əv 'er·ər }

balanced set A set S in a real or complex vector space X such that if x is in S and $|a| \leq 1$, then ax is in S. { 'bal·ənst ‚set }

balance equation An equation expressing a balance of quantities in the sense that the local or individual rates of change are zero. { 'bal·əns i'kwā·zhən }

Banach algebra An algebra which is a Banach space satisfying the property that for every pair of vectors, the norm of the product of those vectors does not exceed the product of their norms. { 'bä‚näk 'al·jə·brə }

Banach's fixed-point theorem A theorem stating that if a mapping f of a metric space E into itself is a contraction, then there exists a unique element x of E such that $f(x) = x$. Also known as Caccioppoli-Banach principle. { ¦bä‚näks ‚fikst ‚póint 'thir·əm }

Banach space A real or complex vector space in which each vector has a non-negative length, or norm, and in which every Cauchy sequence converges to a point of the space. Also known as complete normed linear space. { 'bä‚näk ‚spās }

Banach-Steinhaus theorem If a sequence of bounded linear transformations of a

Banach-Tarski paradox

Banach space is pointwise bounded, then it is uniformly bounded. { ¦bä,näk ¦stīn,haüs ˌthir·əm }

Banach-Tarski paradox A theorem stating that, for any two bounded sets, with interior points in a Euclidean space of dimension at least three, one of the sets can be disassembled into a finite number of pieces and reassembled to form the other set by moving the pieces with rigid motions (translations and rotations). { ¦bä,näk ¦tär·skē 'par·ə,däks }

bar chart See bar graph. { 'bär ˌchärt }

bar graph A diagram of frequency-table data in which a rectangle with height proportional to the frequency is located at each value of a variate that takes only certain discrete values. Also known as bar chart; rectangular graph. { 'bär ˌgraf }

Bartlett's test A method to test for the equalities of variances from a number of independent normal samples by testing the hypothesis. { 'bärt·ləts ˌtest }

barycenter The center of mass of a system of finitely many equal point masses distributed in euclidean space in such a way that their position vectors are linearly independent. { 'bar·ə,sen·tər }

barycentric coordinates The coefficients in the representation of a point in a simplex as a linear combination of the vertices of the simplex. { ˌbar·ə'sen·trik kō'ȯrd·ən,əts }

base 1. A side or face upon which the altitude of a geometric configuration is thought of as being constructed. 2. For a logarithm, the number of which the logarithm is the exponent. 3. For a number system, the number whose powers determine place value. 4. For a topological space, a collection of sets, unions of which form all the open sets of the space. { bās }

base angle Either of the two angles of a triangle that have the base for a side. { 'bās ˌaŋ·gəl }

base for the neighborhood system See local base. { ¦bās fər t̲h̲ə 'nā·bər,hu̇d ˌsis·təm }

base notation See radix notation. { 'bās nō'tā·shən }

base period The period of a year, or other unit of time, used as a reference in constructing an index number. Also known as base year. { 'bās ˌpir·ē·əd }

base space of a bundle The topological space B in the bundle (E,p,B). { ¦bās ˌspās əv ə 'bən·dəl; }

base vector One of a set of linearly independent vectors in a vector space such that each vector in the space is a linear combination of vectors from the set; that is, a member of a basis. { 'bās ˌvek·tər }

base year See base period. { 'bās ˌyir }

base-year method See Laspeyre's index. { ¦bās ˌyir 'meth·əd }

basic solution In bifurcation theory, a simple, explicitly known solution of a nonlinear equation, in whose neighborhood other solutions are studied. { 'bā·sik sə'lü·shən }

basis A set of linearly independent vectors in a vector space such that each vector in the space is a linear combination of vectors from the set. { 'bā·səs }

Bayes decision rule A decision rule under which the strategy chosen from among several available ones is the one for which the expected value of payoff is the greatest. { 'bāz di'sizh·ən ˌrül }

Bayesian statistics An approach to statistics in which estimates are based on a synthesis of a prior distribution and current sample data. { ¦bāz·ē·ən stə'tis·tiks }

Bayesian theory A theory, as of statistical inference or decision making, in which probabilities are associated with individual events or statements rather than with sequences of events. { 'bāz·ē·ən ˌthē·ə·rē }

Bayes rule The rule that the probability $P(E_i|A)$ of some event E_i, given that another event A has been observed, is $P(E_i)P(A|E_i)/P(A)$, where $P(E_i)$ is the prior probability of E_i, determined either objectively or subjectively, and $P(A)$, the probability of A, is given by the sum over all possible events E_j of the quantity $P(E_j)P(A|E_j)$. { 'bāz ˌrül }

Bayes' theorem A theorem stating that the probability of a hypothesis, given the original data and some new data, is proportional to the probability of the hypothesis, given the original data only, and the probability of the new data, given the original data and the hypothesis. Also known as inverse probability principle. { ¦bāz 'thir·əm }

Behrens-Fisher problem The problem of calculating the probability of drawing two random samples whose means differ by some specified value (which may be zero) from normal populations, when one knows the difference of the means of these populations but not the ratio of their variances. { ¦ber·ənz ¦fish·ər ‚präb·ləm }

bei function One of the functions that is defined by $\text{ber}_n(z) \pm i \text{ bei}_n(z) = J_n(ze^{\pm 3\pi i/4})$, where J_n is the nth Bessel function. { 'bī ‚fəŋk·shən }

Bell numbers The numbers, B_n, that count the total number of partitions of a set with n elements. { 'bel ‚nəm·bərz }

bell-shaped curve The curve representing a continuous frequency distribution with a shape having the overall curvature of the vertical cross section of a bell; usually applied to the normal distribution. { ¦bel ¦shāpt 'kərv }

ber function One of the functions defined by $\text{ber}_n(z) \pm i \text{ bei}_n(z) = J_n(ze^{\pm 3\pi i/4})$, where J_n is the nth Bessel function. { 'ber ‚fəŋk·shən }

Bernoulli differential equation See Bernoulli equation. { ber‚nü·lē or ¦ber·nü¦yē ‚dif·ə'ren·chəl i'kwä·zhən }

Bernoulli distribution See binomial distribution. { ber‚nü·lē dis·trə'byü·shən }

Bernoulli equation A nonlinear first-order differential equation of the form $(dy/dx) + yf(x) = y^n g(x)$, where n is a number different from unity and f and g are given functions. Also known as Bernoulli differential equation. { ber‚nü·lē i'kwä·zhən }

Bernoulli experiments See binomial trials. { bər¦nü·lē ik‚sper·ə·məns }

Bernoulli number The numerical value of the coefficient of $x^{2n}/(2n)!$ in the expansion of $xe^x/(e^x-1)$. { ber‚nü·lē ‚nəm·bər }

Bernoulli polynomial The nth such polynomial is

$$\sum_{k=0}^{n} \binom{n}{k} B_k Z^{n-k}$$

where $\binom{n}{k}$ is a binomial coefficient, and B_k is a Bernoulli number. { ber‚nü·lē ‚päl·ə'nō·mē·əl }

Bernoulli's lemniscate A curve shaped like a figure eight whose equation in rectangular coordinates is expressed as $(x^2 + y^2)^2 = a^2(x^2 - y^2)$. { ber‚nü·lēz lem'nis·kət }

Bernoulli theorem See law of large numbers. { ber‚nü·lē 'thir·əm }

Bernoulli trials See binomial trials. { bər'nül·ē ‚trīlz }

Bertrand curve One of a pair of curves having the same principal normals. Also known as associate curve; conjugate curve. { 'ber‚tränd ‚kərv }

Bertrand's postulate The proposition that there exists at least one prime number between any integer greater than three and twice the integer minus two. { 'ber‚tränz 'päs·chə·lət }

Bessel equation The differential equation $z^2 f''(z) + zf'(z) + (z^2 - n^2)f(z) = 0$. { 'bes·əl i'kwä·zhən }

Bessel function A solution of the Bessel equation. Also known as cylindrical function. Symbolized $J_n(z)$. { 'bes·əl ‚fəŋk·shən }

Bessel inequality The statement that the sum of the squares of the inner product of a vector with the members of an orthonormal set is no larger than the square of the norm of the vector. { 'bes·əl ‚in·ē'kwäl·əd·ē }

Bessel transform See Hankel transform. { 'bes·əl 'tranz‚fórm }

best estimate A term applied to unbiased estimates which have a minimum variance. { ¦best 'es·tə·mət }

best fit See goodness of fit. { ¦best 'fit }

beta coefficient Also known as beta weight. **1.** One of the coefficients in a regression equation. **2.** A moment ratio, especially one used to describe skewness and kurtosis. { 'bād·ə kō·ə'fish·ənt }

beta distribution The probability distribution of a random variable with density function

$f(x) = [x^{\alpha-1}(1-x)^{\beta-1}]/B(\alpha,\beta)$, where B represents the beta function, α and β are positive real numbers, and $0 < x < 1$. Also known as Pearson Type I distribution. { 'bād·ə dis·trə'byü·shən }

beta function A function of two positive variables, defined by

$$B(m,n) = \int_0^1 x^{m-1}(1-x)^{n-1}dx$$

{ 'bād·ə ˌfəŋk·shən }

beta random variable A random variable whose probability distribution is a beta distribution. { ¦bād·ə ˌran·dəm 'ver·ē·ə·bəl }

beta weight See beta coefficient. { 'bād·ə ˌwāt }

Betti group See homology group. { 'bät·tē ˌgrüp }

Betti number See connectivity number. { 'bät·tē ˌnəm·bər }

Bézier curve A simple smooth curve whose shape is determined by a mathematical formula from the locations of four points, the two end points of the curve and two interior points. { ¦bāz·yā 'kərv }

Bézout domain An integral domain in which all finitely generated ideals are principal. { ˌbā¸zō dō¸mān }

Bézout's theorem The theorem that the product of the degrees of two algebraic plane curves that lack a common component equals the number of their points of intersection, counted to the degree of their multiplicity, including points of intersection at infinity. { 'bā¸zōz ˌthir·əm }

Bianchi identity A differential identity satisfied by the Riemann curvature tensor: the antisymmetric first covariant derivative of the Riemann tensor vanishes identically. { 'byäŋ·kē ī'den·əd·ē }

bias In estimating the value of a parameter of a probability distribution, the difference between the expected value of the estimator and the true value of the parameter. { 'bī·əs }

biased sample A sample obtained by a procedure that incorporates a systematic error introduced by taking items from a wrong population or by favoring some elements of a population. { ¦bī·əst 'sam·pəl }

biased statistic A statistic whose expected value, as obtained from a random sampling, does not equal the parameter or quantity being estimated. { 'bī·əst stə'tis·tik }

bias error A measurement error that remains constant in magnitude for all observations; a kind of systematic error. { 'bī·əs ˌer·ər }

bicompact set See compact set. { bī'käm¸pakt ¦set }

biconditional operation A logic operator on two statements P and Q whose result is true if P and Q are both true or both false, and whose result is false otherwise. Also known as if and only if operation; match. { ¦bī·kən¸dish·ən·əl ˌäp·ə'rā·shən }

biconditional statement A statement that one of two propositions is true if and only if the other is true. { ˌbī·kən¸dish·ən·əl 'stāt·mənt }

biconnected graph A connected graph in which two points must be removed to disconnect the graph. { ¦bī·kə'nek·təd 'graf }

bicontinuous function See homeomorphism. { ¦bī·kən'tin·yə·wəs 'fəŋk·shən }

bicorn A plane curve whose equation in cartesian coordinates x and y is $(x^2 + 2ay - a^2)^2 = y^2(a^2 - x^2)$, where a is a constant. { 'bī¸kȯrn }

Bieberbach conjecture The proposition, proven in 1984, that if a function $f(z)$ is analytic and univalent in the unit disk, and if it has the power series expansion $z + a_2z^2 + a_z^3 + \cdots$, then, for all n ($n = 2, 3, \ldots$), the absolute value of a_n is equal to or less than n. { 'bē·bə¸bäk kən¸jek·chər }

Bienayme-Chebyshev inequality The probability that the magnitude of the difference between the mean of the sample values of a random variable and the mean of the variable is less than st, where s is the standard deviation and t is any number greater than 1, is equal to or greater than $1 - (1/t^2)$. { ¦bē¸nīm·ə chə·bi'shȯf ¸in·i'kwäl·əd·ē }

bifurcation The appearance of qualitatively different solutions to a nonlinear equation as a parameter in the equation is varied. { bī·fər'kā·shən }

bifurcation theory The study of the local behavior of solutions of a nonlinear equation in the neighborhood of a known solution of the equation; in particular, the study of solutions which appear as a parameter in the equation is varied and which at first approximate the known solution, thus seeming to branch off from it. Also known as branching theory. { ‚bī·fər'kā·shən ‚thē·ə·rē }

bigraded module A collection of modules $E_{s,t}$, indexed by pairs of integers s and t, with each module over a fixed principal ideal domain. { ¦bī‚grād·əd ‚mäj·əl }

biharmonic function A solution to the partial differential equation $\Delta^2 u(x,y,z) = 0$, where Δ is the Laplacian operator; occurs frequently in problems in electrostatics. { ¦bī·här'män·ik 'fəŋk·shən }

bijection A mapping f from a set A onto a set B which is both an injection and a surjection; that is, for every element b of B there is a unique element a of A for which $f(a) = b$. Also known as bijective mapping. { 'bī‚jek·shən }

bijective mapping See bijection. { ‚bī'jek·tiv 'map·iŋ }

bilateral Laplace transform A generalization of the Laplace transform in which the integration is done over the negative real numbers as well as the positive ones. { bī'lad·ə·rəl lə'pläs 'tranz‚fórm }

bilinear concomitant An expression $B(u,v)$, where u, v are functions of x, satisfying $vL(u) - u\bar{L}(v) = (d/dx) \cdot B(u,v)$, where L, $\bar{L}$ are given adjoint differential equations. { bī'lin·ē·ər kən'käm·ə·tənt }

bilinear expression An expression which is linear in each of two variables separately. { bī'lin·ē·ər ik'spresh·ən }

bilinear form 1. A polynomial of the second degree which is homogeneous of the first degree in each of two sets of variables; thus, it is a sum of terms of the form $a_{ij}x_i y_j$, where $x_1, \ldots, x_m$ and $y_1, \ldots, y_n$ are two sets of variables and the a_{ij} are constants. 2. More generally, a mapping $f(x, y)$ from $E \times F$ into R, where R is a commutative ring and $E \times F$ is the Cartesian product of two modules E and F over R, such that for each x in E the function which takes y into $f(x, y)$ is linear, and for each y in F the function which takes x into $f(x, y)$ is linear. { ¦bī‚lin·ē·ər 'fórm }

bilinear transformations See Möbius transformations. { bī'lin·ē·ər tranz·fər'mā·shənz }

billion 1. The number 10^9. 2. In British usage, the number 10^{12}. { 'bil·yən }

bimodal distribution A probability distribution with two different values that are markedly more frequent than neighboring values. { ¦bī·mōd·əl di·strə'byü·shən }

binary notation See binary number system. { 'bīn·ə·rē nō'tā·shən }

binary number A number expressed in the binary number system of positional notation. { 'bīn·ə·rē 'nəm·bər }

binary number system A representation for numbers using only the digits 0 and 1 in which successive digits are interpreted as coefficients of successive powers of the base 2. Also known as binary notation; binary system; dyadic number system. { 'bīn·ə·rē 'nəm·bər ‚sis·təm }

binary numeral One of the two digits 0 and 1 used in writing a number in binary notation. { 'bī‚ner·ē 'nüm·rəl }

binary operation A rule for combining two elements of a set to obtain a third element of that set, for example, addition and multiplication. { 'bīn·ə·rē äp·ə'rā·shən }

binary quantic A quantic that contains two variables. { ‚bīn·ə·rē 'kwän·tik }

binary sequence A sequence, every element of which is 0 or 1. { ‚bīn·ə·rē 'sē·kwəns }

binary system See binary number system. { 'bīn·ə·rē 'sis·təm }

binary-to-decimal conversion The process of converting a number written in binary notation to the equivalent number written in ordinary decimal notation. { 'bīn·ə·rē tə 'des·məl kən'vər·zhən }

binary tree A rooted tree in which each vertex has a maximum of two successors. { 'bīn·ə·rē 'trē }

binomial A polynomial with only two terms. { bī'nō·mē·əl }

binomial array See Pascal's triangle. { bī'nō·mē·əl ə'rā }

binomial coefficient A coefficient in the expansion of $(x + y)^n$, where n is a positive integer; the $(k + 1)$st coefficient is equal to the number of ways of choosing k objects out of n without regard for order. Symbolized $\binom{n}{k}$; $_nC_k$; $C(n,k)$; C_k^n. { bī'nō·mē·əl kō·ə'fish·ənt }

binomial differential A differential of the form $x^p(a + bx^q)^r dx$, where p, q, r are integers. { bī'nō·mē·əl ‚dif·ə'ren·chəl }

binomial distribution The distribution of a binomial random variable; the distribution (n,p) is given by $P(B = r) = \binom{n}{r} p^r q^{n-r}$, $p + q = 1$. Also known as Bernoulli distribution. { bī'nō·mē·əl ‚dis·trə'byü·shən }

binomial equation An equation having the form $x^n - a = 0$. { bī'nō·mē·əl i'kwā·zhən }

binomial expansion *See* binomial series. { bī'nō·mē·əl ik'span·shən }

binomial law The probability of an event occurring r times in n Bernoulli trials is equal to $\binom{n}{r} p^r (1 - p)^{n-r}$, where p is the probability of the event. { bī'nō·mē·əl ‚lȯ }

binomial probability paper Graph paper designed to aid in the analysis of data from a binomial population, that is, data in the form of proportions or as percentages; both axes are marked so that the graduations are square roots of the variable. { ‚bī‚nō·mē·əl ‚prä·bə'bil·əd·ē ‚pā·pər }

binomial random variable A random variable, parametrized by a positive integer n and a number p in the closed interval between 0 and 1, whose range is the set $\{0, 1, \ldots, n\}$ and whose value is the number of successes in n independent binomial trials when p is the probability of success in a single trial. { bī‚nō·mē·əl ‚ran·dəm 'ver·ē·ə·bəl }

binomial series The expansion of $(x + y)^n$ when n is neither a positive integer nor zero. Also known as binomial expansion. { bī'nō·mē·əl 'sir·ēz }

binomial surd A sum of two roots of rational numbers, at least one of which is an irrational number. { bī'nō·mē·əl 'sərd }

binomial theorem The rule for expanding $(x + y)^n$. { bī'nō·mē·əl 'thir·əm }

binomial trials A sequence of trials, each trial offein that a certain result may or may not happen. Also known as Bernoulli experiments; Bernoulli trials. { bī'nō·mē·əl 'trīlz }

binomial trials model A product model in which each factor has two simple events with probabilities p and $q = 1 - p$. { bī'nō·mē·əl 'trīlz ‚mäd·əl }

binormal A vector on a curve at a point so that, together with the positive tangent and principal normal, it forms a system of right-handed rectangular Cartesian axes. { bī'nȯr·məl }

binormal indicatrix For a space curve, all the end points of those radii of a unit sphere that are parallel to the positive directions of the binormals of the curve. Also known as spherical indicatrix of the binormal. { bī'nȯr·məl in'dik·ə‚triks }

biometrician A person skilled in biometry. Also known as biometricist. { bī‚äm·ə'trish·ən }

biometricist *See* biometrician. { ‚bī·ō'me·trə‚sist }

biometrics The use of statistics to analyze observations of biological phenomena. { ‚bī·ō'me·triks }

biometry The use of statistics to calculate the average length of time that a human being lives. { bī'äm·ə·trē }

biostatistics The use of statistics to obtain information from biological data. { ‚bī·ō·stə'tis·tiks }

bipartite cubic The points satisfying the equation $y^2 = x(x - a)(x - b)$. { bī'pär‚tīt 'kyü·bik }

bipartite graph A linear graph (network) in which the nodes can be partitioned into two groups G_1 and G_2 such that for every arc (i,j) node i is in G_1 and node j in G_2. { bī'pär‚tīt 'graf }

bipolar coordinate system 1. A two-dimensional coordinate system defined by the family of circles that pass through two common points, and the family of circles that cut the circles of the first family at right angles. **2.** A three-dimensional coordinate system in which two of the coordinates depend on the x and y coordinates in the same manner as in a two-dimensional bipolar coordinate system and are independent of the z coordinate, while the third coordinate is proportional to the z coordinate. { ¦bī,pō·lər kō'órd·ən·ət ,sis·təm }

biquadratic Any fourth-degree algebraic expression. Also known as quartic. { ¦bī·kwə'drad·ik }

biquadratic equation *See* quartic equation. { ¦bī·kwə'drad·ik i'kwā·zhən }

biquinary abacus An abacus in which the frame is divided into two parts by a bar which separates each wire into two- and five-counter segments. { bī'kwin·ə·rē 'ab·ə·kəs }

biquinary notation A mixed-base notation system in which the first of each pair of digits counts 0 or 1 unit of five, and the second counts 0, 1, 2, 3, or 4 units. Also known as biquinary number system. { bī'kwin·ə·rē nō'tā·shən }

biquinary number system *See* biquinary notation. { bī'kwin·ə·rē 'nəm·bər ,sis·təm }

birectangular Property of a geometrical object that has two right angles. { ¦bī·rek'taŋ·gyə·lər }

Birkhoff-von Neumann theorem The theorem that a matrix is doubly stochastic if and only if it is a convex combination of permutation matrices. { ¦bər,hóf fón 'nói·män ,thir·əm }

birth-death process A method for describing the size of a population in which the population increases or decreases by one unit or remains constant over short time periods. { ¦bərth ¦deth ,prä,səs }

birth process A stochastic process that defines a population whose members may have offspring; usually applied to the case where the population increases by one. { 'bərth ,prä,ses }

bisection algorithm A procedure for determining the root of a function to any desired accuracy by repeatedly dividing a test interval in half and then determining in which half the value of the function changes sign. { 'bī,sek·shən 'al·gə,rith·əm }

bisector The ray dividing an angle into two equal angles. { ,bī'sek·tər }

biserial correlation coefficient A measure of the relationship between two qualities, one of which is a measurable random variable and the other a variable which is dichotomous, classified according to the presence or absence of an attribute; not a product moment correlation coefficient. { ¦bī¦sir·ē·əl ,kär·ə'lā·shən ,kō·ə,fish·ənt }

bit In a pure binary numeration system, either of the digits 0 or 1. Also known as bigit; binary digit. { bit }

bitangent *See* double tangent. { bī'tan·jənt }

biunique correspondence A correspondence that is one to one in both directions. { ¦bī·yü,nēk ,kär·ə'spän·dəns }

bivariate distribution The joint distribution of a pair of variates for continuous or discontinuous data. { bī¦ver·ē·ət ,dis·trə'byü·shən }

Blaschke's theorem The theorem that a bounded closed convex plane set of width 1 contains a circle of radius 1/3. { 'bläsh·kəz ,thir·əm }

blind trial *See* double-blind technique. { 'blīnd 'trīl }

block In experimental design, a homogeneous aggregation of items under observation, such as a group of contiguous plots of land or all animals in a litter. { bläk }

blocking The grouping of sample data into subgroups with similar characteristics. { 'bläk·iŋ }

blurring An operation that decreases the value of the membership function of a fuzzy set if it is greater than 0.5, and increases it if it is less than 0.5. { 'blər·iŋ }

Bochner integral The Bochner integral of a function, f, with suitable properties, from a measurable set, A, to a Banach space, B, is the limit of the integrals over A of a sequence of simple functions, s_n, from A to B such that the limit of the integral over A of the norm of $f - s_n$ approaches zero. { ¦bäk·nər int·i·grəl }

body of revolution A symmetrical body having the form described by rotating a plane curve about an axis in its plane. { 'bäd·ē əv rev·ə'lü·shən }

Bolyai geometry *See* Lobachevski geometry. { 'ból·yī jē'äm·ə·trē }

Bolzano's theorem The theorem that a single-valued, real-valued, continuous function of a real variable is equal to zero at some point in an interval if its values at the end points of the interval have opposite sign. { ,bōl'tsän·ōz ,thir·əm }

Bolzano-Weierstrass property The property of a topological space, each of whose infinite subsets has at least one accumulation point. { bōl¦tsän·ō 'vī·ər,shträs ,präp·ərd·ē }

Bolzano-Weierstrass theorem The theorem that every bounded, infinite set in finite dimensional Euclidean space has a cluster point. { ,bōl'tsän·ō 'vī·ər,shträs ,thir·əm }

Boolean algebra An algebraic system with two binary operations and one unary operation important in representing a two-valued logic. { 'bü·lē·ən 'al·jə·brə }

Boolean calculus Boolean algebra modified to include the element of time. { 'bü·lē·ən 'kal·kyə·ləs }

Boolean determinant A function defined on Boolean matrices which depends on the elements of the matrix in a manner analogous to the manner in which an ordinary determinant depends on the elements of an ordinary matrix, with the operation of multiplication replaced by intersection and the operation of addition replaced by union. { ¦bül·ē·ən di'tər·mə·nənt }

Boolean function A function $f(x,y,\ldots,z)$ assembled by the application of the operations AND, OR, NOT on the variables x, $y,\ldots$, z and elements whose common domain is a Boolean algebra. { 'bü·lē·ən 'fəŋk·shən }

Boolean matrix A rectangular array of elements each of which is a member of a Boolean algebra. { ¦bül·ē·ən 'mā,triks }

Boolean operation table A table which indicates, for a particular operation on a Boolean algebra, the values that result for all possible combination of values of the operands; used particularly with Boolean algebras of two elements which may be interpreted as "true" and "false." { ¦bül·ē·ən ,äp·ə'rā·shən ,tā·bəl }

Boolean operator A logic operator that is one of the operators AND, OR, or NOT, or can be expressed as a combination of these three operators. { ¦bül·ē·ən 'äp·ə,rād·ər }

Boolean ring A commutative ring with the property that for every element a of the ring, $a \times a = a$ and $a + a = 0$; it can be shown to be equivalent to a Boolean algebra. { ¦bül·ē·ən 'riŋ }

bordering For a determinant, the procedure of adding a column and a row, which usually have unity as a common element and all other elements equal to zero. { 'bórd·ər·iŋ }

Borel measurable function 1. A real-valued function such that the inverse image of the set of real numbers greater than any given real number is a Borel set. **2.** More generally, a function to a topological space such that the inverse image of any open set is a Borel set. { bȯ·rel ¦mezh·rə·bəl 'fəŋk·shən }

Borel measure A measure defined on the class of all Borel sets of a topological space such that the measure of any compact set is finite. { bə'rel ,mezh·ər }

Borel set A member of the smallest σ-algebra containing the compact subsets of a topological space. { bȯ·rel ¦set }

Borel sigma algebra The smallest sigma algebra containing the compact subsets of a topological space. { bȯ·rel ¦sig·mə 'al·jə·brə }

borrow An arithmetically negative carry; it occurs in direct subtraction by raising the low-order digit of the minuend by one unit of the next-higher-order digit; for example, when subtracting 67 from 92, a tens digit is borrowed from the 9, to raise the 2 to a factor of 12; the 7 of 67 is then subtracted from the 12 to yield 5 as the units digit of the difference; the 6 is then subtracted from 8, or $9 - 1$, yielding 2 as the tens digit of the difference. { 'bä·rō }

boundary See frontier. { 'baùn·drē }

boundary condition A requirement to be met by a solution to a set of differential equations on a specified set of values of the independent variables. { 'baùn·drē kən'dish·ən }

boundary of a set See frontier. { 'baùn·drē əv ə 'set }

boundary point In a topological space, a point of a set with the property that every neighborhood of the point contains points of both the set and its complement. { 'baún·drē ˌpȯint }

boundary value problem A problem, such as the Dirichlet or Neumann problem, which involves finding the solution of a differential equation or system of differential equations which meets certain specified requirements, usually connected with physical conditions, for certain values of the independent variable. { 'baún·drē ˌval·yü ˌpräb·ləm }

bounded difference For two fuzzy sets A and B, with membership functions m_A and m_B, the fuzzy set whose membership function $m_{A \ominus B}$ has the value $m_A(x) - m_B(x)$ for every element x for which $m_A(x) \geq m_B(x)$, and has the value 0 for every element x for which $m_A(x) \leq m_B(x)$. { 'baúnd·əd 'dif·rəns }

bounded function **1.** A function whose image is a bounded set. **2.** A function of a metric space to itself which moves each point no more than some constant distance. { ˌbaún·dəd 'fəŋk·shən }

bounded growth The property of a function f defined on the positive real numbers which requires that there exist numbers M and a such that the absolute value of $f(t)$ is less than Ma^t for all positive values of t. { ˌbaún·dəd 'grōth }

bounded linear transformation A linear transformation T for which there is some positive number A such that the norm of $T(x)$ is equal to or less than A times the norm of x for each x. { ˌbaún·dəd ˌlin·ē·ər tranz·fər'mā·shən }

bounded product For two fuzzy sets A and B, with membership functions m_A and m_B, the fuzzy set whose membership function $m_{A \odot B}$ has the value $m_A(x) + m_B(x) - 1$ for every element x for which $m_A(x) + m_B(x) \geq 1$, and has the value 0 for every element x for which $m_A(x) + m_B(x) \leq 1$. { 'baúnd·əd 'präd·əkt }

bounded sequence A sequence whose members form a bounded set. { 'baúnd·əd 'sē·kwəns }

bounded set **1.** A collection of numbers whose absolute values are all smaller than some constant. **2.** A set of points, the distance between any two of which is smaller than some constant. { ˌbaún·dəd 'set }

bounded sum For two fuzzy sets A and B, with membership functions m_A and m_B, the fuzzy set whose membership function $m_{A \oplus B}$ has the value $m_A(x) + m_B(x)$ for every element x for which $m_A(x) + m_B(x) \leq 1$, and has the value 1 for every element x for which $m_A(x) + m_B(x) \geq 1$. { ˌbaún·dəd 'səm }

bounded variation A real-valued function is of bounded variation on an interval if its total variation there is bounded. { ˌbaún·dəd ver·ē'ā·shən }

bound variable In logic, a variable that occurs within the scope of a quantifier, and cannot be replaced by a constant. { ˌbaúnd 'ver·ē·ə·bəl }

boxcar function A function whose value is zero except for a finite interval of its argument, for which it has a constant nonzero value. { 'bäks‚kär ˌfəŋk·shən }

braid A braid of order n consists of two parallel lines, sets of n points on each of the lines with a one-to-one correspondence between them, and n nonintersecting space curves, each of which connects one of the n points on one of the parallel lines with the corresponding point on the other; the space curves are configured so that no curve turns back on itself, in the sense that its projection on the plane of the parallel lines lies between the parallel lines and intersects any line parallel to them no more than once, and any two such projections intersect at most a finite number of times. { brād }

branch **1.** A complex function which is analytic in some domain and which takes on one of the values of a multiple-valued function in that domain. **2.** A section of a curve that is separated from other sections of the curve by discontinuities, singular points, or other special points such as maxima and minima. { branch }

branch cut A line or curve of singular points used in defining a branch of a multiple-valued complex function. { 'branch ‚kət }

branching diagram In bifurcation theory, a graph in which a parameter characterizing solutions of a nonlinear equation is plotted against a parameter that appears in the equation itself. { 'branch·iŋ ˌdī·ə‚gram }

branching process A stochastic process in which the members of a population may have offspring and the lines of descent branch out as the new members are born. { 'branch·iŋ 'präs·əs }

branching theory See bifurcation theory. { 'branch·iŋ ˌthē·ə·rē }

branch point 1. A point at which two or more sheets of a Riemann surface join together. **2.** In bifurcation theory, a value of a parameter in a nonlinear equation at which solutions branch off from the basic solution. { 'branch ˌpȯint }

breakdown law The law that if the event E is broken down into the exclusive events $E_1, E_2, \ldots$ so that E is the event E_1 or E_2 or $\ldots$, then if F is any event, the probability of F is the sum of the products of the probabilities of E_i and the conditional probability of F given E_i. { 'brāk‚dau̇n ‚lȯ }

Brianchon's theorem The theorem that if a hexagon circumscribes a conic section, the three lines joining three pairs of opposite vertices are concurrent (or are parallel). { ¦brē·än¦känz ‚thir·əm }

bridge A line whose removal disconnects a component of a graph. Also known as isthmus. { brij }

bridging The operation of carrying in addition or multiplication. { 'brij·iŋ }

Briggsian logarithm See common logarithm. { ¦brigz·ē·ən 'läg·ə‚rith·əm }

Briggs' logarithm See common logarithm. { ¦brigz 'log·ə‚rith·əm }

broken line A line which is composed of a series of line segments lying end to end, and which does not form a continuous line. { ¦brō·kən ‚līn }

Bromwich contour A path of integration in the complex plane running from $c - i\infty$ to $c + i\infty$, where c is a real, positive number chosen so that the path lies to the right of all singularities of the analytic function under consideration. { 'bräm ‚wich ‚kän‚tu̇r }

Brouwer's theorem A fixed-point theorem stating that for any continuous mapping f of the solid n-sphere into itself there is a point x such that $f(x) = x$. { 'brau̇·ərz ‚thir·əm }

Budan's theorem The theorem that the number of roots of an nth-degree polynomial lying in an open interval equals the difference in the number of sign changes induced by n differentiations at the two ends of the interval. { 'bü‚dänz ‚thir·əm }

Buffon's problem The problem of calculating the probability that a needle of specified length, dropped at random on a plane ruled with a series of straight lines a specified distance apart, will intersect one of the lines. { bü'fȯnz ‚präb·ləm }

bullet nose A plane curve whose equation in cartesian coordinates x and y is $(a^2/x^2) - (b^2/y^2) = 1$, where a and b are constants. { 'bu̇l·ət ‚nōz }

bunch-map analysis A graphic technique in confluence analysis; all subsets of regression coefficients in a complete set are drawn on standard diagrams, and the representation of any set of regression coefficients produces a "bunch" of lines; allows the observer to determine the effect of introducing a new variate on a set of variates. { ¦bənch ¦map ə'nal·ə·səs }

bundle A triple (E, p, B), where E and B are topological spaces and p is a continuous map of E onto B; intuitively E is the collection of inverse images under p of points from B glued together by the topology of X. { 'bən·dəl }

bundle of planes See sheaf of planes. { ¦bən·dəl əv 'plānz }

Buniakowski's inequality See Cauchy-Schwarz inequality. { ‚bu̇n·yə'kȯf·skēz ‚in·i'kwäl·əd·ē }

Burali-Forti paradox The order-type of the set of all ordinals is the largest ordinal, but that ordinal plus one is larger. { bu̇'räl·ē 'fȯr·tē 'par·ə‚däks }

Burnside-Frobenius theorem Pertaining to a group of permutations on a finite set, the theorem that the sum over all the permutations, g, of the number of fixed points of g is equal to the product of the number of distinct orbits with respect to the group and the number of permutations in the group. { ¦bərn‚sīd frō'bē·nē· əs ‚thir·əm }

(b,v,r,k,λ)-design See balanced incomplete block design. { ¦bē ¦vē ¦är ¦kä 'lam·də di‚zīn }

C

Caccioppoli-Banach principle *See* Banach's fixed-point theorem. { ˌkä·chē'äp·ə·lē 'bä͵näk ͵prin·sə·pəl }

Calabi conjecture If the volume of a certain type of surface, defined in a higher dimensional space in terms of complex numbers, is known, then a particular kind of metric can be defined on it; the conjecture was subsequently proved to be correct. { kə'lä·bē kən͵jek·chər }

calculus The branch of mathematics dealing with differentiation and integration and related topics. { 'kal·kyə·ləs }

calculus of enlargement *See* calculus of finite differences. { 'kal·kyə·ləs əv in'lärj·mənt }

calculus of finite differences A method of interpolation that makes use of formal relations between difference operators which are, in turn, defined in terms of the values of a function on a set of equally spaced points. Also known as calculus of enlargement. { 'kal·kyə·ləs əv 'fī͵nīt 'dif·rən·səs }

calculus of residues The application of the Cauchy residue theorem and related theorems to compute the residues of a meromorphic function at simple poles, evaluate contour integrals, expand meromorphic functions in series, and carry out related calculations. { 'kal·kyə·ləs əv 'rez·ə͵düz }

calculus of tensors The branch of mathematics dealing with the differentiation of tensors. { 'kal·kyə·ləs əv 'ten·sərs }

calculus of variations The study of problems concerning maximizing or minimizing a given definite integral relative to the dependent variables of the integrand function. { 'kal·kyə·ləs əv ͵ver·ē'ā·shənz }

calculus of vectors That branch of calculus concerned with differentiation and integration of vector-valued functions. { 'kal·kyə·ləs əv 'vek·tərz }

Camp-Meidell condition For determining the distribution of a set of numbers, the guideline stating that if the distribution has only one mode, if the mode is the same as the arithmetic mean, and if the frequencies decline continuously on both sides of the mode, then more than $1 - (1/2.25t^2)$ of any distribution will fall within the closed range $\bar{X} \pm t\sigma$, where t = number of items in a set, $\bar{X}$ = average, and σ = standard deviation. { ¦kamp ͵mī'del kən͵dish·ən }

canal surface The envelope of a family of spheres of equal radii whose centers are on a given space curve. { kə'nal ͵sər·fəs }

cancellation law A rule which allows formal division by common factors in equal products, even in systems which have no division, as integral domains; $ab = ac$ implies that $b = c$. { kan·sə'lā·shən ͵lȯ }

canonical coordinates Any set of generalized coordinates of a system together with their conjugate momenta. { kə'nän·ə·kəl kō'ȯrd·ən·əts }

canonical correlation The maximum correlation between linear functions of two sets of random variables when specific restrictions are imposed upon the coefficients of the linear functions of the two sets. { kə'nän·ə·kəl ͵kȯr·ə'lā·shən }

canonical matrix A member of an equivalence class of matrices that has a particularly simple form, where the equivalence classes are determined by one of the relations defining equivalent, similar, or congruent matrices. { kə'nän·ə·kəl 'mā͵triks }

canonical transformation

canonical transformation Any function which has a standard form, depending on the context. { kə'nän·ə·kəl ,tranz·fər'mā·shən }

Cantor diagonal process A technique of proving statements about infinite sequences, each of whose terms is an infinite sequence by operation on the nth term of the nth sequence for each n; used to prove the uncountability of the real numbers. { 'kän·tȯr dī'ag·ən·əl ,präs·əs }

Cantor function A real-valued nondecreasing continuous function defined on the closed interval [0,1] which maps the Cantor ternary set onto the interval [0,1]. { 'kän·tȯr ,fəŋk·shən }

Cantor's axiom The postulate that there exists a one-to-one correspondence between the points of a line extending indefinitely in both directions and the set of real numbers. { 'kan·tərz 'ak·sē·əm }

Cantor ternary set A perfect, uncountable, totally disconnected subset of the real numbers having Lebesgue measure zero; it consists of all numbers between 0 and 1 (inclusive) with ternary representations containing no ones. { 'kän·tȯr 'tər·nə·rē ,set }

Cantor theorem A theorem that there is no one-to-one correspondence between a set and the collection of its subsets. { 'kän·tȯr 'thir·əm }

cap The symbol ∩, which indicates the intersection of two sets. { kap }

Carathéodory outer measure A positive, countably subadditive set function defined on the class of all subsets of a given set; used for defining measures. { ,kär·ə¦tā·ə'dȯr·ē ¦au̇d·ər 'mezh·ər }

Carathéodory theorem The theorem that each point of the convex span of a set in an n-dimensional Euclidean space is a convex linear combination of points in that set. { ,,kär·ə,tā·ə'dȯr·ē ,thir·əm }

cardinal measurement *See* interval measurement. { 'kärd·nel 'mezh·ər·mənt }

cardinal number The number of members of a set; usually taken as a particular well-ordered set representative of the class of all sets which are in one-to-one correspondence with one another. { 'kärd·nəl 'nəm·bər }

cardioid A heart-shaped curve generated by a point of a circle that rolls without slipping on a fixed circle of the same diameter. { 'kärd·ē,ȯid }

carry An arithmetic operation that occurs in the course of addition when the sum of the digits in a given position equals or exceeds the base of the number system; a multiple m of the base is subtracted from this sum so that the remainder is less than the base, and the number m is then added to the next-higher-order digit. { 'kar·ē }

Cartesian axis One of a set of mutually perpendicular lines which all pass through a single point, used to define a Cartesian coordinate system; the value of one of the coordinates on the axis is equal to the directed distance from the intersection of axes, while the values of the other coordinates vanish. { kär'tē·zhən 'ak·səs }

Cartesian coordinates 1. The set of numbers which locate a point in space with respect to a collection of mutually perpendicular axes. **2.** *See* rectangular coordinates. { kär'tē·zhən kō'ȯrd·nəts }

Cartesian coordinate system A coordinate system in n dimensions where n is any integer made by using n number axes which intersect each other at right angles at an origin, enabling any point within that rectangular space to be identified by the distances from the n lines. Also known as rectangular Cartesian coordinate system. { kär'tē·zhən kō'ȯrd·nət ,sis·təm }

Cartesian geometry *See* analytic geometry. { kär'tē·zhan jē'äm·ə·trē }

Cartesian oval A plane curve consisting of all points P such that $aFP + bF''P = c$, where F and F'' are fixed points and a, b, and c are constants which are not necessarily positive. { kär'tē·zhən 'ō·vəl }

Cartesian plane A plane whose points are specified by Cartesian coordinates. { kär 'tēzh·ən 'plān }

Cartesian product In reference to the product of P and Q, the set $P \times Q$ of all pairs (p,q), where p belongs to P and q belongs to Q. { kär'tē·zhan 'präd·əkt }

Cartesian surface A surface obtained by rotating the curve $n_0(x^2 + y^2)^{1/2} \pm n_1[(x - a)^2 + y^2]^{1/2} = c$ about the x axis. { kär'tē·zhan 'sər·fəs }

Cartesian tensor The aggregate of the functions of position in a tensor field in an n-dimensional Cartesian coordinate system. { kär'tē·zhan 'ten·sər }

Cassinian oval *See* oval of Cassini. { kə'sin·ē·ən 'ō·vəl }

casting-out nines A method of checking the correctness of elementary arithmetical operations, based on the fact that an integer yields the same remainder as the sum of its decimal digits, when divided by 9. { ¦kast·iŋ ¸aùt 'nīnz }

Catalan conjecture The conjecture that the only pair of consecutive positive integers that are powers of smaller integers is the pair (8,9). { 'kä·tə¸län kən¸jek·chər }

Catalan numbers The numbers, c_n, which count the ways to insert parentheses in a string of n terms so that their product may be unambiguously carried out by multiplying two quantities at a time. { 'kat·əl·ən ¸nəm·bərz }

catastrophe theory A theory of mathematical structure in which smooth continuous inputs lead to discontinuous responses. { kə'tas·trə·fē ¸thē·ə·rē }

categorical data Data separable into categories that are mutually exclusive, for example, age groups. { ¸kad·ə¦gör·i·kəl 'dad·ə }

category A class of objects together with a set of morphisms for each pair of objects and a law of composition for morphisms; sets and functions form an important category, as do groups and homomorphisms. { 'kad·ə¸gòr·ē }

catenary The curve obtained by suspending a uniform chain by its two ends; the graph of the hyperbolic cosine function. Also known as alysoid; chainette. { 'kat·ə¸ner·ē }

catenoid The surface of revolution obtained by rotating a catenary about a horizontal axis. { 'kat·ən¸òid }

caterer problem A linear programming problem in which it is required to find the optimal policy for a caterer who must choose between buying new napkins and sending them to either a fast or a slow laundry service. { 'kād·ə·rər ¸präb·ləm }

Cauchy boundary conditions The conditions imposed on a surface in euclidean space which are to be satisfied by a solution to a partial differential equation. { kō·shē 'baùn·drē kən¸dish·ənz }

Cauchy condensation test A monotone decreasing series of positive terms Σa_n converges or diverges as does $\Sigma p^n a_p{}^n$ for any positive integer p. { kō·shē ¸kän·den'sā·shən ¸test }

Cauchy distribution A distribution function having the form $M/[\pi M^2 + (x - a)^2]$, where x is the variable and M and a are constants. Also known as Cauchy frequency distribution. { kō·shē dis·trə'byü·shən }

Cauchy formula An expression for the value of an analytic function f at a point z in terms of a line integral $f(z) = \dfrac{1}{2\pi i} \displaystyle\int_C \dfrac{f(\zeta)}{\zeta - z}\, d\zeta$ where C is a simple closed curve containing z. Also known as Cauchy integral formula. { kō·shē ¸fòr·myə·lə }

Cauchy frequency distribution *See* Cauchy distribution. { kō·shē 'frē·kwən·sē dis·trə'byü·shən }

Cauchy-Hadamard theorem The theorem that the radius of convergence of a Taylor series in the complex variable z is the reciprocal of the limit superior, as n approaches infinity, of the nth root of the absolute value of the coefficient of z^n. { kō·shē 'had·ə·mär ¸thir·əm }

Cauchy inequality The square of the sum of the products of two variables for a range of values is less than or equal to the product of the sums of the squares of these two variables for the same range of values. { kō·shē ¸in·i'kwäl·əd·ē }

Cauchy integral formula *See* Cauchy formula. { kō·shē ¦in·tə·grəl ¦fòr·mya·lə }

Cauchy integral test *See* Cauchy's test for convergence. { kō·shē 'in·tə·grəl ¸test }

Cauchy integral theorem The theorem that if γ is a closed path in a region R satisfying certain topological properties, then the integral around γ of any function analytic in R is zero. { kō·shē 'in·tə·grəl ¸thir·əm }

Cauchy mean The Cauchy mean-value theorem for the ratio of two continuous functions. { kō·shē ¸mēn }

Cauchy mean-value theorem The theorem that if f and g are functions satisfying certain conditions on an interval $[a,b]$, then there is a point x in the interval at which the ratio of derivatives $f'(x)/g'(x)$ equals the ratio of the net change in f, $f(b) - f(a)$, to that of g. { kō·shē ¦mēn ¦val·yü ˌthir·əm }

Cauchy net A net whose members are elements of a topological vector space and which satisfies the condition that for any neighborhood of the origin of the space there is an element a of the directed system that indexes the net such that if b and c are also members of this directed system and $b \geq a$ and $c \geq a$, then $x_b - x_c$ is in this nieghborhood. { 'kō·shē ˌnet }

Cauchy principal value Also known as principal value. **1.** The Cauchy principal value of $\int_{-\infty}^{\infty} f(x)dx$ is $\lim\limits_{s \to \infty} \int_{-s}^{s} f(x)dx$ provided the limit exists. **2.** If a function f is bounded on an interval (a,b) except in the neighborhood of a point c, the Cauchy principal value of $\int_{a}^{b} f(x)dx$ is $\lim\limits_{\Delta \to 0} \left[\int_{a}^{c-\Delta} f(x)dx + \int_{c+\Delta}^{b} f(x)dx \right]$ provided the limit exists. { kō·shē ¦prin·sə·pəl ¦val·yü }

Cauchy problem The problem of determining the solution of a system of partial differential equation of order m from the prescribed values of the solution and of its derivatives of order less than m on a given surface. { kō·shē ˌpräb·ləm }

Cauchy product A method of multiplying two absolutely convergent series to obtain a series which converges absolutely to the product of the limits of the original series: $\left(\sum\limits_{n=0}^{\infty} a_n \right) \left(\sum\limits_{n=0}^{\infty} b_n \right) = \sum\limits_{n=0}^{\infty} c_n$ where $c_n = \sum\limits_{k=0}^{n} a_k b_{n-k}$ { kō·shē ˌpräd·əkt }

Cauchy radical test A test for convergence of series of positive terms: if the nth root of the nth term is less than some number less than unity, the series converges; if it remains equal to or greater than unity, the series diverges. { kō·shē 'rad·i·kəl ˌtest }

Cauchy random variable A random variable that has a Cauchy distribution. { kō·shē ˌran·dəm 'ver·ē·ə·bəl }

Cauchy ratio test A series of nonnegative terms converges if the limit, as n approaches infinity, of the ratio of the $(n + 1)$st to nth term is smaller than 1, and diverges if it is greater than 1; the test fails if this limit is 1. Also known as ratio test. { kō·shē 'rā·shō ˌtest }

Cauchy residue theorem The theorem expressing a line integral around a closed curve of a function which is analytic in a simply connected domain containing the curve, except at a finite number of poles interior to the curve, as a sum of residues of the function at these poles. { kō·shē 'rez·ə,dü ˌthir·əm }

Cauchy-Riemann equations A pair of partial differential equations that is satisfied by the real and imaginary parts of a complex function $f(z)$ if and only if the function is analytic: $\partial u/\partial x = \partial v/\partial y$ and $\partial u/\partial y = - \partial v/\partial x$, where $f(z) = u + iv$ and $z = x + iy$. { kō·shē 'rē,män i'kwā·zhənz }

Cauchy-Schwarz inequality The square of the inner product of two vectors does not exceed the product of the squares of their norms. Also known as Buniakowski's inequality; Schwarz' inequality. { kō·shē 'shwȯrts in·i'kwäl·əd·ē }

Cauchy sequence A sequence with the property that the difference between any two terms is arbitrarily small provided they are both sufficiently far out in the sequence; more precisely stated: a sequence $\{a_n\}$ such that for every $\epsilon > 0$ there is an integer N with the property that, if n and m are both greater than N, then $|a_n - a_m| < \epsilon$. Also known as fundamental sequence; regular sequence. { kō·shē 'sē·kwəns }

Cauchy's mean-value theorem See second mean-value theorem. { kō·shēz ˌmēn ˌval·yü 'thir·əm }

Cauchy's test for convergence **1.** A series is absolutely convergent if the limit as n approaches infinity of its nth term raised to the $1/n$ power is less than unity. **2.** A series a_n is convergent if there exists a monotonically decreasing function f such that $f(n) = a_n$ for n greater than some fixed number N, and if the integral of $f(x)dx$ from N to ∞ converges. Also known as Cauchy integral test; Maclaurin-Cauchy test. { kō·shēz ˌtest fər kən'vər·jəns }

Cauchy transcendental equation An equation whose roots are characteristic values of a certain type of Sturm-Liouville problem: $\tan \sigma\pi = (k + K)/(\sigma^2 - kK)$, where k and K are given, and σ is to be determined. { kō·shē ˌtrans,en'dent·əl i'kwä·zhən }

Cavalieri's theorem The theorem that two solids have the same volume if their altitudes are equal and all plane sections parallel to their bases and at equal distances from their bases are equal. { ˌkav·ə'lyer·ēz ˌthir·əm }

Cayley algebra The nonassociative division algebra consisting of pairs of quaternions; it may be identified with an eight-dimensional vector space over the real numbers. { 'kā·lē ˌal·jə·brə }

Cayley-Hamilton theorem The theorem that a linear transformation or matrix is a root of its own characteristic polynomial. Also known as Hamilton-Cayley theorem. { ˌkāl·ē ˌham·əl·tən ˌthir·əm }

Cayley-Klein parameters A set of four complex numbers used to describe the orientation of a rigid body in space, or equivalently, the rotation which produces that orientation, starting from some reference orientation. { ˌkāl·ē ˌklīn pə,ram·əd·ərz }

Cayley numbers The members of a Cayley algebra. Also known as octonions. { 'kāl·ē ˌnəm·bərz }

Cayley's sextic A plane curve with the equation $r = 4a \cos^3 (\theta/3)$, where r and θ are radial and angular polar coordinates and a is a constant. { 'kā·lēz 'sek·stik }

Cayley's theorem A theorem that any group G is isomorphic to a subgroup of the group of permutations on G. { 'kā,lēz ˌthir·əm }

ceiling The smallest integer that is equal to or greater than a given real number a; symbolized $\lceil a \rceil$. { 'sē·liŋ }

cell 1. The homeomorphic image of the unit ball. 2. One of the $(n - 1)$-dimensional polytopes that enclose a given n-dimensional polytope. { sel }

cell complex A topological space which is the last term of a finite sequence of spaces, each obtained from the previous by sewing on a cell along its boundary. { 'sel ˌkäm,pleks }

cell frequency The number of observations of specified conditional constraints on one or more variables; used mainly in the analysis of data obtained by performing actual counts. { 'sel ˌfrē·kwən·se }

cellular automaton A mathematical construction consisting of a system of entities, called cells, whose temporal evolution is governed by a collection of rules, so that its behavior over time may appear highly complex or chaotic. { 'sel·yə·lər ȯ'täm·ə·tən }

censored data Observations collected by determining in advance whether to record only a specified number of the smallest or largest values, or of the remaining values in a sample of a particular size. { ˌsen·sərd 'dad·ə }

census A complete counting of a population, as opposed to a partial counting or sampling. { 'sen·səs }

center 1. The point that is equidistant from all the points on a circle or sphere. 2. The point (if it exists) about which a curve (such as a circle, ellipse, or hyperbola) is symmetrical. 3. The point (if it exists) about which a surface (such as a sphere, ellipsoid, or hyperboloid) is symmetrical. 4. For a regular polygon, the center of its circumscribed circle. 5. The subgroup consisting of all elements that commute with all other elements in a given group. 6. The subring consisting of all elements a such that $ax = xa$ for all x in a given ring. 7. For a distribution, the expected value of any random variable which has the distribution. { 'sen·tər }

center of area For a plane figure, the center of mass of a thin uniform plate having the same boundaries as the plane figure. Also known as center of figure; centroid. { 'sen·tər əv 'er·ē·ə }

center of curvature At a given point on a curve, the center of the osculating circle of the curve at that point. { 'sen·tər əv 'kər·və·chər }

center of figure See center of area; center of volume. { 'sen·tər əv 'fig·yər }

center of geodesic curvature For a curve on a surface at a given point, the center of curvature of the orthogonal projection of the curve onto a plane tangent to the surface at the point. { ¦sen·tər əv ˌjē·ə'des·ik ˌkərv·ə·chər }

center of inversion The point O with respect to which an inversion is defined, so that every point P is mapped by the inversion into a point Q that is collinear with O and P. { 'sen·tər əv in'vər·zhən }

center of normal curvature For a given point on a surface and for a given direction, the normal section of the surface through the given point and in the given direction. { 'sen·tər əv 'nȯrm·əl 'kər·və·chər }

center of perspective The point specified by Desargues' theorem, at which lines passing through corresponding vertices of two triangles are concurrent. { 'sen·tər əv pər'spek·tiv }

center of principal curvature For a given point on a surface, the center of normal curvature at the point in one of the two principal directions. { 'sen·tər əv 'prin·sə·pəl ˌkər·və·chər }

center of projection The fixed point in a central projection. { ¦sen·tər əv prə'jek·shən }

center of similitude 1. A point of intersection of lines that join the ends of parallel radii of coplanar circles. 2. *See* homothetic center. { 'sen·tər əv si'mil·ə,tüd }

center of spherical curvature The center of the osculating sphere at a specified point on a space curve. { ¦sen·tər əv ¦sfer·ə·kəl 'kər·və·chər }

center of volume For a three-dimensional figure, the center of mass of a homogeneous solid having the same boundaries as the figures.   Also known as center of figure; centroid. { 'sen·tər əv 'väl·yəm }

centrad A unit of plane angle equal to 0.01 radian or to about 0.573 degree. { 'sent,rad }

central angle In a circle, an angle whose sides are radii of the circle. { 'sen·trəl 'aŋ·gəl }

central conic A conic that has a center, namely, a circle, ellipse, or hyperbola. { 'sen·trəl 'kän·ik }

central difference One of a series of quantities obtained from a function whose values are known at a series of equally spaced points by repeatedly applying the central difference operator to these values; used in interpolation or numerical calculation and integration of functions. { 'sen·trəl 'dif·rəns }

central difference operator A difference operator, denoted ∂, defined by the equation $\partial f(x) = f(x + h/2) - f(x - h/2)$, where h is a constant denoting the difference between successive points of interpolation or calculation. { ¦sen·trəl ¦dif·rəns 'äp·ə,räd·ər }

centralizer The subgroup consisting of all elements which commute with a given element of a group. { 'sen·tra,līz·ər }

central-limit theorem The theorem that the distribution of sample means taken from a large population approaches a normal (Gaussian) curve. { ¦sen·trəl ¦lim·ət ˌthir·əm }

central mean operator A difference operator, denoted μ, defined by the equation $\mu f(x) = [f(x + h/2) + f(x - h/2)]/2$, where h is a constant denoting the difference between successive points of interpolation or calculation.   Also known as averaging operator. { ¦sen·trəl ¦mēn 'äp·ə,räd·ər }

central plane For a fixed ruling of a ruled surface, the plane tangent to the surface at the central point of the ruling. { 'sen·trəl 'plān }

central point For a fixed ruling L on a ruled surface, the limiting position, as a variable ruling L' approaches L, of the foot on L of the common perpendicular to L and L'. { 'sen·trəl 'pȯint }

central projection A mapping of a configuration into a plane that associates with any point of the configuration the intersection with the plane of the line passing through the point and a fixed point. { 'sen·trəl prə'jek·shən }

central quadric A quadric surface that has a center, namely, a sphere, ellipsoid, or hyperboloid. { 'sen·trəl 'kwä·drik }

centroid *See* center of area; center of volume. { 'sen,trȯid }

centroids of areas and lines Points positioned identically with the centers of gravity of corresponding thin homogeneous plates or thin homogeneous wires; involved in the analysis of certain problems of mechanics such as the phenomenon of bending. { 'sen,trȯidz əv ˌer·ē·əz ən 'līnz }

Cesáro equation An equation which relates the arc length along a plane curve and the radius of curvature. { chā'zä·rō i,kwā·zhən }

Cesáro summation A method of attaching sums to certain divergent sequences and series by taking averages of the first n terms and passing to the limit. { chā'zä·rō sə'mā·shən }

Ceva's theorem The theorem that if three concurrent straight lines pass through the vertices A, B, and C of a triangle and intersect the opposite sides, produced if necessary, at D, E, and F, then the product $AF \cdot BD \cdot CE$ of the lengths of three alternate segments equals the product $FB \cdot DC \cdot EA$ of the other three. { 'chā·vəz ˌthir·əm }

cevian A straight line that passes through a vertex of a triangle or tetrahedron and intersects the opposite side or face. { 'chāv·ē·ən }

chain See linearly ordered set. { chān }

chain complex A sequence $\{C_n\}$, $-\infty < n < \infty$, of Abelian groups together with a sequence of boundary homomorphisms d_n: $C_n \to C_{n-1}$ such that $d_{n-1} \circ d_n = 0$ for each n. { 'chān ˌkäm,pleks }

chainette See catenary. { chā'net }

chain homomorphism A sequence of homomorphisms f_n: $C_n \to D_n$ between the groups of two chain complexes such that $f_{n-1} d_n = \bar{d}_n f_n$ where d_n and $\bar{d}_n$ are the boundary homomorphisms of $\{C_n\}$ and $\{D_n\}$ respectively. { 'chān ˌhō·mō'mȯr,fiz·əm }

chain index An index number derived by relating the value at any given period to the value in the previous period rather than to a fixed base. { 'chān ˌin,deks }

chain of simplices A member of the free Abelian group generated by the simplices of a given dimension of a simplicial complex. { 'chān əv 'sim·plə,sēz }

chain rule A rule for differentiating a composition of functions: $(d/dx)f(g(x)) = f'(g(x)) \cdot g'(x)$. { 'chān ˌrül }

chance variable See random variable. { ˌchans 'ver·ē·ə·bəl }

character group The set of all continuous homomorphisms of a topological group onto the group of all complex numbers with unit norm. { 'kar·ik·tər ˌgrüp }

characteristic 1. That part of the logarithm of a number which is the integral (the whole number) to the left of the decimal point in the logarithm.   2. For a family of surfaces that depend continuously on a parameter, the limiting curve of intersection of two members of the family as the two values of the parameter determining them approach a common value.   3. For a ring or field, the smallest possible integer whose product with any element of the ring or field equals zero, provided that such an integer exists; otherwise the characteristic is zero. { ˌkar·ik·tə'ris·tik }

characteristic cone A conelike region important in the study of initial value problems in partial differential equations. { ˌkar·ik·tə'ris·tik 'kōn }

characteristic curve 1. One of a pair of conjugate curves in a surface with the property that the directions of the tangents through any point of the curve are the characteristic directions of the surface.   2. A curve plotted on graph paper to show the relation between two changing values.   3. A characteristic curve of a one-parameter family of surfaces is the limit of the curve of intersection of two neighboring surfaces of the family as those surfaces approach coincidence. { ˌkar·ik·tə'ris·tik 'kərv }

characteristic directions For a point P on a surface S, the pair of conjugate directions which are symmetric with respect to the directions of the lines of curvature on S through P. { ˌkar·ik·tə'ris·tik də'rek·shənz }

characteristic equation 1. Any equation which has a solution, subject to specified boundary conditions, only when a parameter occurring in it has certain values.   2. Specifically, the equation $A\mathbf{u} = \lambda\mathbf{u}$, which can have a solution only when the parameter λ has certain values, where A can be a square matrix which multiplies the vector $\mathbf{u}$, or a linear differential or integral operator which operates on the function $\mathbf{u}$, or in general, any linear operator operating on the vector $\mathbf{u}$ in a finite or infinite dimensional vector space. Also known as eigenvalue equation.

3. An equation which sets the characteristic polynomial of a given linear transformation on a finite dimensional vector space, or of its matrix representation, equal to zero. { ‚kar·ik·tə'ris·tik i'kwä·zhən }

characteristic form A means of classifying partial differential equations. { ‚kar·ik· tə'ris·tik 'fȯrm }

characteristic function **1.** The function χ_A defined for any subset A of a set by setting $\chi_A(x) = 1$ if x is in A and $\chi_A = 0$ if x is not in A. Also known as indicator function. **2.** A function that uniquely defines a probability distribution; it is equal to $\sqrt{2\pi}$ times the Fourier transform of the frequency function of the distribution. **3.** *See* eigenfunction. { ‚kar·ik·tə'ris·tik 'fəŋk·shən }

characteristic manifold **1.** A surface used to study the problem of existence of solutions to partial differential equations. **2.** The linear set of eigenvectors corresponding to a given eigenvalue of a linear transformation. { ‚kar·ik·tə'ris·tik 'man·ə‚fōld }

characteristic number *See* eigenvalue. { ‚kar·ik·tə'ris·tik 'nəm·bər }

characteristic point The characteristic point of a one-parameter family of surfaces corresponding to the value u_0 of the parameter is the limit of the point of intersection of the surfaces corresponding to the values u_0, u_1, and u_2 of the parameter as u_1 and u_2 approach u_0 independently. { ‚kar·ik·tə'ris·tik 'pȯint }

characteristic polynomial The polynomial whose roots are the eigenvalues of a given linear transformation on a finite dimensional vector space. { ‚kar·ik·tə'ris·tik ‚päl· ə'nō·mē·əl }

characteristic ray For a differential equation, an integral curve which generates all the others. { ‚kar·ik·tə'ris·tik 'rā }

characteristic root *See* eigenvalue. { ‚kar·ik·tə'ris·tik 'rüt }

characteristic value *See* eigenvalue. { ‚kar·ik·tə'ris·tik 'val·yü }

characteristic vector *See* eigenvector. { ‚kar·ik·tə'ris·tik 'vek·tər }

Charlier polynomials Families of polynomials which are orthogonal with respect to Poisson distributions. { shär¦lyā ‚päl·ə'nō·mē·əlz }

Charpit's method A method for finding a complete integral of the general first-order partial differential equation in two independent variables; it involves solving a set of five ordinary differential equations. { 'chär‚pits ‚meth·əd }

chart An n-chart is a pair (U,h), where U is an open set of a topological space and h is a homeomorphism of U onto an open subset of n-dimensional Euclidean space. { chärt }

Chebyshev approximation *See* min-max technique.

Chebyshev polynomials A family of orthogonal polynomials which solve Chebyshev's differential equation. { 'cheb·ə·shəf ‚päl·i'nō·mē·əlz }

Chebyshev's differential equation A special case of Gauss' hypergeometric second-order differential equation: $(1 - x^2)f''(x) - xf'(x) + n^2 f(x) = 0$. { 'cheb·ə·shəfs dif·ə'ren·chəl i'kwä·zhən }

Chebyshev's inequality Given a nonnegative random variable $f(x)$, and $k > 0$, the probability that $f(x) \geq k$ is less than or equal to the expected value of f divided by k. { 'cheb·ə·shəfs ‚in·i'kwäl·əd·ē }

Chinese remainder theorem The theorem that if the integers m_1, m_2, ..., m_n are relatively prime in pairs and if b_1, b_2, ..., b_n are integers, then there exists an integer that is congruent to b_i modulo m_i for $i=1,2, \ldots, n$. { ¦chī‚nēz ri'mān·der ‚thir·əm }

chirplet A wavelet whose instantaneous frequency drifts upward or downward at a fixed rate throughout its duration. { 'chərp·lət }

chi-square distribution The distribution of the sum of the squares of a set of variables, each of which has a normal distribution and is expressed in standardized units. { 'kī ¦skwer dis·trə'byü·shən }

chi-square statistic A statistic which is distributed approximately in the form of a chi-square distribution; used in goodness-of-fit. { 'kī ¦skwär stə‚tis·tik }

chi-square test A generalization, and an extension, of a test for significant differences between a binomial population and a multinomial population, wherein each observation may fall into one of several classes and furnishes a comparison among several samples instead of just two.   { 'kī ¦skwer 'test }

Choquet theorem Let K be a compact convex set in a locally convex Hausdorff real vector space and assume that either (1) the set of extreme points of K is closed or (2) K is metrizable; then for every point x in K there is at least one Radon probability measure m on X, concentrated on the set of extreme points of K, such that x is the centroid of m.   { shō'kā ‚thir·əm }

chord A line segment which intersects a curve or surface only at the endpoints of the segment.   { kȯrd }

Christoffel symbols Symbols that represent particular functions of the coefficients and their first-order derivatives of a quadratic form.   Also known as three-index symbols.   { 'kris·tȯf·əl ‚sim·bəlz }

chromatic number For a specified surface, the smallest number n such that for any decomposition of the surface into regions the regions can be colored with n colors in such a way that no two adjacent regions have the same color.   { krō'mad·ik 'nəm·bər }

Church-Rosser theorem If for a lambda expression there is a terminating reduction sequence yielding a reduced form B, then the leftmost reduction sequence will yield a reduced form that is equivalent to B up to renaming.   { ¦chərch ¦rȯs·ər ¦thir·əm }

Church's thesis The claim that a function is computable in the intuitive sense if and only if it is computable by a Turing machine.   Also known as Turing's thesis.   { ¦chərch·əz ¦thē·səs }

circle 1. The set of all points in the plane at a given distance from a fixed point.   2. A unit of angular measure, equal to one complete revolution, that is, to 2π radians or 360°.   Also known as turn.   { 'sər·kəl }

circle graph See pie chart.   { 'sər·kəl ‚graf }

circle of convergence The region in which a power series possesses a limit.   { 'sər·kəl əv kən'vər·jəns }

circle of curvature The circle tangent to a curve on the concave side and having the same curvature at the point of tangency as does the curve.   { 'sər·kəl əv 'kər·və·chər }

circle of inversion A circle with respect to which two specified curves are inverse curves.   { 'sər·kəl əv in'vər·zhən }

circuit See cycle.   { 'sər·kət }

circulant determinant A determinant in which the elements of each row are the same as those of the previous row moved one place to the right, with the last element put first.   { 'sər·kyə·lənt də'tər·mə·nənt }

circulant matrix A matrix in which the elements of each row are those of the previous row moved one place to the right.   { 'sər·kyə·lənt 'mā‚triks }

circular arc See arc.   { 'sər·kyə·lər 'ärk }

circular argument An argument that is not valid because it uses the theorem to be proved or a consequence of that theorem that is not proven.   { ¦sər·kyə·lər 'är·gyə·mənt }

circular cone A cone whose base is a circle.   { 'sər·kyə·lər 'kōn }

circular conical surface The lateral surface of a right circular cone.   { ¦sər·kyə·lər ¦kän·ə·kəl 'sər·fəs }

circular cylinder A solid bounded by two parallel planes and a cylindrical surface whose intersections with planes perpendicular to the straight lines forming the surface are circles.   { 'sər·kyə·lər 'sil·ən·dər }

circular functions See trigonometric functions.   { 'sər·kyə·lər 'fəŋk·shənz }

circular helix A curve that lies on a right circular cylinder and intersects all the elements of the cylinder at the same angle.   { 'sər·kyə·lər 'hē‚liks }

circular nomograph A chart with concentric circular scales for three variables, laid out so that any straight line passes through values of the variables satisfying a given equation.   { 'sər·kyə·lər 'nō·mə‚graf }

circular permutation An arrangement of objects around a circle. { ¦sər·kyə·lər ˌpər·myə'tā·shən }

circular point A point on a surface at which the normal curvature is the same in all directions. { 'sər·kyə·lər 'pȯint }

circular point at infinity In projective geometry, one of two points at which every circle intersects the ideal line. { ¦sər·kyə·lər ¦pȯint at in'fin·əd·ē }

circular segment Portion of circle cut off from the main body of the circle by a straight line (chord) through the circle. { 'sər·kyə·lər 'seg·mənt }

circular slide rule A slide rule in a circular form whose advantages over a straight slide rule are its precision, because it is equivalent to a straight slide rule many times longer than the circular slide rule's diameter, and ease of multiplication, because the scale is continuous. { 'sər·kyə·lər 'slīd ˌrül }

circular word A sequence of elements arranged clockwise around a circle. { ¦sər·kyə·lər 'wərd }

circulation For the circulation of a vector field around a closed path, the line integral of the field vector around the path. { ˌsər·kyə·'lā·shən }

circumcenter For a triangle or a regular polygon, the center of the circle that is circumscribed about the triangle or polygon. { ¦sər·kəm¦sen·tər }

circumcircle A circle that passes through all the vertices of a given polygon, if such a circle exists. { 'sər·kəm,sər·kəl }

circumference 1. The length of a circle. **2.** For a sphere, the length of any great circle on the sphere. { sər'kəm·fə·rəns }

circumradius The radius of a circle that is circumscribed about a polygon. { ¦sər·kəm'rād·ē·əs }

circumscribed 1. A closed curve (or surface) is circumscribed about a polygon (or polyhedron) if every vertex of the polygon (or polyhedron) is incident upon the curve (or surface) and the polygon (or polyhedron) is contained in the curve (or surface). **2.** A polygon (or polyhedron) is circumscribed about a closed curve (or surface) if every side of the polygon (or face of the polyhedron) is tangent to the curve (or surface) and the curve (or surface) is contained within the polygon (or polyhedron). { 'sər·kəm,skrībd }

cissoid A plane curve consisting of all points which lie on a variable line passing through a fixed point, and whose distance from the fixed point is equal to the distance between the intersections of the line with two given curves. { 'sis,ȯid }

cissoid of Diocles The cissoid of a circle and a tangent line with respect to a fixed point on the circumference of the circle diametrically opposite the point of tangency. { 'si,sȯid əv 'dī·ə·klēz }

class 1. A set that consists of all the sets having a specified property. **2.** The class of a plane curve is the largest number of tangents that can be drawn to the curve from any point in the plane that is not on the curve. A collection of adjacent values of a random variable. { klas }

class C^n The class of all functions that are continuous on a given domain and have continuous derivatives of all orders up to and including the nth. { ˌklas 'sē 'en }

class formula A formula which states that the order of a finite group G is equal to the sum, over a set of representatives x_i of the distinct conjugacy classes of G, of the index of the normalizer of x_i in G. { 'klas ˌfȯr·myə·lə }

class frequency The frequency with which a random variable assumes the values included in a given class interval. { 'klas ¦frē·kwən·sē }

classical canonical matrix A form to which any matrix can be reduced by a collineatory transformation, with zeros except for a sequence of Jordan matrices siutated along the principal diagonal. { 'klas·ə·kəl kə'nän·ə·kəl 'mā·triks }

class interval One of several convenient intervals into which the values of the variate of a frequency distribution may be grouped. { ¦klas 'int·ər·vəl }

class limits The lower and upper limits of a class interval. { 'klas ¦lim·its }

class mark The mid-value of a class interval, or the integral value nearest the midpoint of the interval. { 'klas ˌmärk }

clique In a graph, a complete subgraph of that graph. { klēk }

34

cloithoid *See* Cornu's spiral. { 'klȯi‚thȯid }

closed ball In a metric space, a closed set about a point x which consists of all points that are equal to or less than a fixed distance from x. { ¦klōzd 'bȯl }

closed braid A modification of a braid in which plane curves are added that connect each of the n points on one of the parallel lines specified in the definition of the braid to one of the n points on the other in such a way that no two of these curves intersect or terminate at the same point, and the parallel lines themselves are deleted. { ¦klōzd 'brād }

closed circular region The union of the interior of a circle with the circle itself. { ¦klōzd ¦sər·kyə·lər 'rē·jən }

closed covering A closed covering of a set S in a topological space is a collection of closed sets whose union contains S. { ¦klōzd 'kəv·ər·iŋ }

closed curve A curve that has no end points. { ¦klōzd 'kərv }

closed dipath A directed path whose initial and final vertices are the same. { ¦klōzd 'dī‚path }

closed disk A circle and its interior. Also known as disk. { ¦klōzd 'disk }

closed graph theorem If T is a linear transformation on Banach space X to Banach space Y whose domain $D(T)$ is closed and whose graph, that is, the set of pairs (x,Tx) for x in $D(T)$, is closed in $X \times Y$, then T is bounded (and hence continuous). { ¦klōzd ¦graf 'thir·əm }

closed half plane A half plane that includes the line that bounds it. { ¦klōzd ¦haf 'plān }

closed half space A half space that includes the plane that bounds it. { ¦klōzd ¦half 'spās }

closed intervals A closed interval of real numbers, denoted by $[a,b]$, consists of all numbers equal to or greater than a and equal to or less than b. { ¦klōzd 'in·tər·vəlz }

closed linear manifold A topologically closed vector subspace of a topological vector space. { ¦klōzd ¦lin·ē·ər 'man·ə·fōld }

closed linear transformation A linear transformation T such that the set of points of the form $[x,T(x)]$ is closed in the Cartesian product $D \times R$ of the closure of the domain D and the closure of the range R of T. { ¦klōzd ¦lin·ē·ər ‚tranz·fər'mā·shən }

closed map A function between two topological spaces which sends each closed set of one into a closed set of the other. { ¦klōzd 'map }

closed-mapping theorem The theorem that a linear, surjective mapping between two Banach spaces is continuous if and only if it is closed. { ¦klōzd 'map·iŋ ‚thir·əm }

closed n-cell A set that is homeomorphic with the set of points in n-dimensional Euclidean space ($n = 1, 2, \ldots$) whose distance from the origin is equal to or less than unity. { ¦klōzd 'en ‚sel }

closed operator A linear transformation f whose domain A is contained in a normed vector space X satisfying the condition that if $\lim x_n = x$ for a sequence x_n in A, and $\lim f(x_n) = y$, then x is in A and $f(x) = y$. { ¦klōzd 'äp·ə‚rād·ər }

closed orthonormal set *See* complete orthonormal set. { ¦klōzd ¦ȯr·thō¦nȯr·məl 'set }

closed path In a graph, a path whose initial and final vertices are the same. { ¦klōzd 'path }

closed polygonal region The union of the interior of a polygon with the polygon itself. { ¦klōzd pə¦lig·ən·əl 'rē·jən }

closed pyramidal surface A surface generated by a line passing through a fixed point and moving along a polygon in a plane not containing that point. { ¦klōzd ‚pir·ə¦mid·əl 'sər·fəs }

closed rectangular region The union of the interior of a rectangle with the rectangle itself. { ¦klōzd rek¦taŋ·gyə·lər 'rē·jən }

closed region The closure of an open, connected set. { ¦klōzd 'rē·jən }

closed set A set of points which contains all its cluster points. Also known as topologically closed set. { ¦klōzd 'set }

closed surface A surface that has no bounding curve. { ¦klōzd 'sər·fəs }

closed triangular region The union of the interior of a triangle with the triangle itself. { ¦klōzd trī¦aŋ·gyə·lər 'rē·jən }

closure **1.** The union of a set and its cluster points; the smallest closed set containing the set. **2.** Property of a mathematical set such that a specified mathematical operation that is applied to elements of the set produces only elements of the same set { 'klō·zhər }

clothoid *See* Cornu's spiral. { 'klȯth·ȯid }

cluster analysis A general approach to multivariate problems whose aim is to determine whether the individuals fall into groups or clusters. { 'kləs·tər ə'nal·ə·səs }

cluster point A cluster point of a set in a topological space is a point p whose neighborhoods all contain at least one point of the set other than p. Also known as accumulation point; limit point. { 'kləs·tər ˌpȯint }

cluster sampling A random sampling plan in which the population is subdivided into groups called clusters so that there is small variability within clusters and large variability between clusters. { 'kləs·tər ˌsam·pliŋ }

clutter *See* Sperner set. { 'kləd·ər }

coarser A partition P of a set is coarser than another partition Q of the same set if each member of Q is a subset of a member of P. { 'kȯrs·ər }

coaxial circles Family of circles such that any pair have the same radical axis. { kō'ak·sē·əl 'sər·kəlz }

coaxial cylinders Two cylinders whose cylindrical surfaces consist of the lines that pass through concentric circles in a given plane and are perpendicular to this plane. { kō'ak·sē·əl 'sil·ən·dərz }

coaxial planes Planes that pass through the same straight line. Also known as collinear planes. { kō'ak·sē·əl 'planz }

coboundary An image under the coboundary operator. { kō'baůn·drē }

coboundary operator If {C^n} is a sequence of Abelian groups, coboundary operators are homomorphisms {δ^n} such that $\delta^n: C^n \to C^{n+1}$ and $\delta^{n+1} - \delta^n = 0$. { kō'baůn·drē 'äp·ə,rād·ər }

cochain complex A sequence of Abelian groups C^n, $-\infty < n < \infty$, together with coboundary homomorphisms $\delta^n: C^n \to C^{n+1}$ such that $\delta^{n+1} \circ \delta^n = 0$. { 'kō,chān 'käm,pleks }

cochleoid A plane curve whose equation in polar coordinates is $r\theta = a \sin \theta$. { 'käk·lē,ȯid }

Cochran's test A test used when one estimated variance appears to be very much larger than the remainder of the estimated variances; based on the ratio of the largest estimate of the variance to the total of all the estimates. { 'käk·rənz ˌtest }

cocycle A chain of simplices whose coboundary is 0. { 'kō,sī·kəl }

coefficient A factor in a product. { ˌkō·ə'fish·ənt }

coefficient of alienation A statistic that measures the lack of linear association between two variables; computed by taking the square root of the difference between 1 and the square of the correlation coefficient. { ˌkō·ə'fish·ənt əv ˌā·lē·ə'nā·shən }

coefficient of association A statistic used as a measure of the association of data grouped in a 2×2 table; the value of the statistic ranges from -1 to $+1$, with the former indicating perfect negative association and the latter perfect positive association. Usually designated as Q. { ˌkō·ə'fish·ənt əv ə,sō·sē'ā·shən }

coefficient of concordance A statistic that measures the agreement among sets of rankings by two or more judges. { ˌkō·ə'fish·ənt əv kən'kȯrd·əns }

coefficient of contingency A measure of the strength of dependence between two statistical variables, based on a contingency table. { ˌkō·ə'fish·ənt əv kən'tin·jən·sē }

coefficient of determination A statistic which indicates the strength of fit between two variables implied by a particular value of the sample correlation coefficient r. Designated by r^2. { ˌkō·ə'fish·ənt əv di,tər·mə'nā·shən }

coefficient of multiple correlation A measure used as an index of the strength of a relationship between a variable y and a set of one or more variables x_i; computed by deriving the square root of the ratio of the explained variation to the total variation. { ˌkō·ə'fish·ənt əv ˌməl·tə·pəl ˌkär·ə'lā·shən }

coefficient of nondetermination The coefficient of alienation squared; represents that part of the dependent variable's total variation not accounted for by linear association with the independent variable. { ¦kō·ə'fish·ənt əv ¸nän·di¸tər·mə'nā·shən }

coefficient of strain Multiplier used in transformations to elongate or compress configurations in a direction parallel to an axis. { ¦kō·ə'fish·ənt əv 'strān }

coefficient of variation The ratio of the standard deviation of a distribution to its arithmetic mean. { ¦kō·ə'fish·ənt əv ¸ver·ē'ā·shən }

cofactor *See* minor. { 'kō¸fak·tər }

cofinal A subset *C* of a directed set *D* is cofinal if for each element of *D* there is a larger element in *C*. { kō'fīn·əl }

cohomology group One of a series of Abelian groups $H^n(K)$ that are used in the study of a simplicial complex *K* and are closely related to homology groups, being associated with cocycles and coboundaries in the same manner as homology groups are associated with cycles and boundaries. { 'kō·hə'mäl·ə·jē ¸grüp }

cohomology theory A theory which uses algebraic groups to study the geometric properties of topological spaces; closely related to homology theory. { kō·hō'mäl·ə·jē 'thē·ə·rē }

cohort A group of individuals who experience a significant event, such as birth, during the same period of time. { 'kō¸hòrt }

collinear Lying on a single straight line. { kə'lin·ē·ər }

collinear planes *See* coaxial planes. { ¸kō'lin·ē·ər 'plānz }

collinear vectors Two vectors, one of which is a non-zero scalar multiple of the other. { kə'lin·ē·ər 'vek·tərz }

collineation A mapping which transforms points into points, lines into lines, and planes into planes. Also known as collineatory transformation. { kə¸lin·ē'ā·shən }

collineatory transformation *See* collineation. { kə'lin·yə¸tòr·ē ¸tranz·fər'mā·shən }

colog *See* cologarithm. { 'kō¸läg }

cologarithm The cologarithm of a number is the logarithm of the reciprocal of that number. Abbreviated colog. { ¦kō'läg·ə¸ri<u>th</u>·əm }

color class In a given coloring of a graph, the set of vertices which are assigned the same color. { 'kəl·ər ¸klas }

coloring An assignment of colors to the vertices of a graph so that adjacent vertices are assigned different colors. { 'kəl·ər·iŋ }

column *See* place. { 'käl·əm }

column matrix *See* column vector. { 'käl·əm ¸mā·triks }

column operations A set of rules for manipulating the columns of a matrix so that the image of the corresponding linear transformation remains unchanged. { 'käl·əm ¸äp·ə'rā·shənz }

column rank The number of linearly independent columns of a matrix; the dimension of the image of the corresponding linear transformation. { 'käl·əm ¸raŋk }

column space The vector space spanned by the columns of a matrix. { 'käl·əm ¸spās }

column vector A matrix consisting of only one column. Also known as column matrix. { 'käl·əm ¸vek·tər }

Combescure transformation A one-to-one continuous mapping of one space curve onto another space curve so that tangents to corresponding points are parallel. { 'kōm·bes¸kyùr tranz·fər'mā·shən }

combination A selection of one or more of the elements of a given set without regard to order. { ¸käm·bə'nā·shən }

combinatorial analysis 1. The determination of the number of possible outcomes in ideal games of chance by using formulas for computing numbers of combinations and permutations. **2.** The study of large finite problems. { kəm¸bī·nə'tòr·ē·əl ə'nal·ə·səs }

combinatorial proof A proof that uses combinatorial reasoning instead of calculation. { ¸käm·bə·nə¸tòr·ē·l prüf }

combinatorial theory The branch of mathematics which studies the arrangements of elements into sets. { kəm¸bī·nə'tòr·ē·əl 'thē·ə·rē }

combinatorial topology

combinatorial topology The study of polyhedrons, simplicial complexes, and generalizations of these. Also known as piecewise linear topology. { kəm,bī·nə'tȯr·ē· əl tə'päl·ə·jē }

combinatorics Combinatorial topology which studies geometric forms by breaking them into simple geometric figures. { ,kəm·bə·nə'tȯr·iks }

common denominator Any common multiple of the denominators of a collection of fractions. { ¦käm·ən də'näm·ə,nād·ər }

common difference The fixed difference between any term in an arithmetic progression and the preceding term. { 'käm·ən 'dif·rəns }

common divisor For a set of integers, an integer c such that each of the integers in the set is divisible by c. Also known as common factor. { 'käm·ən di'vīz·ər }

common factor *See* common divisor { ¦käm·ən 'fak·tər }

common fraction A fraction whose numerator and denominator are both integers. Also known as simple fraction; vulgar fraction. { 'käm·ən 'frak·shən }

common logarithm The exponent in the representation of a number as a power of 10. Also known as Briggsian logarithm; Briggs' logarithm. { ¦käm·ən 'läg·ə,rith·əm }

common multiple A quantity (polynomial number) divisible by all quantities in a given set. { ¦käm·ən 'məl·tə·pəl }

common tangent A common tangent of two circles is a line that is tangent to both circles. { ,käm·ən 'tan·jənt }

commutative algebra An algebra in which the multiplication operation obeys the commutative law. { ¦käm·yə,tād·iv 'al·jə·brə }

commutative diagram A diagram in which any two mappings between the same pair of sets, formed by composition of mappings represented by arrows in the diagram, are equal. { ¦käm·yə,tād·iv 'dī·ə,gram }

commutative group *See* Abelian group. { ¦käm·yə,tād·iv ,grüp }

commutative law A rule which requires that the result of a binary operation be independent of order; that is, $ab = ba$. { ¦käm·yə,tād·iv ,lȯ }

commutative operation A binary operation that obeys a commutative law, such as addition and multiplication on the real or complex numbers. Also known as Abelian operation. { ¦käm·yə,tād·iv ,äp·ə'rā·shən }

commutative ring A ring in which the multiplication obeys the commutative law. Also known as Abelian ring. { ¦käm·yə,tād·iv ,riŋ }

commutator The commutator of a and b is the element c of a group such that $bac = ab$. { 'käm·yə,tād·ər }

commutator subgroup The subgroup of a given group G consisting of all products of the form $g_1g_2 \ldots g_n$, where each g_i is the commutator of some pair of elements in G. { 'käm·yə,tād·ər 'səb,grüp }

compactification For a topological space X, a compact topological space that contains X. { käm'pak·tə·fe,kā·shən }

compact-open topology A topology on the space of all continuous functions from one topological space into another; a subbase for this topology is given by the sets $W(K,U) = \{f : f(K) \subset U\}$, where K is compact and U is open. { ¦käm,pakt ¦ō·pən tə'päl·ə·jē }

compact operator A linear transformation from one normed vector space to another, with the property that the image of every bounded set has a compact closure. { ¦käm,pakt 'äp·ə,rād·ər }

compact set A set in a topological space with the property that every open cover has a finite subset which is also a cover. Also known as bicompact set. { ¦käm ,pakt 'set }

compact space A topological space which is a compact set. { ¦käm,pakt 'spās }

compact support The property of a function whose support is a compact set. { 'käm ,pak sə,pȯrt }

compactum A topological space that is metrizable and compact. { käm'pak·təm }

comparable functions Two real-valued functions with a common domain of definition such that the values of one of the functions are equal to or greater than the values of the other for all the points in this domain. { 'käm·prə·bəl 'fəŋk·shənz }

comparable pair A pair of elements, x and y, of a partially ordered set such that either $x \leq y$ or $y \leq x$. { ¦käm·prə·bəl 'per }

comparative experiments Experiments conducted to determine statistically whether one procedure is better than another. { kəm'par·əd·iv ik'sper·ə·məns }

comparison property *See* trichotomy property. { kəm'par·ə·sən ˌpräp·ərd·ē }

comparison test A simple test for the convergence of an infinite series, according to which a series converges if the absolute values of each of its terms are equal to or less than the corresponding term of a series that is known to converge, and diverges if each of its terms is equal to or greater than the absolute value of the corresponding term of a series that is known to diverge. { kəm'par·ə·sən ˌtest }

complement 1. The complement of a number A is another number B such that the sum $A + B$ will produce a specified result. **2.** For a subset of a set, the collection of all members of the set which are not in the given subset. **3.** For a fuzzy set A with membership function m_A, the complement of A is the fuzzy set $\bar{A}$ whose membership function $m_{\bar{A}}$ has the value $1 - m_A(x)$ for every element x. **4.** The complement of a simple graph, G, is the graph, G with the same vertices as G, in which there is an edge between two vertices if and only if there is no edge between those vertices in G. **5.** The complement of an angle A is another angle B such that the sum $A + B$ equals 90°. **6.** *See* radix complement. { 'käm·plə·mənt }

complementary angle One of a pair of angles whose sum is 90°. { ˌkäm·plə'men·trē 'aŋ·gəl }

complementary function Any solution of the equation obtained from a given linear differential equation by replacing the inhomogeneous term with zero. { ˌkäm·plə'men·trē 'fəŋk·shən }

complementary minor *See* minor. { ˌkäm·plə'men·trē 'mī·nər }

complementary operation An operation on a Boolean algebra of two elements (labeled "true" and "false") whose result is the negation of a given operation; for example, NAND is complementary to the AND function. { ˌkäm·plə'men·trē ˌäp·ə'rā·shən }

complementation The act of replacing a set by its complement. { ˌkäm·plə·mən'tā·shən }

complementation law The law that the probability of an event E is 1 minus the probability of the event not E. { ˌkäm·plə·mən'tā·shən ˌlò }

complemented lattice A lattice with distinguished elements a and b, and with the property that corresponding to each point x of the lattice, there is a y such that the greatest lower bound of x and y is a, and the least upper bound of x and y is b. { 'käm·pləˌment·əd 'lad·əs }

complete bipartite graph A graph whose vertices can be partitioned into two sets such that every edge joins a vertex in one set with a vertex in the other, and each vertex in one set is joined to each vertex in the other by exactly one edge. { kəm'plēt bī'pär,tīt ˌgraf }

complete class of decision functions A concept in decision theory which states that for a class of decision functions to be complete it must include a uniformly better decision function, which is a decision function that is sometimes better but never worse (according to some criterion) than each decision function not in the class. { kəm'plēt ¦klas əv di'sizh·ən ˌfəŋk·shənz }

complete four-point *See* four-point. { kəm'plēt 'fòr ˌpòint }

complete graph A graph with exactly one edge connecting each pair of distinct vertices and no loops. { kəm'plēt 'graf }

complete induction *See* mathematical induction. { kəm'plēt in'dək·shən }

complete integral 1. A solution of an nth order ordinary differential equation which depends on n arbitrary constants as well as the independent variable. Also known as complete primitive. **2.** A solution of a first-order partial differential equation with n independent variables which depends upon n arbitrary parameters as well as the independent variables. { kəm'plēt 'in·tə·grəl }

complete lattice A partially ordered set in which every subset has both a supremum and an infimum. { kəm'plēt 'lad·əs }

complete limit *See* limit superior. { kəm'plēt 'lim·ət }

complete linear topological space A topological vector space in which each Cauchy net undergoes Moore-Smith convergence to some point in the space. { kəm¦plēt ¦lin·ē·ər ‚täp·ə¦läj·ə·kəl 'spās }

completely additive set function *See* countably additive set function. { kəm¦plēt·le ¦ad·əd·iv 'set ‚faŋk·shən }

completely ordered set *See* linearly ordered set. { kəm¦plēt·lē ‚órd·ərd 'set }

completely normal space A topological space with the property that any pair of sets with disjoint closures can be separated by open sets. { kəm'plēt·lē ¦nór·məl 'spās }

completely reducible representation A representation of a group as a family of linear operators of a vector space V such that V is the direct sum of subspaces $V_1, \ldots,$ V_n which are invariant under these operators, but $V_1, \ldots, V_n$ do not have any proper closed subspaces which are also invariant under these operators. Also known as semisimple representation. { kəm¦plēt·lē ri¦düs·ə·bəl ‚rep·ri·zen'tā·shən }

completely regular space A topological space X where for every point x and neighborhood U of x there is a continuous function from X to [0,1] with $f(x) = 1$ and $f(y) = 0$, if y is not in U. { kəm'plēt·lē ¦reg·yə·lər 'spās }

completely separable space *See* perfectly separable space. { kəm¦plēt·lē ¦sep·rə·bəl 'spās }

complete matching A subset of the edges of a bipartite graph that consists of edges joining each of the vertices in one of the sets of vertices defining the bipartite structure with distinct vertices in the other such set. { kəm¦plēt 'mach·iŋ }

complete metric space A metric space in which every Cauchy sequence converges to a point of the space. Also known as complete space. { kəm'plēt ¦me·trik 'spās }

complete normed linear space *See* Banach space. { kəm'plēt ¦nórmd ¦lin·ē·ər 'spas }

complete order *See* linear order. { kəm'plēt 'órd·ər }

complete ordered field An ordered field in which every nonempty set that has an upper bound also has a least upper bound. { kəm¦plēt ¦órd·ərd 'fēld }

complete orthonormal set A set of mutually orthogonal unit vectors in a (possibly infinite dimensional) vector space which is contained in no larger such set, that is no nonzero vector is perpendicular to all the vectors in the set. Also known as closed orthonormal set. { kəm'plēt ¦ór·thō¦nór·məl 'set }

complete primitive *See* complete integral. { kəm'plēt 'prim·əd·iv }

complete quadrangle A plane figure consisting of a quadrangle and its two diagonals. Also known as complete quadrilateral. { kəm'plēt 'kwä‚draŋ·gəl }

complete quadrilateral *See* complete quadrangle. { kəm'plēt ‚kwä·drə'lad·ə·rəl }

complete residue system modulo n A set of integers that includes one and only one member of each number class modulo n. { kəm¦plēt ¦rez·ə·dü ¦sis·təm ¦mäj·ə‚lō 'en }

complete space *See* complete metric space. { kəm'plēt 'spās }

complete system of representations A set of representations of a group by matrices (or operators) such that, for any member of the group other than the identity, there is at least one representation for which this member does not correspond to the identity matrix (or the identity operator). { kəm¦plēt ¦sis·təm əv ‚rep·ri·zen'tā·shənz }

completing the square A method of solving quadratic equations, consisting of moving all terms to the left side of the equation, dividing through by the coefficient of the square term, and adding to both sides a number sufficient to make the left side a perfect square. { kəm'plēd·iŋ thə 'skwer }

completion For a metric space X, a complete metric space obtained from X by formally adding limits to Cauchy sequences. { kəm'plē·shən }

complex A space which is represented as a union of simplices which intersect only on their faces. { 'käm‚pleks }

complex conjugate **1.** One of a pair of complex numbers with identical real parts and with imaginary parts differing only in sign. Also known as conjugate. **2.** The matrix whose elements are the complex conjugates of the corresponding elements of a given matrix. { 'käm‚pleks 'kän·jə·gət }

complex fourier series For a function $f(x)$, the series $\sum_{n=\infty}^{\infty} c_n e^{inx}$ with $c_n =$ $\frac{1}{2\pi} \int_{-\pi}^{\pi} f(x) e^{-inx} dx$ { ‚käm·pleks 'für·yä ‚sir·ēz }

complex fraction A fraction whose numerator or denominator is a fraction. { 'käm ‚pleks 'frak·shən }

complex integer *See* Gaussian integer. { ¦käm‚pleks 'int·ə·jər }

complex measure A function whose domain is a sigma algebra of subsets of a particular set, whose range is in the complex numbers, whose value on the empty set is 0, and whose value on a countable union of pairwise disjoint sets is the sum of its values on each of these sets. { ¦käm‚pleks 'mezh·ər }

complex number Any number of the form $a + bi$, where a and b are real numbers, and $i^2 = -1$. { 'käm‚pleks 'nəm·bər }

complex number system The field of complex numbers. { ¦käm‚pleks 'nəm·bər ‚sis·təm }

complex plane A plane whose points are assigned the real and imaginary parts of complex numbers for coordinates. { ¦käm‚pleks 'plān }

complex sphere *See* Riemann sphere. { 'käm‚pleks 'sfir }

complex unit Any complex number, $x + iy$, whose absolute value, $\sqrt{(x^2 + y^2)}$, equals 1. { 'käm‚pleks 'yü·nət }

complex variable A variable which assumes complex numbers for values. { 'käm ‚pleks 'ver·ē·ə·bəl }

component 1. In a graph system, a connected subgraph which is not a subgraph of any other connected subgraph. 2. For a set S, a connected subset of S that is not a subset of any other connected subset of S. 3. The projection of a vector in a given direction of a coordinate system. { kəm'pō·nənt }

component bar chart A bar chart which shows within each bar the components that make up the bar; each component is represented by a section proportional in size to its representation in the total of each bar. { kəm¦pō·nənt 'bär ‚chärt }

component vectors Vectors parallel to specified (usually perpendicular) axes whose sum equals a given vector. { kəm'pō·nənt ‚vek·tərz }

composite function A function of one or more independent variables that are themselves functions of one or more other independent variables. { kəm'päz·ət 'fəŋk·shən }

composite group A group that contains normal subgroups other than the identity element and the whole group. { kəm'päz·ət 'grüp }

composite hypothesis A hypothesis that specifies a range of values for the distribution of the observed random variables. { kəm'päz·ət hī'päth·ə·səs }

composite number Any positive integer which is not prime. Also known as composite quantity. { kəm'päz·ət 'nəm·bər }

composite quantity *See* composite number. { kəm'päz·ət 'kwän·əd·ē }

composition 1. The composition of two mappings, f and g, denoted $g \circ f$, where the domain of g includes the range of f, is the mapping which assigns to each element x in the domain of f the element $g(y)$, where $y = f(x)$. 2. *See* addition. { ‚käm·pə'zish·ən }

composition series A normal series G_1, G_2, . . ., of a group, where each G_i is a proper normal subgroup of G_{i-1} and no further normal subgroups both contain G_i and are contained in G_{i-1}. { ‚käm·pə'zish·ən 'sir‚ēz }

compositum Let E and F be fields, both contained in some field L; the compositum of E and F, denoted EF, is the smallest subfield of L containing E and F. { kəm'päz·əd·əm }

compound curve A curve made up of two arcs of differing radii whose centers are on the same side, connected by a common tangent; used to lay out railroad curves because curvature goes from nothing to a maximum gradually, and vice versa. { 'käm‚paünd 'kərv }

compound distribution A frequency distribution resulting from the combining of two or more separate distributions of the same general type. { ¦käm‚pu̇nd ‚dis·trə'byü·shən }

compound event 1. An event whose probability of occurrence depends upon the probability of occurrence of two or more independent events. **2.** An event that consists of two or more events that are not mutually exclusive. { 'käm‚pau̇nd i'vent }

compound number A quantity which is expressed as the sum of two or more quantities in terms of different units, for example, 3 feet 10 inches, or 2 pounds 5 ounces. { 'käm‚pau̇nd 'nəm·bər }

computable function A function whose value can be calculated by some Turing machine in a finite number of steps. Also known as effectively computable function. { kəm¦pyüd·ə·bəl 'fəŋk·shən }

computation 1. The act or process of calculating. **2.** The result so obtained. { ‚käm·pyə'tā·shən }

computational statistics The conversion of statistical algorithms into computer code that can retrieve useful information from large, complex data sets. Also known as statistical computing. { ‚käm·pyü'tā·shən·əl stə'tis·tiks }

concave function A function $f(x)$ is said to be concave over the interval a,b if for any three points x_1, x_2, x_3 such that $a < x_1 < x_2 < x_3 < b$, $f(x_2) \geq L(x_2)$, where $L(x)$ is the equation of the straight line passing through the points $[x_1, f(x_1)]$ and $[x_3, f(x_3)]$. { 'kän‚kāv 'fəŋk·shən }

concave polygon A polygon at least one of whose angles is greater than 180°. { 'kän ‚kāv 'päl·ə‚gän }

concave polyhedron A polyhedron for which there is at least one plane that contains a face of the polyhedron and that is such that parts of the polyhedron are on both sides of the plane. { 'kän¦kāv ‚päl·ə'hē·drən }

concentrated A measure (or signed measure) m is concentrated on a measurable set A if any measurable set B with nonzero measure has a nonnull intersection with A. { 'kän·sən‚träd·əd }

concentration An operation that provides a relatively sharp boundary to a fuzzy set; for a fuzzy set A with membership function m_A, a concentration of A is a fuzzy set whose membership function has the value $[m_A(x)]^\alpha$ for every element x, where α is a fixed number that is greater than 1. { ‚kän·sən'trä·shən }

concentric circles A family of coplanar circles with the same center. { kən'sen·trik 'sər·kəlz }

conchoid A plane curve consisting of the locus of both ends of a line segment of constant length on a line which rotates about a fixed point, while the midpoint of the segment remains on a fixed curve which does not contain the fixed point. { 'käŋ‚kȯid }

conchoid of Nicomedes The conchoid of a straight line with respect to a fixed point that does not lie on the line. { 'käŋ‚kȯid əv ‚nik·ə'mē·dēz }

concurrent line One of two or more lines that have a point in common. { kən'kər· ənt ‚līn }

concurrent plane One of three or more planes that have a point in common. { kən'kər· ənt 'plān }

concyclic points Points that are located on a common circle. { kən¦sīk·lik 'pȯins }

condensation point For a set in a topological space, a point whose neighborhoods all contain uncountably many points of the set. { ‚kän·dən'sā·shən ‚pȯint }

condition The product of the norm of a matrix and of its inverse. { kən'dish·ən }

conditional convergence The property of a series that is convergent but not absolutely convergent. { kən'dish·ən·əl kən'vər·jəns }

conditional distribution If W and Z are random variables with discrete values w_1, w_2, ..., and $z_1, z_2, \ldots$, the conditional distribution of W given $Z = z$ is the distribution which assigns to w_i, $i = 1, 2, \ldots$, the conditional probability of $W = w_i$ given $Z = z$. { kən'dish·ən·əl dis·trə'byü·shən }

conditional expectation 1. If X is a random variable on a probability space (Ω, F, P), the conditional expectation of X with respect to a given sub σ-field F' of F is an

F'-measurable random variable whose expected value over any set in F' is equal to the expected value of X over this set. **2.** The expected value of a conditional distribution. { kən'dish·ən·əl ,ek,spek'tā·shən }

conditional frequency If r and s are possible outcomes of an experiment which is performed n times, the conditional frequency of s given that r has occurred is the ratio of the number of times both r and s have occurred to the number of times r has occurred. { kən'dish·ən·əl 'frē·kwən·sē }

conditional implication *See* implication. { kən¦dish·ən·əl ,im·plə'kā·shən }

conditional inequality An inequality which fails to hold true for some of the values of the variable involved. { kən¦dish·ən·əl ,in·i'kwäl·əd·ē }

conditionally compact set A set whose closure is compact. Also known as relatively compact set. { kən'dish·ən·əl·ē ¦käm,pakt ,set }

conditional probability The probability that a second event will be B if the first event is A, expressed as $P(B/A)$. { kən'dish·ən·əl ,präb·ə'bil·əd·ē }

cone A solid bounded by a region enclosed in a closed curve on a plane and a surface formed by the segments joining each point of the closed curve to a point which is not in the plane. { kōn }

cone of revolution The surface obtained by rotating a line around another line which it intersects, using the intersection point as a pivot. { 'kōn əv rev·ə'lü·shən }

confidence The degree of assurance that a specified failure rate is not exceeded. { 'kän·fə·dəns }

confidence coefficient The probability associated with a confidence interval; that is, the probability that the interval contains a given parameter or characteristic. Also known as confidence level. { 'kän·fə·dəns ,kō·i'fish·ənt }

confidence interval An interval which has a specified probability of containing a given parameter or characteristic. { 'kän·fə·dəns ,in·tər·vəl }

confidence level *See* confidence coefficient. { 'kän·fə·dəns ,lev·əl }

confidence limit One of the end points of a confidence interval. { 'kän·fə·dəns ,lim·ət }

configuration An arrangement of geometric objects. { kən,fig·yə'rā·shən }

confluent hypergeometric function A solution to differential equation $z(d^2w/dz^2) + (\rho - z)(dw/dz) - \alpha w = 0$. { kən'flü·ənt ¦hī·pər,jē·ə¦me,trik 'fəŋk·shən }

confocal conics 1. A system of ellipses and hyperbolas that have the same pair of foci. **2.** A system of parabolas that have the same focus and the same axis of symmetry. { kän'fō·kəl 'kän·iks }

confocal coordinates Coordinates of a point in the plane with norm greater than 1 in terms of the system of ellipses and hyperbolas whose foci are at $(1,0)$ and $(-1,0)$. { kän'fō·kəl ,kō'örd·ən·əts }

confocal quadrics Quadrics that have the same principal planes and whose sections by any one of these planes are confocal conics. { kän'fō·kəl 'kwäd·riks }

conformable matrices Two matrices which can be multiplied together; this is possible if and only if the number of columns in the first matrix equals the number of rows in the second. { kən'för·mə·bəl 'mā·trə,sēz }

conformal mapping An angle-preserving analytic function of a complex variable. { kən'för·məl 'map·iŋ }

confounding Method used in design of factorial experiments in which some information about higher-order interaction is sacrificed so that estimates of main effects in lower-order interactions can be more precise. { kən'faùnd·iŋ }

congruence 1. The property of geometric figures that can be made to coincide by a rigid transformation. Also known as superposability. **2.** The property of two integers having the same remainder on division by another integer. { kən'grü·əns }

congruence transformation 1. Also known as transformation. **2.** A mapping which associates with each real quadratic form on a set of coordinates the quadratic form that results when the coordinates are subjected to a linear transformation. **3.** A mapping which associates with each square matrix A the matrix $B = SAT$, where S and T are nonsingular matrices, and T is the transpose of S; if A represents the coefficients of a quadratic form, then this definition is equivalent to definition 1. { kən'grü·əns ,tranz·fər,mā·shən }

congruent figures Two geometric figures (plane or solid), one of which can be made to coincide with the other by a rigid motion in space. { kən¦grü·ənt 'fig·yərz }

congruent matrices Two matrices A and B related by the transformation $B = SAT$, where S and T are nonsingular matrices and T is the transpose of S. { kən'grü·ənt 'mā·trə,sēz }

congruent numbers Two numbers having the same remainder when divided by a given quantity called the modulus. { kən'grü·ənt 'nəm·bərz }

conic A curve which may be represented as the intersection of a cone with a plane; the four types of conics are circle, ellipse, parabola, and hyperbola. Also known as conic section. { 'kän·ik }

conical helix A curve that lies on a cone and cuts all the elements of the cone at the same angle. { 'kän·ə·kəl 'hē·liks }

conical projection A projection which associates with each point P in a plane Q the point p in a second plane q which is collinear with O and P, where O is a fixed point lying outside Q. { 'kän·ə·kəl prə'jek·shən }

conical surface A surface formed by the lines which pass through each of the points of a closed plane curve and a fixed point which is not in the plane of the curve. { 'kän·ə·kəl 'sər·fəs }

conicoid A quadric surface (ellipsoid, paraboloid, or hyperboloid) other than a limiting (degenerate) case of such a surface. { 'kän·ə,kòid }

conic section See conic. { ¦kän·ik 'sek·shən }

conjugate 1. An element y of a group related to a given element x by $y = z^{-1}xz$ or $zy = xz$, where z is another element of the group. Also known as transform. **2.** For a quaternion, $x = x_0 + x_1 i + x_2 j + x_3 k$, the quaternion $\bar{x} = x_0 - x_1 i - x_2 j - x_3 k$. **3.** See complex conjugate. { 'kän·jə·gət }

conjugate angles Two angles whose sum is 360° or 2π radians. Also known as explementary angles. { 'kän·jə·gət 'aŋ·gəlz }

conjugate arcs Two arcs of a circle whose sum is the complete circle. { 'kän·jə·gət 'ärks }

conjugate axis For a hyperbola whose equation in cartesian coordinates has the standard form $(x^2/a^2) - (y^2/b^2) = 1$, the portion of the y axis from $(0,-b)$ to $(0,b)$. { 'kän·jə·gət 'ak·səs }

conjugate binomial surds See conjugate radicals. { 'kän·jə·gət bī'nōm·ē·əl 'sərdz }

conjugate convex functions Two functions $f(x)$ and $g(y)$ are conjugate convex functions if the derivative of $f(x)$ is 0 for $x = 0$ and constantly increasing for $x > 0$, and the derivative of $g(y)$ is the inverse of the derivative of $f(x)$. { 'kän·jə·gət 'kän,veks 'fəŋk·shənz }

conjugate curve 1. A member of one of two families of curves on a surface such that exactly one member of each family passes through each point P on the surface, and the directions of the tangents to these two curves at P are conjugate directions. **2.** See Bertrand curve. { 'kän·jə·gət 'kərv }

conjugate diameters 1. For a conic section, any pair of straight lines either of which bisects all the chords that are parallel to the other. **2.** For an ellipsoid or hyperboloid, any three lines passing through the point of symmetry of the surface such that the plane containing the conjugate diameters (first definition) of one of the lines also contains the other two lines. { 'kän·jə·gət dī'am·əd·ərz }

conjugate diametral planes A pair of diametral planes, each of which is parallel to the chords that define the other. { 'kän·jə·gət ¸dī·ə'me·trəl 'plānz }

conjugate directions For a point on a surface, a pair of directions, one of which is the direction of a curve on the surface through the point, while the other is the direction of the characteristic of the planes tangent to the surface at points on the curve. { 'kän·jə·gət di'rek·shənz }

conjugate elements 1. Two elements a and b in a group G for which there is an element x in G such that $ax = xb$. **2.** Two elements of a determinant that are interchanged if the rows and columns of the determinant are interchanged. { 'kän·jə·gət 'el·ə·mənts }

conjugate foci See conjugate points. { 'kän·jə·gət 'fō,sī }

conjugate hyperbolas Two hyperbolas having the same asymptotes with semiaxes interchanged. { 'kän·jə·gət hī'pər·bə·ləz }

conjugate lines 1. For a conic section, two lines each of which passes through the intersection of the tangents to the conic at its points of intersection with the other line. **2.** For a quadric surface, two lines each of which intersects the polar line of the other. { 'kän·jə·gət 'līnz }

conjugate partition If P is a partition, a conjugate partition of P is a partition that is obtained from P by interchanging the rows and columns in its star diagram. { ¦kän·jə·gət pär'tish·ən }

conjugate planes For a quadric surface, two planes each of which contains the pole of the other. { 'kän·jə·gət 'plānz }

conjugate points For a conic section, two points either of which lies on the line that passes through the points of contact of the two tangents drawn to the conic from the other. { 'kän·jə·gət 'póins }

conjugate quaternion One of a pair of quaternions that can be expressed as $q = s + ia + jb + kc$ and $\bar{q} = s - (ia + jb + kc)$, where s, a, b, and c are real numbers and i, j, and k are generators of the quaternions. { ¦kän·ji·gət kwə'tər·nē·ən }

conjugate radicals Binomial surds that are of the type $a\sqrt{b} + c\sqrt{d}$ and $a\sqrt{b} - c\sqrt{d}$, where a, b, c, d are rational but $\sqrt{b}$ and $\sqrt{d}$ are not both rational. Also known as conjugate binomial surds. { 'kän·jə·gət 'rad·ə·kəlz }

conjugate roots Conjugate complex numbers which are roots of a given equation. { 'kän·jə·gət 'rüts }

conjugate ruled surface The ruled surface whose rulings are the lines that are tangent to a given ruled surface at the points of its line of striction and are perpendicular to the rulings of the given ruled surface at these points. { 'kän·jə·gət ¦rüld 'sər·fəs }

conjugate space The set of all continuous linear functionals defined on a normed linear space. { 'kän·jə·gət 'spās }

conjugate subgroups Two subgroups A and B of a group G for which there exists an element x in G such that B consists of the elements of the form xax^{-1}, where a is in A. { 'kän·jə·gət 'səb,grüps }

conjugate system of curves Two one-parameter families of curves on a surface such that a unique curve of each family passes through each point of the surface, and the directions of the tangents to these two curves at any point on the surface are the conjugate directions at that point. { 'kän·jə·gət ¦sis·təm əv 'kərvz }

conjugate triangles Two triangles in which the poles of the sides of each with respect to a given curve are the vertices of the other. { 'kän·jə·gət 'trī,aŋ·gəlz }

conjunction The connection of two statements by the word "and." { kən'jəŋk·shən }

conjunctive matrices Two matrices A and B related by the transformation $B = SAT$, where S and T are nonsingular matrices and S is the Hermitian conjugate of T. { kən'jəŋk·tiv 'mā·trə,sēz }

conjunctive transformation The transformation $B = SAT$, where S is the Hermitian conjugate of T, and matrices A and B are equivalent. { kən'jəŋk·tiv ,tranz·fər'mā·shən }

connected graph A graph in which each pair of points is connected by a path. { kə'nek·təd 'graf }

connected relation A relation such that for any two distinct elements a and b, either (a,b) or (b,a) is a member of the relation. { kə,nek·təd ri'lā·shən }

connected set A set in a topological space which is not the union of two nonempty sets A and B for which both the intersection of the closure of A with B and the intersection of the closure of B with A are empty; intuitively, a set with only one piece. { kə'nek·təd 'set }

connected space A topological space which cannot be written as the union of two nonempty disjoint open subsets. { kə'nek·təd 'spās }

connected surface A surface between any two points of which there is a continuous path that does not cross the surface's boundary. { kə'nek·təd 'sər·fəs }

connectivity number 1. The number of points plus 1 which can be removed from a curve without separating the curve into more than one piece. **2.** The number of

45

closed cuts or cuts joining points of previous cuts (or joining points on the boundary) plus 1 which can be made on a surface without separating the surface. Also known as Betti number. **3.** In general, the n-dimensional connectivity number of a topological space X is the number of infinite cyclic groups whose direct sum with the torsion group $G_n(X)$ forms the homology group $H_n(X)$. { kə,nek'tiv·əd·ē ,nəm·bər }

consecutive Immediately following one another in a sequence. { kən'sek·yəd·iv }

consecutive angles Two angles of a polygon that have a common side. { kən,sek·yəd·iv 'aŋ·gəlz }

consecutive sides Two sides of a polygon that have a common angle. { kən,sek·yəd·iv 'sīdz }

consequent 1. The second term or denominator of a ratio. **2.** The second of the two statements in an implication. **3.** *See* successor. { 'kän·sə·kwənt }

consistency condition The requirement that a mathematical theory be free from contradiction. { kən'sis·tən·sē kən'dish·ən }

consistent equations Two or more equations that are all satisfied by at least one set of values of the variables. { kən'sis·tənt i'kwā·zhənz }

consistent estimate A method of estimation which has the property that the estimate is practically certain to fall very close to a parameter being estimated, provided there are sufficient observations. { kən'sis·tənt 'es·tə·mət }

constant-effect model A model of a test in which the effect of a treatment is the same for all subjects. { ¦kän·stənt i'fekt ,mäd·əl }

constant function A function whose value is the same number for all elements of the function's domain. { 'kän·stənt ,fəŋk·shən }

constant of integration An arbitrary constant that must be added to an indefinite integral of a function to obtain all the indefinite integrals of that function. Also known as integration constant. { 'kän·stənt əv ,in·tə'grā·shən }

constant term A term that does not contain a variable. Also known as absolute term. { 'kän·stənt 'tərm }

constrained optimization problem A nonlinear programming problem in which there are constraint functions. { kən'strānd äp·tə·mə'zā·shən ,präb·ləm }

constraint function A function defining one of the prescribed conditions in a nonlinear programming problem. { kən'strānt ,fəŋk·shən }

construction The process of drawing with suitable instruments a geometrical figure satisfying certain specified conditions. { kən'strək·shən }

contact transformation *See* canonical transformation. { 'kän,takt ,tranz·fər'mā·shən }

contagious distribution A probability distribution which is dependent on a parameter that itself has a probability distribution. { kən'tā·jəs dis·trə'byü·shən }

content *See* Jordan content. { 'kän,tent }

contiguous functions Any pair of hypergeometric functions in which one of the parameters differs by unity and the other two are equal. { kən'tig·yə·wəs 'fəŋk·shənz }

contingency table A table for classifying elements of a population according to two variables, the rows corresponding to one variable and the columns to the other. { kən'tin·jən·sē ,tā·bəl }

continuant The determinant of a continuant matrix. { kən'tin·yə·wənt }

continuant matrix A square matrix all of whose nonzero elements lie on the principal diagonal or the diagonals immediately above and below the principal diagonal. Also known as triple-diagonal matrix. { kən'tin·yə·wənt 'mā·triks }

continued equality An expression in which three or more quantities are set equal by means of two or more equality signs. { kən'tin·yüd i'kwäl·əd·ē }

continued fraction The sum of a number and a fraction whose denominator is the sum of a number and a fraction, and so forth; it may have either a finite or an infinite number of terms. { kən'tin·yüd 'frak·shən }

continued-fraction expansion 1. An expansion of a driving-point function about infinity (or zero) in a continued fraction, in which the terms are alternately constants and multiples of the complex frequency (or multiples of the reciprocal of the complex frequency). **2.** A representation of a real number by a continued fraction, in a manner similar to the representation of real numbers by a decimal expansion. { kən'tin·yüd 'frak·shən ik'span·shən }

continued product A product of three or more factors, or of an infinite number of factors. { kən'tin·yüd 'präd·əkt }

continuous at a point A function f is continuous at a point x, if for every sequence $\{x_n\}$ whose limit is x, the sequence $f(x_n)$ converges to $f(x)$; in a general topological space, for every neighborhood W of $f(x)$, there is a neighborhood N of x such that $f^{-1}(W)$ is contained in N. { kən'tin·yə·wəs ad ə 'pȯint }

continuous deformation A transformation of an object that magnifies, shrinks, rotates, or translates portions of the object in any manner without tearing. { kən'tin·yə·wəs ,dē·fȯr'mā·shən }

continuous distribution Distribution of a continuous population, which is a class of pairs such that the second member of each pair is a value, and the first member of the pair is a proportion density for that value. { kən'tin·yə·wəs ,dis·trə'byü·shən }

continuous extension A continuous function which is equal to another continuous function defined on a smaller domain. { kən'tin·yə·wəs ik'sten·shən }

continuous function A function which is continuous at each point of its domain. Also known as continuous transformation. { kən'tin·yə·wəs 'fəŋk·shən }

continuous geometry A generalization of projective geometry. { kən'tin·yə·wəs jē'äm·ə·trē }

continuous image The image of a set under a continuous function. { kən'tin·yə·wəs 'im·ij }

continuous operator A linear transformation of Banach spaces which is continuous with respect to their topologies. { kən'tin·yə·wəs 'äp·ə,rād·ər }

continuous population A population in which a random variable is measuring a continuous characteristic. { kən'tin·yə·wəs ,päp·yə'lā·shən }

continuous set In an infinite number of outcomes of an experiment, those outcomes in which any value in a given interval can occur. { kən'tin·yə·wəs 'set }

continuous spectrum The portion of the spectrum of a linear operator which is a continuum. { kən'tin·yə·wəs 'spek·trəm }

continuous surface The range of a continuous function from a plane or a connected region in a plane to three-dimensional Euclidean space. { kən'tin·yə·wəs 'sər·fəs }

continuous transformation *See* continuous function. { kən'tin·yə·wəs tranz·fər'mā·shən }

continuum A compact, connected set. { kən'tin·yə·wəm }

continuum hypothesis The conjecture that every infinite subset of the real numbers can be put into one-to-one correspondence with either the set of positive integers or the entire set of real numbers. { kən'tin·yü·əm hī,päth·ə·səs }

contour integral A line integral of a complex function, usually over a simple closed curve. { 'kän,túr ,in·tə·grəl }

contracted curvature tensor A symmetric tensor of second order, obtained by summation on two indices of the Riemann curvature tensor which are not antisymmetric. Also known as contracted Riemann-Christoffel tensor; Ricci tensor. { kən'trak·təd 'kər·və·chər ,ten·sər }

contracted Riemann-Christoffel tensor *See* contracted curvature tensor. { kən'trak·təd ¦rē·män kris'tȯf·əl ,ten·sər }

contraction A function f from a metric space to itself for which there is a constant K that is less than 1 such that, for any two elements in the space, a and b, the distance between $f(a)$ and $f(b)$ is less than K times the distance between a and b. { kən'trak·shən }

contraction semigroup A strongly continuous semigroup all of whose elements have norms which are equal to or less than a constant which is, in turn, less than 1. { kən'trak·shən 'sem·i,grüp }

contrapositive The contrapositive of the statement "if p, then q" is the equivalent statement "if not q, then not p." { ¦kän·trə'päz·əd·iv }

contravariant functor A functor which reverses the sense of morphisms. { ¦kän·trə'ver·ē·ənt 'fəŋk·tər }

contravariant index A tensor index such that, under a transformation of coordinates, the procedure for obtaining a component of the transformed tensor for which this index has the value p involves taking a sum over q of the product of a component of the original tensor for which the index has the value q times the partial derivative of the pth transformed coordinate with respect to the qth original coordinate; it is written as a superscript. { ¦kän·trə'ver·ē·ənt 'in¸deks }

contravariant tensor A tensor with only contravariant indices. { ¦kän·trə'ver·ē·ənt 'ten·sər }

contravariant vector A contravariant tensor of degree 1, such as the tensor whose components are differentials of the coordinates. { ¦kän·trə'ver·ē·ənt 'vek·tər }

control 1. A test made to determine the extent of error in experimental observations or measurements. 2. A procedure carried out to give a standard of comparison in an experiment. 3. Observations made on subjects which have not undergone treatment, to use in comparison with observations made on subjects which have undergone treatment. { kən'trōl }

control group A sample in which a factor whose effect is being estimated is absent or is held constant, in order to provide a comparison. { kən'trōl ¸grüp }

convergence The property of having a limit for infinite series, sequences, products, and so on. { kən'vər·jəns }

convergence in measure A sequence of functions $f_n(x)$ converges in measure to $f(x)$ if given any $\epsilon > 0$, the measure of the set of points at which $|f_n(x) - f(x)| > \epsilon$ is less than ϵ, provided n is sufficiently large. { kən'vər·jəns in 'mezh·ər }

convergent One of the continued fractions that is obtained from a given continued fraction by terminating after a finite number of terms. { kən'vər·jənt }

convergent integral An improper integral which has a finite value. { kən'vər·jənt 'in·tə·grəl }

convergent sequence A sequence which has a limit. { kən'vər·jənt 'sē·kwəns }

convergent series A series whose sequence of partial sums has a limit. { kən'vər·jənt 'sir¸ēz }

converse The converse of the statement "if p, then q" is the statement "if q, then p." { 'kän¸vərs }

conversion factor The numerical factor by which one must multiply (or divide) a quantity that is expressed in terms of a certain unit to express the quantity in terms of another unit. Also known as conversion ratio; unit conversion factor. { kən'vər·zhən ¸fak·tər }

conversion ratio *See* conversion factor. { kən'vər·zhən ¸rā·shō }

convex angle A polyhedral angle that lies entirely on one side of each of its faces. { 'kän¸veks 'aŋ·gəl }

convex body A convex set that has at least one interior point. { 'kän¸veks 'bäd·ē }

convex combination A linear combination of vectors in which the sum of the coefficients is 1. { 'kän¸veks ¸käm·bə'nā·shən }

convex curve A plane curve for which any straight line that crosses the curve crosses it at just two points. { 'kän¸veks 'kərv }

convex function A function $f(x)$ is considered to be convex over the interval a,b if for any three points x_1, x_2, x_3 such that $a < x_1 < x_2 < x_3 < b, f(x_2) \leq L(x_2)$, where $L(x)$ is the equation of the straight line passing through the points $[x_1, f(x_1)]$ and $[x_3, f(x_3)]$. { 'kän¸veks 'fəŋk·shən }

convex function in the sense of Jensen A function $f(x)$ over an interval a, b such that, for any two points x_1 and x_2 satisfying $a < x_1 < x_2$ then $f[(x_1 + x_2)/2] \leq (1/2) [f(x_1) + f(x_2)]$. { ¦kän¸veks ¦fəŋk·shən in thə ¸sens əv 'jen·sən }

convex hull The smallest convex set containing a given collection of points in a real linear space. Also known as convex linear hull. { 'kän¸veks 'həl }

convex linear combination A linear combination in which the scalars are nonnegative real numbers whose sum is 1. { ¦kän¸veks ¸lin·ē·ər ¸käm·bə'nā·shən }

convex linear hull *See* convex hull. { 'kän¸veks 'lin·ē·ər ¸həl }

convex polygon A polygon all of whose interior angles are less than or equal to 180°. { 'kän¸veks 'päl·i¸gän }

convex polyhedron A polyhedron in the plane which is a convex set, for example, any regular polyhedron. { 'kän,veks ¦päl·i¦hē·drən }

convex polytope A bounded, convex subset of an n-dimensional space enclosed by a finite number of hyperplanes. { ¦kän,veks 'päl·i,tōp }

convex programming Nonlinear programming in which both the function to be maximized or minimized and the constraints are appropriately chosen convex or concave functions of the independent variables. { 'kän,veks 'prō,gram·iŋ }

convex sequence A sequence of numbers, $a_1, a_2, \ldots,$ such that $a_{i+1} \leqq (1/2)(a_i + a_{i+2})$ for all $i \geqq 1$ (or for all i satisfying $1 \leqq i < n - 2$ if the sequence is a finite sequence with n terms). { 'kän,veks 'sē·kwəns }

convex set A set which contains the entire line segment joining any pair of its points. { 'kän,veks 'set }

convex span For a set A, the intersection of all convex sets that contain A. { ¦kn,veks 'span }

convolution 1. The convolution of the functions f and g is the function F, defined by $F(x) = \displaystyle\int_0^x f(t)g(x - t) \, dt$. **2.** A method for finding the distribution of the sum of two or more random variables; computed by direct integration or summation as contrasted with, for example, the method of characteristic functions. { ,kän·və'lü·shən }

convolution family See faltung. { ,kän·və'lü·shən ,fam·lē }

convolution rule The statement that $C(p + q, r)$ is the sum over the index j from $j = 0$ to $j = r$ of the quantity $C(p, j) \, C(q, r - j)$, where, in general, $C(n, r)$ is the number of distinct subsets of r elements in a set of n elements (the binomial coefficient). Also known as Vandermonde's identity. { ,kän·və,lü·shən ,rül }

convolution theorem A theorem stating that, under specified conditions, the integral transform of the convolution of two functions is equal to the product of their integral transforms. { ,kän·və'lü·shən ,thir·əm }

coordinate axes One of a set of lines or curves used to define a coordinate system; the value of one of the coordinates uniquely determines the location of a point on the axis, while the values of the other coordinates vanish on the axis. { kō'órd·ən·ət 'ak,sēz }

coordinate basis A basis for tensors on a manifold induced by a set of local coordinates. { kō'órd·ən·ət 'bā·səs }

coordinates A set of numbers which locate a point in space. { kō'órd·ən·əts }

coordinate systems A rule for designating each point in space by a set of numbers. { kō'órd·ən·ət ,sis·təmz }

coordinate transformation A mathematical or graphic process of obtaining a modified set of coordinates by performing some nonsingular operation on the coordinate axes, such as rotating or translating them. { kō'órd·ən·ət tranz·fər'mā·shən }

Cornu's spiral A plane curve whose curvature is proportional to its arc length, and whose Cartesian coordinates are given in parametric form by the Fresnel integrals. Also known as clothoid; Euler's spiral. { 'kór·nüz ¦spī·rəl }

correction for attenuation A method used to adjust correlation coefficients upward because of errors of measurement when two measured variables are correlated; the errors always serve to lower the correlation coefficient as compared with what it would have been if the measurement of the two variables had been perfectly reliable. { kə'rek·shən fór ə,ten·yə'wā·shən }

correlation The interdependence or association between two variables that are quantitative or qualitative in nature. { ,kär·ə'lā·shən }

correlation coefficient A measurement, which is unchanged by both addition and multiplication of the random variable by positive constants, of the tendency of two random variables X and Y to vary together; it is given by the ratio of the covariance of X and Y to the square root of the product of the variance of X and the variance of Y. { ,kär·ə'lā·shən ,kō·i'fish·ənt }

correlation curve See correlogram. { ,kär·ə'lā·shən ,kərv }

correlation ratio A measure of the nonlinear relationship between two variables; in a two-way frequency table it may be regarded as the ratio of the variance between arrays to the total variance. { ˌkär·ə'lā·shən ˌrā·shō }

correlation table A table designed to categorize paired quantitative data; used to calculate correlation coefficients. { ˌkär·ə'lā·shən ˌtā·bəl }

correlogram A curve showing the assumed correlation between two mathematical variables. Also known as correlation curve. { kə'rel·əˌgram }

corresponding angles For two lines, l_1 and l_2, cut by a transversal t, a pair of angles such that (1) one of the angles has sides l_1 and t while the other has sides l_2 and t; (2) both angles are on the same side of t; and (3) the angles are on the same sides of l_1 and l_2, respectively. { 'kär·əˌspänd·iŋ 'aŋ·gəlz }

cos *See* cosine function.

cosecant The reciprocal of the sine. Denoted csc. { kō'sēˌkant }

coset For a subgroup of a group, a set consisting of all elements of the form xh or of all elements of the form hx, where h is an element of the subgroup and x is a fixed element of the group. { 'kōˌset }

cosh *See* hyperbolic cosine.

cosine function In a right triangle with an angle θ, the cosine function gives the ratio of adjacent side to hypotenuse; more generally, it is the function which assigns to any real number θ the abscissa of the point on the unit circle obtained by moving from (1,0) counterclockwise θ units along the circle, or clockwise |θ| units if θ is less than 0. Denoted cos. { 'kōˌsīn ˌfəŋk·shən }

cosine series A Fourier series that contains only terms that are even in the independent variable, that is, the constant term and terms involving the cosine function. { 'kōˌsīn ˌsir·ēz }

cot *See* cotangent.

cotangent The reciprocal of the tangent. Denoted cot; ctn. { kō'tan·jənt }

coterminal angles Two angles that have the same initial line and the same terminal line and therefore differ by a multiple of 2π radians or 360°. { ¦kō¦tərm·ən·əl 'aŋ·gəlz }

coth *See* hyperbolic cotangent.

count 1. To name a set of consecutive positive integers in order of size, usually starting with 1. 2. To associate consecutive positive integers, starting with 1, with the members of a finite set in order to determine the cardinal number of the set. { kaunt }

countability axioms Two conditions which are satisfied by a euclidean space and one or the other of which is often assumed in the study of a general topological space; the first states that any point in the topological space has a countable local base, while the second states that the topological space has a countable base. { ˌkaun·tə'bil·əd·ē ˌax·sē·əmz }

countable Either finite or denumerable. Also known as enumerable. { 'kaunt·ə·bəl }

countably additive Given a measure m, and a sequence of pairwise disjoint measurable sets, the property that the measure of the union is equal to the sum of the measures of the sets. { 'kaunt·ə·blē 'ad·əd·iv }

countably additive set function A set function with the properties that (1) the union of any finite or countable collection of sets in the range of the function is also in this range, and (2) the value of the function at the union of a finite or countable collection of sets that are in the range of the set function and are pairwise disjoint is equal to the sum of the values at each set in the collection. Also known as completely additive set function. { ¦kaun·tə·blē ¦ad·əd·iv 'set ˌfəŋk·shən }

countably compact set A set with the property that every cover with countably many open sets contains a finite number of sets which is also a cover. { 'kaunt·ə·blē ˌkäm,pakt 'set }

countably infinite set *See* denumerable set. { 'kaunt·ə·blē ˌin·fə·nət 'set }

countably metacompact space A topological space with the property that every open covering F is associated with a point-finite open covering G, such that every element of G is a subset of an element of F. { ˌkaunt·ə·blē ˌmed·ə,käm,pakt 'späs }

countably paracompact space A topological space with the property that every countable open covering F is associated with a locally finite open covering G, such that every element of G is a subset of an element of F. { ˌkau̇nt·ə·blē ˌpar·ə͵käm ͵pakt 'spās }

countably subadditive A set function m is countably subadditive if, given any sequence of sets, the measure of the union is less than or equal to the sum of the measures of the sets. { kau̇nt·ə·blē səb'ad·əd·iv }

countably subadditive set function A real-valued function defined on a class of sets such that the value of the function on the union of any sequence of sets is equal to or less than the sum of the sequence of the values of the function on the sets. { ͵kau̇n·tə·blē səb͵ad·əd·iv 'set ͵fəŋk·shən }

counting number One of the numbers used in counting objects, either the set of positive integers or the set of positive integers and the number 0. { 'kau̇nt·iŋ ͵nəm·bər }

covariance A measurement of the tendency of two random variables, X and Y, to vary together, given by the expected value of the variable $(X - \bar{X})(Y - \bar{Y})$, where $\bar{X}$ and $\bar{Y}$ are the expected values of the variables X and Y respectively. { kō'ver·ē·əns }

covariance analysis An extension of the analysis of variance which combines linear regression with analysis of variance; used when members falling into classes have values of more than one variable. { kō'ver·ē·əns ə͵nal·ə·səs }

covariant components Vector or tensor components which, in a transformation from one set of basis vectors to another, transform in the same manner as the basis vectors. { kō'ver·ē·ənt kəm'pō·nəns }

covariant derivative For a tensor field at a point P of an affine space, a new tensor field equal to the difference between the derivative of the original field defined in the ordinary manner and the derivative of a field whose value at points close to P are parallel to the value of the original field at P as specified by the affine connection. { kō'ver·ē·ənt də'riv·əd·iv }

covariant functor A functor which does not change the sense of morphisms. { kō'ver· ē·ənt 'fəŋk·tər }

covariant index A tensor index such that, under a transformation of coordinates, the procedure for obtaining a component of the transformed tensor for which this index has value p involves taking a sum over q of the product of a component of the original tensor for which the index has the value q times the partial derivative of the qth original coordinate with respect to the pth transformed coordinate; it is written as a subscript. { kō'ver·ē·ənt 'in͵deks }

covariant tensor A tensor with only covariant indices. { kō'ver·ē·ənt 'ten·sər }

covariant vector A covariant tensor of degree 1, such as the gradient of a function. { kō'ver·ē·ənt 'vek·tər }

cover 1. An element, x, of a partially ordered set covers another element y if x is greater than y, and the only elements that are both greater than or equal to y and less than or equal to x are x and y themselves. **2.** *See* covering. { 'kəv·ər }

covering For a set A, a collection of sets whose union contains A. Also known as cover. { 'kəv·riŋ }

covers *See* coversed sine.

coversed sine The coversed sine of A is $1 - \sin A$. Denoted covers. Also known as coversine; versed cosine. { ͵kō͵vərst 'sīn }

coversine *See* coversed sine. { ˌkō͵vər'sīn }

cracovian An object which is the same as a matrix except that the product of cracovians A and B is equal to the matrix product $A'B$, where A' is the transpose of A. { krə'kō·vē·ən }

Cramér-Rao inequality An inequality that is the basis of a method for determining a lower bound to the variance of an estimator of a parameter. { krə͵mä 'rau̇ ͵in· i͵kwäl·əd·ē }

Cramer's rule The method of solving a system of linear equations by means of determinants. { 'krā·mərz ͵rül }

crisp set A conventional set, wherein the degree of membership of any object in the set is either 0 or 1. { ¦krisp 'set }

critical function A function satisfying the Euler equations in the calculus of variations. { 'krid·ə·kəl 'fəŋk·shən }

critical point A point at which the first derivative of a function is either 0 or does not exist. { 'krid·ə·kəl 'pȯint }

critical ratio The ratio of a particular deviation from the mean value to the standard deviation. { 'krid·ə·kəl 'rā·shō }

critical region In testing hypotheses, the set of sample values leading to rejection of the null hypothesis. { 'krid·ə·kəl 'rē·jən }

critical table A table, usually for a function that varies slowly, which gives only values of the argument near which changes in the value of the function, as rounded to the number of decimal places displayed in the table, occur. { 'krid·ə·kəl ‚tā·bəl }

critical value The value of the dependent variable at a critical point of a function. A number which causes rejection of the null hypothesis if a given test statistic is this number or more, and acceptance of the null hypothesis if the test statistic is smaller than this number. { 'krid·ə·kəl 'val·yü }

cross-cap The self-intersecting surface that results when a Möbius band is deformed so that its boundary is a circle. { 'krȯs ‚kap }

cross-correlation 1. Correlation between corresponding members of two or more series: if $q_1, \ldots, q_n$ and $r_1, \ldots, r_n$ are two series, correlation between q_i and r_i, or between q_i and r_{i+j} (for fixed j), is a cross correlation. **2.** Correlation between or expectation of the inner product of two series of random variables, where the difference in indices between the corresponding values of the two series is fixed. { 'krȯs kär·ə'lā·shən }

cross curve A plane curve whose equation in cartesian coordinates x and y is $(a^2/x^2) + (b^2/y^2) = 1$, where a and b are constants. Also known as cruciform curve. { 'krȯs ‚kərv }

cross multiplication Multiplication of the numerator of each of two fractions by the denominator of the other, as when eliminating fractions from an equation. { ¦krȯs ‚məl·tə·plə'kā·shən }

crossover length A length characteristic of a fractal network such that at scales which are small compared with this length the fractal nature of the structure is manifest in its dynamics, whereas at scales which are large compared with this length the dynamics resemble those of a crystalline structure. { 'krȯs‚ō·vər ‚leŋkth }

cross product 1. An anticommutative multiplication on the vectors of Euclidean three-dimensional space. Also known as vector product. **2.** The product of the two mean terms of a proportion, or the product of the two extreme terms; in the proportion $a/b = c/d$, it is ad or bc. { 'krȯs ‚prä·dəkt }

cross ratio For four collinear points, A, B, C, and D, the ratio $(AB)(CD)/(AD)(CB)$, or one of the ratios obtained from this quantity by a permutation of A, B, C, and D. { 'krȯs ‚rä·shō }

cross section 1. The intersection of an n-dimensional geometric figure in some Euclidean space with a lower dimensional hyperplane. **2.** A right inverse for the projection of a fiber bundle. { 'krȯs ‚sek·shən }

Crout reduction Modification of the Gauss procedure for numerical solution of simultaneous linear equations; adapted for use on desk calculators and digital computers. { 'kraȯt ri'dək·shən }

cruciform curve *See* cross curve. { 'krü·sə‚fȯrm ‚kərv }

crunode A point on a curve through which pass two branches of the curve with different tangents. Also known as node. { ¦krü¦nōd }

csc *See* cosecant.

csch *See* hyperbolic cosecant.

ctn *See* cotangent.

cubature The numerical integration of a function of two variables. { 'kyüb·ə·chər }

cube 1. Regular polyhedron whose faces are all square. **2.** For a number a, the new number obtained by taking the threefold product of a with itself: $a \times a \times a$. { kyüb }

cube root Another number whose cube is the original number. { 'kyüb 'rüt }

cubical parabola A plane curve whose equation in Cartesian coordinates x and y is $y = x^3$. { 'kyüb·ə·kəl pə'rab·ə·lə }

cubic curve A plane curve which has an equation of the form $f(x,y) = 0$, where $f(x,y)$ is a polynomial of degree three in x and y. { 'kyü·bik 'kərv }

cubic determinant A mathematical form analogous to an ordinary determinant, with the elements forming a cube instead of a square. { 'kyü·bik di'tər·mə·nənt }

cubic equation A polynomial equation with no exponent larger than 3. { 'kyü·bik i'kwā·zhən }

cubic polynomial A polynomial in which all exponents are no greater than 3. { 'kyü·bik ˌpäl·ə'nō·mē·əl }

cubic quantic A quantic of the third degree. { ˌkyüb·ik 'kwän·tik }

cubic spline One of a collection of cubic polynomials used in interpolating a function whose value is specified at each of a collection of distinct ordered values, X_i ($i = 1, \ldots, n$), and whose slope is specified at X_1 and X_n; one cubic polynomial is found for each interval, such that the interpolating system has the prescribed values at each of the X_i, the prescribed slope at X_n and X_n, and a continuous slope at each of the X_i. { 'kyü·bik 'splīn }

cubic surd A cube root of a rational number that is itself an irrational number. { 'kyü·bik 'sərd }

cuboctahedron A polyhedron whose faces consist of six equal squares and eight equal equilateral triangles, and which can be formed by cutting the corners off a cube; it is one of the 13 Archimedean solids. Also spelled cubooctahedron. { ˌkyü¦bäk·tə'hē·drən }

cuboid See rectangular parallelepiped. { 'kyü,bóid }

cubooctahedron See cuboctahedron. { ¦kyü·bo¸äk·tə'hē·drən }

Cullen number A number having the form $C_n = (n \cdot 2^n) + 1$ for $n = 0, 1, 2, \ldots$ { 'kəl·ən ˌnəm·bər }

cumulants A set of parameters k_h ($h = 1, \ldots r$) of a one-dimensional probability distribution defined by $\ln \chi_x(q) = \sum_{h=1}^{r} k_h[(iq)^h/h!] + o(q^r)$ where $\chi_x(q)$ is the characteristic function of the probability distribution of x. Also known as semi-invariants. { 'kyü·myə·ləns }

cumulative error An error whose magnitude does not approach zero as the number of observations increases. Also known as accumulative error. { 'kyü·myə·ləd·iv 'er·ər }

cumulative frequency distribution The frequency with which a variable assumes values less than or equal to some number, obtained by summing the values in a frequency distribution. { 'kyü·myə·ləd·əv 'frē·kwən·sē ˌdi·strə¸byü·shən }

cup The symbol ∪, which indicates the union of two sets. { kəp }

cup product A multiplication defined on cohomology classes; it gives cohomology a ring structure. { 'kəp ˌpräd·əkt }

curl The curl of a vector function is a vector which is formally the cross product of the del operator and the vector. Also known as rotation (rot). { kərl }

curtate cycloid A trochoid in which the distance from the center of the rolling circle to the point describing the curve is less than the radius of the rolling circle. { 'kər¸tät 'sī¸klóid }

curvature The reciprocal of the radius of the circle which most nearly approximates a curve at a given point; the rate of change of the unit tangent vector to a curve with respect to arc length of the curve. { 'kər·və·chər }

curvature tensor See Riemann-Christoffel tensor. { 'kər·və·chər ˌten·sər }

curve The continuous image of the unit interval. { kərv }

curved surface A surface having no part that is a plane surface. { 'kərvd 'sər·fəs }

curve fitting The calculation of a curve of some particular character (as a logarithmic curve) that most closely approaches a number of points in a plane. { 'kərv ˌfid·iŋ }

curve tracing The method of graphing a function by plotting points and analyzing symmetries, derivatives, and so on. { 'kərv ˌträs·iŋ }

curvilinear coordinates Any linear coordinates which are not Cartesian coordinates; frequently used curvilinear coordinates are polar coordinates and cylindrical coordinates. { 'kər·və'lin·ē·ər kō'órd·ən·əts }

curvilinear regression Regression study of jointly distributed random variables where the function measuring their statistical dependence is analyzed in terms of curvilinear coordinates. Also known as nonlinear regression. { 'kər·və'lin·ē·ər ri'gresh·ən }

curvilinear solid A solid whose surfaces are not planes. { ‚kər·və'lin·ē·ər 'säl·əd }

curvilinear transformation A transformation from one coordinate system to another in which the coordinates in the new system are arbitrary twice-differentiable functions of the coordinates in the old system. { 'kər·və'lin·ē·ər tranz·fər'mā·shən }

curvilinear trend A nonlinear trend which may be expressed as a polynomial or a smooth curve. { ‚kər·və'lin·ē·ər 'trend }

cusp A singular point of a curve at which the limits of the tangents of the portions of the curve on either side of the point coincide. Also known as spinode. { kəsp }

cuspidal cubic A cubic curve that has one cusp, one point of inflection, and no node. { 'kəs·pəd·əl 'kyü·bik }

cuspidal locus A curve consisting of the cusps of some family of curves. { 'kəs·pəd·əl 'lō·kəs }

cusp of the first kind A cusp such that the two portions of the curve adjacent to the cusp lie on opposite sides of the limiting tangent to the curve at the cusp. Also known as simple cusp. { 'kəsp əv t͟hə 'fərst ‚kīnd }

cusp of the second kind A cusp such that the two portions of the curve adjacent to the cusp lie on the same side of the limiting tangent to the curve at the cusp. { 'kəsp əv t͟hə 'sek·ənd kīnd }

cut A subset of a given set whose removal from the original set leaves a set that is not connected. { kət }

cut capacity For a network whose points have been partitioned into two specified classes, C_1 and C_2, the sum of the capacities of all the segments directed from a point in C_1 to a point in C_2. Also known as cut value. { 'kət kə'pas· əd·ē }

cut point A point in a component of a graph whose removal disconnects that component. Also known as articulation point. { 'kət ‚póint }

cut value See cut capacity. { 'kət ‚val·yü }

cycle 1. A member of the kernel of a boundary homomorphism. **2.** A closed path in a graph that does not pass through any vertex more than once and passes through at least three vertices. Also known as circuit. **3.** See cyclic permutation. A periodic movement in a time series. { 'sī·kəl }

cyclic curve 1. A curve (such as a cycloid, cardioid, or epicycloid) generated by a point of a circle that rolls (without slipping) on a given curve. **2.** The intersection of a quadric surface with a sphere. Also known as spherical cyclic curve. **3.** The stereographic projection of a spherical cyclic curve. Also known as plane cyclic curve. { 'sīk·lik 'kərv }

cyclic extension A Galois extension whose Galois group is cyclic. { 'sīk·lik ik'sten·chən }

cyclic graph A graph whose vertices correspond to the vertices of a regular polygon and whose edges correspond to the sides of the polygon. { ‚sī·klik 'graf }

cyclic group A group that has an element a such that any element in the group can be expressed in the form a^n, where n is an integer. { 'sīk·lik ‚grüp }

cyclic identity The principle that the sum of any component of the Riemann-Christoffel tensor and two other components obtained from it by cyclic permutation of any three indices, while the fourth is held fixed, is zero. { 'sīk·lik Ī‚den·təd·ē }

cyclic left module A left module over a ring A that has a member x such that any member of the module has the form ax, where a is a member of A. { ‚sī·klik ‚left 'mäj·əl }

cyclic permutation A permutation of an ordered set of symbols which sends the first to the second, the second to the third, . . ., the last to the first. Also known as cycle. { 'sīk·lik pər·myə'tā·shən }

cyclic polygon A polygon whose vertices are located on a common circle. { ¦sī·klik 'päl·i̧gän }

cyclindroid 1. A cylindrical surface generated by the lines perpendicular to a plane that pass through an ellipse in the plane. **2.** A surface that is generated by a straight line that moves so as to intersect two curves and remain parallel to a given plane. { si'kliņdróid }

cycloid The curve traced by a point on the circumference of a circle as the circle rolls along a straight line. { 'sī̧klóid }

cyclomatic number For a graph, the number $e - n + 1$, where e is the number of edges and n is the number of nodes. { ¦sī·klə̧mad·ik 'nəm·bər }

cyclosymmetric function A function whose value is unchanged under a cyclic permutation of its variables. { ¦si·klō·si¦me·trik 'fəŋk·shən }

cyclotomic equation An equation which has the form $x^{n-1} + x^{n-2} + \cdots + x + 1 = 0$, where n is a prime number. { ¦sī·klō¦täm·ik i'kwā·zhən }

cyclotomic field The extension field of a given field K which is the smallest extension field of K that includes the nth roots of unity for some integer n. { ̧sī·klə¦täm· ik 'fēld }

cyclotomic integer A number of the form $a_0 + a_1 z + a_2 z^2 + \cdots + a_{n-1} z^{n-1}$, where z is a primitive nth root of unity and each a_i is an ordinary integer. { ̧sī·klə̧täm· ik 'in·ə·jər }

cyclotomic polynomial The nth cyclotomic polynomic is the monic polynomial of degree $\phi(n)$ [where ϕ represents Euler's phi function] whose zeros are the primitive nth roots of unity. { sī·klə̧täm·iķpäl·ə'nōmē·əl }

cyclotomy Theory of dividing the circle into equal parts or constructing regular polygons or, analytically, of finding the nth roots of unity. { sī'kläd·ə·mē }

cylinder 1. A solid bounded by a cylindrical surface and two parallel planes, or the surface of such a solid. **2.** *See* cylindrical surface. { 'sil·ən·dər }

cylinder function Any solution of the Bessel equation, including Bessel functions, Neumann functions, and Hankel functions. { 'sil·ən·dər ̧faŋk·shən }

cylindrical coordinates A system of curvilinear coordinates in which the position of a point in space is determined by its perpendicular distance from a given line, its distance from a selected reference plane perpendicular to this line, and its angular distance from a selected reference line when projected onto this plane. { sə'lin· drə·kəl ̧kō'órd·ən·əts }

cylindrical function *See* Bessel function. { sə'lin·dri·kəl ̧faŋk·shən }

cylindrical helix A curve lying on a cylinder which intersects the elements of the cylinder at a constant angle. { sə'lin·drə·kəl 'hȩ̄liks }

cylindrical surface A surface consisting of each of the straight lines which are parallel to a given straight line and pass through a given curve. Also known as cylinder. { sə'lin·drə·kəl 'sər·fəs }

D

d'Alembertian A differential operator in four-dimensional space, $\dfrac{\partial^2}{\partial x^2} + \dfrac{\partial^2}{\partial y^2} + \dfrac{\partial^2}{\partial z^2} - \dfrac{1}{c^2}\dfrac{\partial^2}{\partial t^2}$, which is used in the study of relativistic mechanics.   { ¦dal·əm¦bər·shən }

d'Alembert's test for convergence A series Σa_n converges if there is an N such that the absolute value of the ratio a_n/a_{n-1} is always less than some fixed number smaller than 1, provided n is at least N, and diverges if the ratio is always greater than 1.   Also known as generalized ratio test.   { ¦dal·əm¦bərz ‚test fər kən'vər·jəns }

damped regression analysis *See* ridge regression analysis.   { ¦dampt ri'gresh·ən ə‚nal·ə·səs }

Dandelin sphere For a conic that is represented as the intersection of a plane and a circular cone, a sphere that is a tangent to both the plane and the cone.   { 'dänd ‚lan ‚sfir }

Darboux's monodromy theorem The proposition that, if the function $f(z)$ of the complex variable z is analytic in a domain D bounded by a simple closed curve C, and $f(z)$ is continuous in the union of D and C and is injective for z on C, then $f(z)$ is injective for z in D.   { 'där·büz ¦män·ə‚drä·mē ‚thir·əm }

data reduction The conversion of all information in a data set into fewer dimensions for a particular purpose, as, for example, a single measure such as a reliability measure.   { 'dad·ə ri‚dək·shən }

decagon A 10-sided polygon.   { 'dek·ə‚gän }

decahedron A polyhedron that has 10 faces.   { ‚dek·ə'hē·drən }

decidable predicate A predicate for which there exists an algorithm which, for any given value of its independent variables, provides a definite answer as to whether or not it is true.   { di'sīd·ə·bəl 'pred·ə·kət }

decile Any of the points which divide the total number of items in a frequency distribution into 10 equal parts.   { 'des‚īl }

decimal A number expressed in the scale of tens.   { 'des·məl }

decimal fraction Any number written in the form: an integer followed by a decimal point followed by a (possibly infinite) string of digits.   { ¦des·məl ¦frak·shən }

decimal number A number signifying a decimal fraction by a decimal point to the left of the numerator with the number of figures to the right of the point equal to the power of 10 of the denominator.   { ¦des·məl ¦nəm·bər }

decimal number system A representational system for the real numbers in which place values are read in powers of 10.   { ¦des·məl 'nəm·bər ‚sis·təm }

decimal place Reference to one of the digits following the decimal point in a decimal fraction; the kth decimal place registers units of 10^{-k}.   { ¦des·məl ¦plās }

decimal point A dot written either on or slightly above the line; used to mark the point at which place values change from positive to negative powers of 10 in the decimal number system.   { 'des·məl ‚pȯint }

decimal system A number system based on the number 10; in theory, each unit is 10 times the next smaller one.   { 'des·məl ‚sis·təm }

decision-making under uncertainty The process of drawing conclusions from limited information or conjecture.   { di¦sizh·ən ¦māk·iŋ ‚ən·dər ən'sərt·ən·tē }

decomino One of the 4655 plane figures that can be formed by joining 10 unit squares along their sides.   { ‚dek·ə'mē·nō }

decomposable process A process which can be reduced to several basic events. { dē·kəm'pō·zə·bəl 'präs·əs }

decomposition 1. The expression of a fraction as a sum of partial fractions. **2.** The representation of a set as the union of pairwise disjoint subsets. { dē,käm·pə'zish·ən }

decreasing function A function, f, of a real variable, x, whose value gets smaller as x gets larger; that is, if $x < y$ then $f(x) > f(y)$. Also known as strictly decreasing function. { di'krēs·iŋ ,fəŋk·shən }

decreasing sequence A sequence of real numbers in which each term is less than the preceding term. { di¦krēs·iŋ 'sē·kwəns }

decrement The quantity by which a variable is decreased. { 'dek·rə·mənt }

Dedekind cut A set of rational numbers satisfying certain properties, with which a unique real number may be associated; used to define the real numbers as an extension of the rationals. { 'dā·də·kint ,kət }

Dedekind test If the series $\sum_i (b_i - b_{i+1})$ converges absolutely, the b_i converge to zero, and the series $\sum_i a_i$ has bounded partial sums, then the series $\sum_i a_i b_i$ converges. { 'dā·də·kint ,test }

deduction The process of deriving a statement from certain assumed statements by applying the rules of logic. { di'dək·shən }

defective equation An equation that has fewer roots than another equation from which it has been derived. { di'fekt·iv i'kwā·zhən }

defective number *See* deficient number. { di'fek·tiv 'nəm·bər }

deficiency index For a curve or equation involving two complex variables this is the genus of the Riemann surface associated to the equation. { də'fish·ən·sē ,in,deks }

deficient number A positive integer the sum of whose divisors, including 1 but excluding itself, is less than itself. Also known as defective number. { də'fish·ənt 'nəm·bər }

definite Riemann integral A number associated with a function defined on an interval $[a,b]$ which is $\lim_{N \to \infty} \sum_{k=0}^{N-1} f\left(a + \frac{k}{N}\right) \cdot \frac{b-a}{N}$ if f is bounded and continuous; denoted by $\int_a^b f(x)dx$; if f is a positive function, the definite integral measures the area between the graph of f and the x axis. { ¦def·ə·nət 'rē,män ,in·tə·grəl }

deformation A homotopy of the identity map to some other map. { ,def·ər'mā·shən }

degeneracy The condition in which two characteristic functions of an operator have the same characteristic value. { di'jen·ə·rə·sē }

degenerate conic A straight line, a pair of straight lines, or a point, which is a limiting form of a conic. { di'jen·ə·rət 'kän·ik }

degenerate simplex A modification of a simplex in which the points $p_0, \ldots, p_n$ on which the simplex is based are linearly dependent. { di'jen·ə·rət 'sim,pleks }

degree 1. A unit for measurement of plane angles, equal to 1/360 of a complete revolution, or 1/90 of a right angle. Symbolized °. **2.** For a term in one variable, the exponent of that variable. **3.** For a term in several variables, the sum of the exponents of its variables. **4.** For a polynomial, the degree of the highest-degree term. **5.** For a differential equation, the greatest power to which the highest-order derivative occurs. **6.** For an algebraic curve defined by the polynomial equation $f(x,y) = 0$, the degree of the polynomial $f(x,y)$. **7.** For a vertex in a graph, the number of arcs which have that vertex as an end point. **8.** For an extension of a field, the dimension of the extension field as a vector space over the original field. { di'grē }

degree of degeneracy The number of characteristic functions of an operator having the same characteristic value. Also known as order of degeneracy. { di'grē əv di'jen·ə·rə·sē }

degree of freedom A number one less than the number of frequencies being tested with a chi-square test. { di'grē əv 'frē·dəm }

degree vector The sequence of degrees of the vertices of a simple graph, arranged in nonincreasing order. { di'grē ,vek·tər }

de Gua's rule The rule that if, in a polynomial equation $f(x) = 0$, a group of r consecutive terms is missing, then the equation has at least r imaginary roots if r is even, or the equation has at least $r + 1$ or $r - 1$ imaginary roots if r is odd (depending on whether the terms immediately preceding and following the group have like or unlike signs). { də'gwäz ,rül }

Delambre analogies *See* Gauss formulas. { də'lam·brə ə,nal·ə·jēz }

del operator The rule which replaces the function f of three variables, x, y, z, by the vector valued function whose components in the x, y, z directions are the respective partial derivatives of f. Written ∇f. Also known as nabla. { 'del ,äp·ə,rād·ər }

delta function A distribution δ such that $\int_{-\infty}^{\infty} f(t)\Delta(x - t)dt$ is $f(x)$. Also known as Dirac delta function; Dirac distribution; unit impulse. { 'del·tə ,fəŋk·shən }

deltahedron Any polyhedron whose faces are congruent equilateral triangles. { ,del·tə'hē·drən }

deltoid 1. The plane curve traced by a point on a circle while the circle rolls along the inside of another circle whose radius is three times as great. **2.** A concave quadrilateral with two pairs of adjacent equal sides. Also known as Steiner's hypocycloid tricuspid. { 'del,tȯid }

De Moivre's theorem The nth power of the quantity $\cos\theta + i\sin\theta$ is $\cos n\theta + i\sin n\theta$ for any integer n. { də'mwäv·rəz ,thir·əm }

De Morgan's rules The complement of the union of two sets equals the intersection of their respective complements; the complement of the intersection of two sets equals the union of their complements. { də'mȯr·gənz ,rülz }

De Morgan's test A series with term u_n, for which $|u_{n+1}/u_n|$ converges to 1, will converge absolutely if there is $c > 0$ such that the limit superior of $n(|u_{n+1}/u_n| - 1)$ equals $-1 - c$. { də'mȯr·gənz ,test }

denial *See* negation. { di'nī·əl }

denominator In a fraction, the term that divides the other term (called the numerator), and is written below the line. { də'näm·ə,nād·ər }

dense-in-itself set A set every point of which is an accumulation point; a set without any isolated points. { 'dens in it'self ,set }

dense subset A subset of a topological space whose closure is the entire space. { ¦dens 'səb,set }

density For an increasing sequence of integers, the greatest lower bound of the quantity $F(n)/n$, where $F(n)$ is the number of integers in the sequence (other than zero) equal to or less than n. { 'den· səd·ē }

density function 1. A density function for a measure m is a function which gives rise to m when it is integrated with respect to some other specified measure. **2.** *See* probability density function. { 'den·səd·ē ,fəŋk·shən }

denumerable set A set which may be put in one-to-one correspondence with the positive integers. Also known as countably infinite set. { də'nüm·rə·bəl 'set }

dependence The existence of a relationship between frequencies obtained from two parts of an experiment which does not arise from the direct influence of the result of the first part on the chances of the second part but indirectly from the fact that both parts are subject to influences from a common outside factor. { di'pen·dəns }

dependent equation 1. An equation is dependent on one or more other equations if it is satisfied by every set of values of the unknowns that satisfy all the other equations. **2.** A set of equations is dependent if any member of the set is dependent on the others. { di¦pen·dənt i'kwā·zhən }

dependent events Two events such that the occurrence of one affects the probability of the occurrence of the other. { di'pen·dənt i'vens }

dependent variable If y is a function of x, that is, if the function assigns a single value of y to each value of x, then y is the dependent variable. { di'pen·dənt 'ver·ē·ə·bəl }

depressed equation An equation that results from reducing the number of roots in a given equation with one unknown by dividing the original equation by the difference of the unknown and a root. { di'prest i'kwā·zhən }

derangement A permutation of a finite set of elements that carries no element of the set into itself. { di'rānj·mənt }

derangement numbers The numbers D_n, $n = 1, 2, 3, \ldots$, giving the number of permutations of a set of n elements that carry no element of the set into itself. { di'rānj·mənt ‚nəm·bərz }

derivation 1. The process of deducing a formula. **2.** A function D on an algebra which satisfies the equation $D(uv) = uD(v) + vD(u)$. { ‚der·ə'vā·shən }

derivative The slope of a graph $y = f(x)$ at a given point c; more precisely, it is the limit as h approaches zero of $f(c + h) - f(c)$ divided by h. Also known as differential coefficient; rate of change. { də'riv·əd·iv }

derived curve A curve whose ordinate, for each value of the abscissa, is equal to the slope of some given curve. Also known as first derived curve. { də'rīvd 'kərv }

derived set The set of cluster points of a given set. { də'rīvd 'set }

derogatory matrix A matrix whose order is greater than that of its reduced characteristic equation. { də'räg·ə‚tȯr·ē 'mā·triks }

Desarguesian plane Any projective plane in which points and lines satisfy Desargues' theorem. Also known as Arguesian plane. { dā·zär‚gā·zē·ən 'plān }

Desargues' theorem If the three lines passing through corresponding vertices of two triangles are concurrent, then the intersections of the three pairs of corresponding sides lie on a straight line, and conversely. { dā'zärgz ‚thir·əm }

Descartes' rule of signs A polynomial with real coefficients has at most k real positive roots, where k is the number of sign changes in the polynomial. { dā'kärts 'rül əv 'sīnz }

descending chain condition The condition on a ring that every descending sequence of left ideals (or right ideals) has only a finite number of distinct members. { di‚send·iŋ 'chān kən‚dish·ən }

descending sequence 1. A sequence of elements in a partially ordered set such that each member of the sequence is equal to or less than the preceding one. **2.** In particular, a sequence of sets such that each member of the sequence is a subset of the preceding one. { di‚send·iŋ 'sē·kwəns }

descriptive geometry The application of graphical methods to the solution of three-dimensional space problems. { di'skrip·tiv jē'äm·ə·trē }

descriptive statistics Presentation of data in the form of tables and charts or summarization by means of percentiles and standard deviations. { di'skrip·tiv stə'tis·tiks }

determinant A certain real-valued function of the column vectors of a square matrix which is zero if and only if the matrix is singular; used to solve systems of linear equations and to study linear transformations. { də'tər·mə·nənt }

determinant tensor A tensor whose components are each equal to the corresponding component of the Levi-Civita tensor density times the square root of the determinant of the metric tensor, and whose contravariant components are each equal to the corresponding component of the Levi-Civita density divided by the square root of the metric tensor. Also known as permutation tensor. { də'tər·mə·nənt 'ten·sər }

developable surface A surface that can be obtained from a plane sheet by deformation, without stretching or shrinking. { di‚vel·əp·ə·bəl 'sər·fəs }

deviation The difference between any given number in a set and the mean average of those numbers. { ‚dēv·ē'ā·shən }

devil on two sticks *See* devil's curve. { 'dev·əl ȯn ‚tü 'stiks }

devil's curve A plane curve whose equation in Cartesian coordinates x and y is $y^4 - a^2y^2 = x^4 - b^2x^2$, where a and b are constants. Also known as devil on two sticks. { 'dev·əlz 'kərv }

dextrorse curve *See* right-handed curve. { 'dek‚strȯrs ‚kərv }

dextrorsum *See* right-handed curve. { dek'strȯr·səm }

diagonal 1. The set of points all of whose coordinates are equal to one another in an n-dimensional coordinate system. **2.** A line joining opposite vertices of a polygon with an even number of sides. { dī'ag·ən·əl }

diagonalize To convert a square matrix to a diagonal matrix, usually by multiplying it on the left by a second matrix A of the same order, and on the right by the inverse of A. { dī'ag·ən·ə,līz }

diagonal Latin square A Latin square in which each of the symbols appears exactly once in each diagonal. { di¦ag·ən·əl ‚lat·ən 'skwer }

diagonally dominant matrix A matrix in which the absolute value of each diagonal element is either greater than the sum of the absolute values of the off-diagonal elements of the same row or greater than the sum of the off-diagonal elements in the same column. { dī'ag·ən·əl·ē 'däm·ə·nənt 'mā‚triks }

diagonal matrix A matrix whose nonzero entries all lie on the principal diagonal. { dī'ag·ən·əl 'mā·triks }

diagram A picture in which sets are represented by symbols and mappings between these sets are represented by arrows. { 'dī·ə‚gram }

diakoptics A piecewise approach to the solution of large-scale interconnected systems, in which the large system is first broken up into several small pieces or subdivisions, the subdivisions are solved separately, and finally the effect of interconnection is determined and added to each subdivision to yield the complete solution of the system. { ‚dī·ə'käp·tiks }

diameter 1. A line segment which passes through the center of a circle, and whose end points lie on the circle. **2.** The length of such a line. **3.** For a conic, any straight line that passes through the midpoints of all the chords of the conic that are parallel to a given chord. **4.** For a set, the smallest number that is greater than or equal to the distance between every pair of points of the set. { dī'am·əd·ər }

diametral curve A curve that passes through the midpoints of a family of parallel chords of a given curve. { dī'am·ə·trəl 'kərv }

diametral plane 1. A plane that passes through the center of a sphere. **2.** A plane that passes through the mid-points of a family of parallel chords of a quadric surface that are parallel to a given chord. { dī'am·ə·trəl 'plān }

diametral surface A surface that passes through the midpoints of a family of parallel chords of a given surface that are parallel to a given chord. { dī'am·ə·trəl 'sər·fəs }

dicycle A simple closed dipath. Also known as directed cycle. { 'dī‚sī·kəl }

Dido's problem The problem of finding the curve, with a given perimeter, that encloses the greatest possible area; the curve is a circle. { 'dē‚dōz ‚präb·ləm }

diffeomorphic sets Sets in Euclidean space such that there is a diffeomorphism between them. { ‚dif·ē·ə‚mȯr·fik 'sets }

diffeomorphism A bijective function, with domain and range in the same or different Euclidean spaces, such that both the function and its inverse have continuous mixed partial derivatives of all orders in neighborhoods of each point of their respective domains. { ‚dif·ē·ə'mȯr·fiz·əm }

difference 1. The result of subtracting one number from another. **2.** The difference between two sets A and B is the set consisting of all elements of A which do not belong to B; denoted $A - B$. { 'dif·rəns }

difference equation An equation expressing a functional relationship of one or more independent variables, one or more functions dependent on these variables, and successive differences of these functions. { 'dif·rəns i'kwā·zhən }

difference methods Versions of the predictor-corrector methods of calculating numerical solutions of differential equations in which the prediction and correction formulas express the value of the solution function in terms of finite differences of a derivative of the function. { 'dif·rəns ‚meth·ədz }

difference operator One of several operators, such as the displacement operator, forward difference operator, or central mean operator, which can be used to conveniently express formulas for interpolation or numerical calculation or integration of functions and can be manipulated as algebraic quantities. { 'dif·rəns ‚äp·ə‚rād·ər }

difference quotient The increment of the value of a function divided by the increment of the independent variable; for the function $y = f(x)$, it is $\Delta y/\Delta x = [f(x + \Delta x) - f(x)]\Delta x$, where Δx and Δy are the increments of x and y. { 'dif·rəns ‚kwō·shənt }

differentiable atlas

differentiable atlas A family of embeddings $h_i{:}E^n \to M$ of Euclidean space into a topological space M with the property that $h_i^{-1}h_j{:}E^n \to E^n$ is a differentiable map for each pair of indices, i, j. { ‚dif·ə'ren·chə·bəl 'at·ləs }

differentiable function A function which has a derivative at each point of its domain. { ‚dif·ə'ren·chə·bəl 'fəŋk·shən }

differentiable manifold A topological space with a maximal differentiable atlas; roughly speaking, a smooth surface. { ‚dif·ə'ren·chə·bəl 'man·ə‚fōld }

differential 1. The differential of a real-valued function $f(x)$, where x is a vector, evaluated at a given vector c, is the linear, real-valued function whose graph is the tangent hyperplane to the graph of $f(x)$ at $x = c$; if x is a real number, the usual notation is $df = f'(c)dx$. **2.** See total differential. { ‚dif·ə'ren·chəl }

differential calculus The study of the manner in which the value of a function changes as one changes the value of the independent variable; includes maximum-minimum problems and expansion of functions into Taylor series. { ‚dif·ə'ren·chəl 'kal·kyə·ləs }

differential coefficient See derivative. { ‚dif·ə·ren·chəl ‚kō·i'fish·ənt }

differential equation An equation expressing a relationship between functions and their derivatives. { ‚dif·ə'ren·chəl i'kwā·zhən }

differential form A homogeneous polynomial in differentials. { ‚dif·ə'ren·chəl 'fȯrm }

differential game A game in which the describing equations are differential equations. { ‚dif·ə'ren·chəl 'gām }

differential geometry The study of curves and surfaces using the methods of differential calculus. { ‚dif·ə'ren·chəl jē'äm·ə·trē }

differential operator An operator on a space of functions which maps a function f into a linear combination of higher-order derivatives of f. { ‚dif·ə'ren·chəl 'äp·ə‚rād·ər }

differential selection A biased selection of a conditioned sample. { ‚dif·ə'ren·chəl si'lek·shən }

differential topology The branch of mathematics dealing with differentiable manifolds. { ‚dif·ə'ren·chəl tə'päl·ə·jē }

differentiation The act of taking a derivative. { ‚dif·ə‚ren·chē'ā·shən }

digamma function The derivative of the natural logarithm of the gamma function. { 'dī‚gam·ə ‚fəŋk·shən }

digit A character used to represent one of the nonnegative integers smaller than the base of a system of positional notation. Also known as numeric character. { 'dij·ət }

digit place See digit position. { 'dij·ət ‚plās }

digit position The position of a particular digit in a number that is expressed in positional notation, usually numbered from the lowest significant digit of the number. Also known as digit place. { 'dij·ət pə‚zish·ən }

digraph See directed graph. { 'dī‚graf }

dihedral See dihedron. { dī'hē·drəl }

dihedral angle The angle between two planes; it is said to be zero if the planes are parallel; if the planes intersect, it is the plane angle between two lines, one in each of the planes, which pass through a point on the line of intersection of the two planes and are perpendicular to it. { dī'hē·drəl ‚aŋ·gəl }

dihedral group The group of rotations of three-dimensional space that carry a regular polygon into itself. { dī'hē·drəl ‚grüp }

dihedron A geometric figure formed by two half planes that are bounded by the same straight line. Also known as dihedral. { dī'hē·drən }

dilation 1. A transformation which changes the size, and only the size, of a geometric figure. **2.** An operation that provides a relatively flexible boundary to a fuzzy set; for a fuzzy set A with membership function m_A, a dilation of A is a fuzzy set whose membership function has the value $[m_A(x)]^\beta$ for every element x, where β is a fixed number that is greater than 0 and less than 1. { də'lā·shən }

Dilworth's theorem The theorem that, in a finite partially ordered set, the maximum cardinality of an antichain is equal to the minimum number of disjoint chains into which the partially ordered set can be partitioned. { 'dil,wərths ,thir·əm }

dimension 1. The number of coordinates required to label the points of a geometrical object. **2.** For a vector space, the number of vectors in any basis of the vector space. **3.** For a simplex, one less than the number of vertices of the simplex. **4.** For a simplicial complex, the largest of the dimensions of the simplices that make up the complex. **5.** The length of one of the sides of a rectangle. **6.** The length of one of the edges of a rectangular parallelepiped. { də'men·chən }

dimensionless number A ratio of various physical properties (such as density or heat capacity) and conditions (such as flow rate or weight) of such nature that the resulting number has no defining units of weight, rate, and so on. Also known as nondimensional parameter. { də'men·chən·ləs 'nəm·bər }

dimension theory The study of abstract notions of dimension, which are topological invariants of a space. { də'men·chən ,thē·ə·rē }

Dini condition A condition for the convergence of a Fourier series of a function f at a number x, namely, that the limits of f at x on the left and right, $f(x-)$ and $f(x+)$, both exist, and that the function given by the absolute value of $[f(x + t) - f(x+) + f(x - t) - f(x-)]/t$ be integrable on some closed interval, $-d \le t \le d$, where d is a positive number. { 'dē·nē kən¦didh·ən }

Dini theorem The theorem that, if a monotone sequence of continuous real-valued functions converges to a continuous function f on a compact set C, this convergence is uniform; that is, the sequence converges uniformly to f on C. { 'dē·nē ,thir·əm }

dioctahedral Having 16 faces. { ,dī,äk·tə'hē·drəl }

diophantine analysis A means of determining integer solutions for certain algebraic equations. { ¦dī·ə¦fant·ən ə'nal·ə·səs }

diophantine equations Equations with more than one independent variable and with integer coefficients for which integer solutions are desired. { ¦dī·ə¦fant·ən i'kwā·zhənz }

dipath *See* directed path. { 'dī,path }

Dirac delta function *See* delta function. { di'rak 'del·tə ,fəŋk·shən }

Dirac distribution *See* delta function. { də¦rak di·strə'byü·shən }

Dirac spinor *See* spinor. { di'rak 'spin·ər }

directed angle An angle for which one side is designated as initial, the other as terminal. { də'rek·təd 'aŋ·gəl }

directed cycle *See* dicycle. { də¦rek·təd 'sī·kəl }

directed graph A graph in which a direction is shown for every arc. Also known as digraph. { də'rek·təd 'graf }

directed line A line on which a positive direction has been specified. { də'rek·təd 'līn }

directed network A directed graph in which each arc is assigned a unique nonnegative integer called its weight. { də¦rek·təd 'net,wərk }

directed number A number together with a sign. { də'rek·təd 'nəm·bər }

directed path A sequence of vertices, $v_1, v_2, \ldots, v_n$, in a directed graph such that there is an arc from v_i to v_{i+1} for $i = 1, 2, \ldots, n - 1$. Also known as dipath. { də¦rek·təd 'path }

directed set A partially ordered set with the property that for every pair of elements a,b in the set, there is a third element which is larger than both a and b. Also known as directed system; Moore-Smith set. { də'rek·təd 'set }

directed system *See* directed set. { də¦rek·təd 'sis·təm }

directional derivative The rate of change of a function in a given direction; more precisely, if f maps an n-dimensional Euclidean space into the real numbers, and $\mathbf{x} = (x_1, \ldots, x_n)$ is a vector in this space, and $\mathbf{u} = (u_1, \ldots, u_n)$ is a unit vector in the space (that is, $u_1^2 + \cdots + u_n^2 = 1$), then the directional derivative of f at $\mathbf{x}$ in the direction of $\mathbf{u}$ is the limit as h approaches zero of $[f(\mathbf{x} + h\mathbf{u}) - f(\mathbf{x})]/h$. { də'rek·shən·əl də'riv·əd·iv }

direction angles The three angles which a line in space makes with the positive x, y, and z axes. { də'rek·shən 'aŋ·gəlz }

direction cosine The cosine of one of the direction angles of a line in space. { də'rek·shən 'kō,sīn }

direction numbers Any three numbers proportional to the direction cosines of a line in space. Also known as direction ratios. { di'rek·shən ,nəm·bərz }

direction ratios *See* direction numbers. { di'rek·shən ,rā·shōz }

directly congruent figures Two solid geometric figures, one of which can be made to coincide with the other by a rigid motion in space, without reflection. { də,rek·lē kən¦grü·ənt 'fig·yərz }

director circle A circle consisting of the points of intersection of pairs of perpendicular tangents to an ellipse or hyperbola. { di'rek·tər 'sər·kəl }

direct product Given a finite family of sets $A_1, \ldots, A_n$, the direct product is the set of all n-tuples $(a_1, \ldots, a_n)$, where a_i belongs to A_i for $i = 1, \ldots, n$. { də'rekt 'präd·əkt }

direct proof An argument that establishes the truth of a statement by making direct use of the hypotheses, as opposed to a proof by contradiction. { də,rekt 'prüf }

direct proportion A statement that the ratio of two variable quantities is equal to a constant. { də'rekt prə'pòr·shən }

directrix 1. A fixed line used in one method of defining a conic; the distance from this line divided by the distance from a fixed point (called the focus) is the same for all points on the conic. **2.** A curve through which a line generating a given ruled surface always passes. { də'rek·triks }

direct sum If each of the sets in a finite direct product of sets has a group structure, this structure may be imposed on the direct product by defining the composition "componentwise"; the resulting group is called the direct sum. { də¦rekt 'səm }

direct variation 1. A relationship between two variables wherein their ratio remains constant. **2.** An equation or function expressing such a relationship. { də'rekt ,ver·ē'ā·shən }

Dirichlet conditions The requirement that a function be bounded, and have finitely many maxima, minima, and discontinuities on the closed interval $[-\pi, \pi]$. { ,dē·rē'klā kən,dish·ənz }

Dirichlet drawer principle *See* pigeonhole principle. { ,dē·rē'klā 'drò·ər ,prin·sə·pəl }

Dirichlet problem To determine a solution to Laplace's equation which satisfies certain conditions in a region and on its boundary. { ,dē·rē'klā ,präb·ləm }

Dirichlet series A series whose nth term is a complex number divided by n to the zth power. { ,dē·rē'klā ,sir·ēz }

Dirichlet test for convergence If Σb_n is a series whose sequence of partial sums is bounded, and if $\{a_n\}$ is a monotone decreasing null sequence, then the series

$$\sum_{n=1}^{\infty} a_n b_n \text{ converges.}$$ { ,dē·rē·klā ,test fər kən'vər·jəns }

Dirichlet theorem The theorem that, if a and b are relatively prime numbers, there are infinitely many prime numbers of the form $a + nb$, where n is an integer. { dē·rē'klā ,thir·əm }

Dirichlet transform For a function $f(x)$, this is the integral of $f(x) \cdot \sin(kx)/x$; its convergence determines the convergence of the Fourier series of $f(x)$. { ,dē·rē'klā ,tranz,fòrm }

disc *See* disk. { disk }

disconnected set A set in a topological space that is the union of two nonempty sets A and B for which both the intersection of the closure of A with B and the intersection of the closure of B with A are empty. { ,dis·kə¦nek·təd 'set }

discontinuity A point at which a function is not continuous. { dis,känt·ən'ü·əd·ē }

discrete mathematics *See* finite mathematics. { di,skrēt ,math·ə'mat·iks }

discrete Fourier transform A generalization of the Fourier transform to finite sets of data; for a function f defined at N data values, $0, 1, 2, \ldots, N - 1$, the discrete Fourier transform is a function, F, also defined on the set $(0, 1, 2, \ldots, N - 1)$, whose value at n is the sum over the variable r, from 0 through $N-1$, of the quantity $N^{-1} f(r) \exp(-i2\pi nr/N)$. { di¦skrēt für·yā 'tranz,fòrm }

discrete set A set with no cluster points. { di'skrēt 'set }

discrete topology For a set A, the set of all subsets of A. { di,skrēt tə'päl·ə·jē }

discrete variable A variable for which the possible values form a discrete set. { di¦skrēt 'ver·ē·ə·bəl }

discretization A procedure in the numerical solution of partial differential equations in which the domain of the independent variable is subdivided into cells or elements and the equations are expressed in discrete form at each point by finite difference, finite volume, or finite element methods. { dis¸krēd·ə'zā·shən }

discretization error The error in the numerical calculation of an integral that results from using an approximate expression for the true mathematical function to be integrated. { ¸dis·krə·də'zā·shən ¸er·ər }

discriminant 1. The quantity $b^2 - 4ac$, where a, b, c are coefficients of a given quadratic polynomial: $ax^2 + bx + c$. **2.** More generally, for the polynomial equation $a_0x^n + a_1x^{n-1} + \cdots + a_nx_0 = 0$, a_0^{2n-2} times the product of the squares of all the differences of the roots of the equation, taken in pairs. { di'skrim·ə·nənt }

discriminant function A linear combination of a set of variables that will classify events or items for which the variables are measured with the smallest possible proportion of misclassifications. { di¦skrim·ə·nənt 'fəŋk·shən }

disintegration of measure The representation of a measure as an integral of a family of positive measures. { dis¸in·tə'grā·shən əv 'mezh·ər }

disjoint sets Sets with no elements in common. { dis'jȯint 'sets }

disjunction The connection of two statements by the word "or." Also known as alternation. { dis'jəŋk·shən }

disk Also spelled disc. **1.** The region in the plane consisting of all points with norm less than 1 (sometimes less than or equal to 1). **2.** *See* closed disk. { disk }

disk method A method of computing the volume of a solid of revolution, by integrating over the volumes of infinitesimal disk-shaped slices bounded by planes perpendicular to the axis of revolution. { 'disk ¸meth·əd }

dispersion The degree of spread shown by observations in a sample or a population. { də'spər·zhən }

dispersion index Statistics used to determine the homogeneity of a set of samples. { di'spər·zhən ¸in¸deks }

displacement operator A difference operator, denoted E, defined by the equation $Ef(x) = f(x + h)$, where h is a constant denoting the difference between successive points of interpolation or calculation. Also known as forward shift operator. { dis'plās·mənt ¸äp·ə¸rād·ər }

dissimilar terms Terms that do not contain the same unknown factors or that do not contain the same powers of these factors. { di¦sim·ə·lər 'tərmz }

distance 1. A nonnegative number associated with pairs of geometric objects. **2.** The spatial separation of two points, measured by the length of a hypothetical line joining them. **3.** For two parallel lines, two skew lines, or two parallel planes, the length of a line joining the two objects and perpendicular to both. **4.** For a point and a line or plane, the length of the perpendicular from the point to the line or plane. { 'dis·təns }

distribution 1. An abstract object which generalizes the idea of function; used in applied mathematics, quantum theory, and probability theory; the delta function is an example. Also known as generalized function. **2.** For a discrete random variable, a function (or table) which assigns to each possible value of the random variable the probability that this value will occur; for a continuous random variable x, the monotone nondecreasing function which assigns to each real t the probability that x is less than or equal to t. Also known as distribution function; probability distribution; statistical distribution. { ¸dis·trə'byü·shən }

distribution curve The graph of the distribution function of a random variable. { ¸dis·trə'byü·shən 'kərv }

distribution-free method Any method of inference that does not depend on the characteristics of the population from which the samples are obtained. { ¸dis·trə'byü·shən ¸frē ¸meth·əd }

distribution function *See* distribution. { ¸dis·trə'byü·shən ¸fəŋk·shən }

distributive lattice A lattice in which "greatest lower bound" obeys a distributive law with respect to "least upper bound," and vice versa. { di'strib·yəd·iv 'lad·əs }

distributive law A rule which stipulates how two binary operations on a set shall behave with respect to one another; in particular, if $+$, $\circ$ are two such operations then $\circ$ distributes over $+$ means $a \circ (b + c) = (a \circ b) + (a \circ c)$ for all a,b,c in the set. { di'strib·yəd·iv 'lȯ }

divergence For a vector-valued function, the sum of the diagonal entries of the Jacobian matrix; it is the scalar product of the del operator and the vector. { də'vər·jəns }

divergence theorem *See* Gauss' theorem. { də'vər·jəns ,thir·əm }

divergent integral An improper integral which does not have a finite value. { də'vər·jənt 'in·tə·grəl }

divergent sequence A sequence which does not converge. { də'vər·jənt 'sē·kwəns }

divergent series An infinite series whose sequence of partial sums does not converge. { də'vər·jənt 'sir·ēz }

divide One object (integer, polynomial) divides another if their quotient is an object of the same type. { də'vīd }

divide-and-conquer relation A recurrence relation which expresses the value of a number-theoretic function for an argument n in terms of its value for an argument n/b, where b is an integer greater than 1. { di¦vīd ən 'käŋ·kər ri,lā·shən }

divided differences Quantities which are used in the interpolation or numerical calculation or integration of a function when the function is known at a series of points which are not equally spaced, and which are formed by various operations on the difference between the values of the function at successive points. { də'vīd·əd 'dif·rən·səs }

dividend A quantity which is divided by another quantity in the operation of division. { 'div·ə,dend }

divine proportion *See* golden section. { di,vīn prə'pȯr·shən }

division The inverse operation of multiplication; the number a divided by the number b is the number c such that b multiplied by c is equal to a. { də'vizh·ən }

division algebra A hypercomplex system that is also a skew field. { də'vizh·ən ,al·jə·brə }

division algorithm The theorem that, for any integer m and any positive integer n, there exist unique integers q and r such that $m = qn + r$ and r is equal to or greater than 0 and less than n. { di¦vizh·ən 'al·gə,rith·əm }

division modulo p Division in the finite field with p elements, where p is a prime number. { də'vizh·ən ¦mäj·ə·lō 'pē }

division ring 1. A ring in which the set of nonzero elements form a group under multiplication. **2.** More generally, a nonassociative ring with nonzero elements in which, for any two elements a and b, there are elements x and y such that $ax = b$ and $ya = b$. { di'vizh·ən ,riŋ }

division sign 1. The symbol $\div$, used to indicate division. **2.** The diagonal $/$, used to indicate a fraction. { di'vizh·ən ,sīn }

divisor 1. The quantity by which another quantity is divided in the operation of division. **2.** An element b in a commutative ring with identity is a divisor of an element a if there is an element c in the ring such that $a = bc$. { də'vīz·ər }

divisor of zero A nonzero element x of a commutative ring such that $xy = 0$ for some nonzero element y of the ring. Also known as zero divisor. { di¦vī·zər əv 'zir·ō }

Dobinski's equality A formula which expresses a Bell number as the sum of an infinite series. { dō¦bin·skēz ē'kwäl·əd·ē }

dodecagon A 12-sided polygon. { dō'dek·ə,gän }

dodecahedron A polyhedron with 12 faces. { dō,dek·ə'hē,drən }

dodecomino One of the 63,600 plane figures that can be formed by joining 12 unit squares along their sides. { ,dō·dek·əmē·nō }

domain 1. For a function, the set of values of the independent variable. **2.** A nonempty open connected set in Euclidean space. Also known as open region; region. **3.** *See* Abelian field. { dō'mān }

domain of dependence For an initial-value problem for a partial differential equation, a portion of the range such that the initial values on this portion determine the solution over the entire range.   { ¦dō¦mān əv di'pen·dəns }

dominant strategy Relative to a given pure strategy for one player of a game, a second pure strategy for that player that has at least as great a payoff as the given strategy for any pure strategy of the opposing player.   { 'däm·ə·nənt 'strad·ə·jē }

dominated convergence theorem If a sequence $\{f_n\}$ of Lebesgue measurable functions converges almost everywhere to f and if the absolute value of each f_n is dominated by the same integrable function, then f is integrable and $\lim \int f_n dm = \int f dm$.   { 'däm·ə¸nād·əd kən'vər·jəns ¸thir·əm }

dominating edge set A set of edges of a graph such that every edge is either a member of this set or has a vertex in common with a member of this set.   { ¦däm·ə¸nād·iŋ 'ej ¸set }

dominating integral An improper integral whose nonnegative, nonincreasing integrand function has the property that its value for all sufficiently large positive integers n is no smaller than the nth term of a given series of positive terms; used in the integral test for convergence.   { 'däm·ə¸nad·iŋ 'in·tə·grəl }

dominating series A series, each term of which is larger than the respective term in some other given series; used in the comparison test for convergence of series.   { 'däm·ə¸nad·iŋ 'sir·ēz }

dominating vertex set A set of vertices in a simple graph such that every vertex of the graph is either a member of this set or is adjacent to a member of this set. Also known as external dominating set.   { ¦däm·ə¸nād·iŋ 'vər¸teks ¸set }

domino The plane figure formed by joining two unit squares along a common side; a rectangle whose length is twice its width.   { 'däm·ə¸nō }

dot product *See* inner product.   { 'dät ¸präd·əkt }

double angle formula An equation that expresses a trigonometric function of twice an angle in terms of trigonometric functions of the angle.   { ¦dəb·əl 'aŋ·gəl ¸fór·myə·lə }

double-blind technique An experimental procedure in which neither the subjects nor the experimenters know the makeup of the test and control group during the actual course of the experiments.   Also known as blind trial.   { ¦dəb·əl 'blīnd ¸tek¸nēk }

double cusp A point on a curve through which two branches of the curve with the same tangent pass, and at which each branch extends in both directions of the tangent.   Also known as point of osculation; tacnode.   { 'dəb·əl ¸kəsp }

double integral The Riemann integral of functions of two variables.   { ¦dəb·əl 'in·tə·grəl }

double law of the mean *See* second mean-value theorem.   { ¦dəb·əl ¦lò əv thə 'mēn }

double minimal surface A minimal surface that is also a one-sided surface.   { ¦dəb·əl ¦min·ə·məl 'sər·fəs }

double point A point on a curve at which a curve crosses or touches itself, or has a cusp; that is, a point at which the curve has two tangents (which may be coincident).   { ¦dəb·əl 'póint }

double root For an algebraic equation, a number a such that the equation can be written in the form $(x - a)^2 p(x) = 0$ where $p(x)$ is a polynomial of which a is not a root.   { ¦dəb·əl 'rüt }

double series A two-dimensional array of numbers whose sum is the limit of $S_{m,n}$, the sum of the terms in the rectangular array formed by the first n terms in each of the first m rows, as m and n increase.   { ¦dəb·əl 'sir¸ēz }

double tangent A line which is tangent to a curve at two distinct noncoincident points.   Also known as bitangent. Two coincident tangents to branches of a curve at a given point, such as the tangents to a cusp.   { ¦dəb·əl 'tan·jənt }

doubly ruled surface A ruled surface that can be generated by either of two distinct moving straight lines; quadric surfaces are the only surfaces of this type.   { ¦dəb·lē ¦rüld 'sər·fəs }

doubly stochastic matrix A matrix of nonnegative real numbers such that every row sum and every column sum are equal to 1.   { ¦dəb·lē stō¦kas·tik 'mā·triks }

dual basis 1. For a finite-dimensional vector space with basis $x_1, x_2, \ldots, x_n$, the dual basis of the conjugate space is the set of linear functionals $f_1, f_2, \ldots, f_n$ with $f_i(x_i) = 1$ and $f_i(x_j) = 0$ for i not equal to j. **2.** For a Banach space with basis $x_1, x_2, \ldots,$ the dual basis of the conjugate space is the sequence of continuous linear functionals, $f_1, f_2, \ldots,$ defined by $f_i(x_i) = 1$ and $f_i(x_j) = 0$ for i not equal to j, provided that the conjugate space is shrinking. { ¦dü·əl 'bā·ses }

dual coordinates Point coordinates and plane coordinates are dual in geometry since an equation about one determines an equation about the other. { ¦dü·əl kō'ȯrd·ən·əts }

dual graph A planar graph corresponding to a planar map obtained by replacing each country with its capital and each common boundary by an arc joining the two countries. { ¦dü·əl 'graf }

dual group The group of all homomorphisms of an Abelian group G into the cyclic group of order n, where n is the smallest integer such that g^n is the identity element of G. { ¦dü·əl 'grüp }

duality principle Also known as principle of duality. A principle that if a theorem is true, it remains true if each object and operation is replaced by its dual; important in projective geometry and Boolean algebra. { dü'al·əd·ē ¸prin·sə·pəl }

duality theorem 1. A theorem which asserts that for a given n-dimensional space, the $(n - p)$ dimensional homology group is isomorphic to a p-dimensional cohomology group for each $p = 0, \ldots, n$, provided certain conditions are met. **2.** Let G be either a compact group or a discrete group, let X be its character group, and let G' be the character group of X; then there is an isomorphism of G onto G' so that the groups G and G' may be identified. **3.** If either of two dual linear-programming problems has a solution, then so does the other. { dü'al·əd·ē ¸thir·əm }

dual linear programming Linear programming in which the maximum and minimum number are the same number. { 'dü·əl ¦lin·ē·ər 'prō¸gram·iŋ }

dual operation In projective geometry, an operation that is obtained from a given operation by replacing points with lines, lines with points, the drawing of a line through a point with the marking of a point on a line, and so forth. { ¦dül äp·ə'rā·shən }

dual space The vector space consisting of all linear transformations from a given vector space into its scalar field. { 'dü·əl 'spās }

dual tensor The product of a given tensor, covariant in all its indices, with the contravariant form of the determinant tensor, contracting over the indices of the given tensor. { 'dü·əl 'ten·sər }

dual theorem In projective geometry, the theorem that is obtained from a given theorem by replacing points with lines, lines with points, and operations with their dual operations. Also known as reciprocal theorem. { ¦dül 'thir·əm }

dual variables Mutually dependent variables. { 'dü·əl 'ver·ē·ə·bəlz }

Duhamel's theorem If f and g are continuous functions, then

$$\lim_{|\Delta x| \to 0} \sum_{i=1}^{n} f(x_i')g(x_i'')\Delta x_i = \int_{a}^{b} f(x)g(x)dx$$

where x_i' and x_i'' are between x_{i-1} and x_i, $i = 1, \ldots, n$, and $|\Delta x| = \max \{x_i - x_{i-1}\}$ for a partition $a = x_0 < x_1 < \cdots < x_n = b$. { dyə'melz ¸thir·əm }

dummy suffix A suffix which has no true mathematical significance and is used only to facilitate notation; usually an index which is summed over. { ¦dəm·ē 'səf·iks }

dummy variable A variable which has no true mathematical significance and is used only to facilitate notation; usually a variable which is integrated over. { ¦dəm·ē 'ver·ē·ə·bəl }

duodecimal number system A representation system for real numbers using 12 as the base. { ¸dü·ə¦des·məl 'nəm·bər ¸sis·təm }

Dupin's theorem The proposition that, given three families of mutually orthogonal surfaces, the line of intersection of any two surfaces of different families is a line of curvature for both the surfaces. { dyü'paz ¸thir·əm }

Durer's conchoid A plane curve consisting of points that lie on a variable line passing through points Q and R and are a constant distance a from Q, where Q and R have

Cartesian coordinates (*q*,0) and (0,*r*) and *q* and *r* satisfy the equation *q* + *r* = *b*, where *b* is a constant.   { 'dùr·ərz 'kän,kȯid }

Durfee square The largest square that is filled with asterisks in the star diagram of a particular partition.   { 'dər·fē ,skwer }

dyad An abstract object which is a pair of vectors **AB** in a given order on which certain operations are defined.   { 'dī,ad }

dyadic expansion The representation of a number in the binary number system.   { dī'ad·ik ik'span·chən }

dyadic number system *See* binary number system.   { dī¦ad·ik 'nəm·bər ,sis·təm }

dyadic operation An operation that has only two operands.   { dī'ad·ik ,äp·ə'rā·shən }

dyadic rational A fraction whose denominator is a power of 2.   { dī'ad·ik 'rash·ən·əl }

dynamical system An abstraction of the concept of a family of solutions to an ordinary differential equation; namely, an action of the real numbers on a topological space satisfying certain "flow" properties.   { dī¦nam·ə·kəl 'sis·təm }

dynamic programming A mathematical technique, more sophisticated than linear programming, for solving a multidimensional optimization problem, which transforms the problem into a sequence of single-stage problems having only one variable each.   { dī¦nam·ik 'prō·grə·miŋ }

E

e The base of the natural logarithms; the number defined by the equation $\int_1^e \frac{1}{x}\,dx = 1$; approximately equal to 2.71828.

eccentric angle 1. For an ellipse having semimajor and semiminor angles of lengths a and b respectively, lying along the x and y axes of a coordinate system respectively, and for a point (x,y) on the ellipse, the angle $\text{arc}\cos\frac{x}{a} = \text{arc}\sin\frac{y}{b}$. **2.** For a hyperbola having semitransverse and semiconjugate axes of lengths a and b respectively, lying along the x and y axes of a coordinate system respectively, and for a point (x,y) on the hyperbola, the angle $\text{arc}\sec\frac{x}{a} = \text{arc}\tan\frac{y}{b}$. { ek¦sen·trik 'ang·əl }

eccentric circles 1. For an ellipse, two circles whose centers are at the center of the ellipse and whose diameters are, respectively, the major and minor axes of the ellipse. **2.** For a hyperbola, two circles whose centers are at the center of symmetry of the hyperbola and whose diameters are, respectively, the transverse and conjugate axes of the hyperbola. { ek¦sen·trik 'sərk·əlz }

eccentricity The ratio of the distance of a point on a conic from the focus to the distance from the directrix. { ‚ek·sən'tris·əd·ē }

echelon matrix A matrix in which the rows whose terms are all zero are below those with some nonzero terms, the first nonzero term in a row is 1, and this 1 appears to the right of the first nonzero term in any row above it. { 'esh·ə‚län ‚mā·triks }

edge 1. A line along which two plane faces of a solid intersect. **2.** A line segment connecting nodes or vertices in a graph (a geometric representation of the relation among situations). **3.** The edge of a half plane is the line that bounds it. Also known as arc. { ej }

edge cover A set of edges in a graph such that every vertex of positive degree is the vertex of at least one of the edges in this set. { 'ej ‚kəv·ər }

edge-covering number For a graph, the sum of the number of edges in a minimum edge cover and the number of isolated vertices. { 'ej ‚kəv·ər·iŋ ‚nəm·bər }

edge domination number For a graph, the smallest possible number of edges in a dominating edge set. { ¦ej ‚däm·ə'nā·shən ‚nəm·bər }

edge independence number For a graph, the largest possible number of edges in a matching. { ¦ej ‚in·də'pen·dəns ‚nəm·bər }

edge-induced subgraph A subgraph whose vertices consist of all the vertices in the original graph that are incident on at least one edge in the subgraph. { ¦ej in‚düst 'səb‚graf }

edge of regression The curve swept out by the characteristic point of a one-parameter family of surfaces. { ¦ej əv rē'gresh·ən }

effectively computable function Any function that can be computed on the natural numbers by means of an effective procedure. { ə¦fek·tiv·lē kəm¦pyüd·ə·bəl 'fəŋk·shən }

effective procedure A procedure or process determined by a finite list of precise instructions. { i¦fek·div prə'sē·jər }

effective transformation group A transformation group in which the identity element is the only element to leave all points fixed. { ə¦fek·tiv ‚tranz·fər'mā·shən ‚grüp }

efficiency Abbreviated eff. **1.** An estimator is more efficient than another if it has a smaller variance.   **2.** An experimental design is more efficient than another if the same level of precision can be obtained in less time or with less cost.   { ə'fish·ən·sē }

efficient estimator A statistical estimator that has minimum variance.   { ə¦fish·ənt 'es·tə,mād·ər }

Egerov's theorem If a sequence of measurable functions converges almost everywhere on a set of finite measure to a real-valued function, then given any $\epsilon > 0$ there is a set of measure smaller than ϵ on whose complement the sequence converges uniformly.   { 'eg·ə,räfs ,thir·əm }

eigenfunction Also known as characteristic function.   **1.** An eigenvector for a linear operator on a vector space whose vectors are functions.   Also known as proper function.   **2.** A solution to the Sturm-Liouville partial differential equation.   { 'ī·gən,fəŋk·shən }

eigenfunction expansion By using spectral theory for linear operators defined on spaces composed of functions, in certain cases the operator equals an integral or series involving its eigenvectors; this is known as its eigenfunction expansion and is particularly useful in studying linear partial differential equations.   { 'ī·gən,fəŋk·shən ik'span·chən }

eigenmatrix Corresponding to a diagonalizable matrix or linear transformation, this is the matrix all of whose entries are 0 save those on the principal diagonal where appear the eigenvalues.   { 'ī·gən,mā·triks }

eigenvalue One of the scalars λ such that $T(v) = \lambda v$, where T is a linear operator on a vector space, and v is an eigenvector.   Also known as characteristic number; characteristic root; characteristic value; latent root; proper value.   { 'ī·gən,val·yü }

eigenvalue equation See characteristic equation.   { 'ī·gən,val·yü i,kwā·zhən }

eigenvalue problem See Sturm-Liouville problem.   { 'ī·gən,val·yü ,präb·ləm }

eigenvector A nonzero vector v whose direction is not changed by a given linear transformation T; that is, $T(v) = \lambda v$ for some scalar λ.   Also known as characteristic vector.   { 'ī·gən,vek·tər }

eight curve A plane curve whose equation in Cartesian coordinates x and y is $x^4 = a^2(x^2 - y^2)$, where a is a constant.   Also known as lemniscate of Gerono.   { 'āt ,kərv }

Einstein space A Riemannian space in which the contracted curvature tensor is proportional to the metric tensor.   { ¦īn¦stīn 'spās }

Einstein's summation convention A notational convenience used in tensor analysis whereupon it is agreed that any term in which an index appears twice will stand for the sum of all such terms as the index assumes all of a preassigned range of values.   { 'īn,stīnz sə'mā·shən kən,ven·chən }

Eisentein irreducibility criterion The proposition that a polynomial with integer coefficients is irreducible in the field of rational numbers if there is a prime p that does not divide the coefficient of x^n but divides all the other coefficients, and if p^2 does not divide the coefficient of x^0.

element See component.   { 'el·ə·mənt }

elementary event A single outcome of an experiment.   Also known as simple event.   { ,el·ə¦men·trē i'vent }

elementary function Any function which can be formed from algebraic functions and the exponential, logarithmic, and trigonometric functions by a finite number of operations consisting of addition, subtraction, multiplication, division, and composition of functions.   { ,el·ə'men·trē 'fəŋk·shən }

elementary symmetric functions For a set of n variables, a set of n functions, σ_1, σ_2, $\ldots$, σ_n, where σ_k is the sum of all products of k of the n variables.   { el·ə'men·trē si¦me·trik 'fəŋk·shənz }

eliminant See resultant.   { i'lim·ə·nənt }

elimination A process of deriving from a system of equations a new system with fewer variables, but with precisely the same solutions.   { ə,lim·ə'nā·shən }

ellipse The locus of all points in the plane at which the sum of the distances from a fixed pair of points, the foci, is a given constant.   { ə'lips }

ellipsoid A surface whose intersection with every plane is an ellipse (or circle). { ə'lip,sȯid }

ellipsoidal coordinates Coordinates in space determined by confocal quadrics. { ə,lip-'sȯid·əl kō'ȯrd·ən·əts }

ellipsoidal harmonics Lamé functions that play a role in potential problems on an ellipsoid analogous to that played by spherical harmonics in potential problems on a sphere. { ə|lip,sȯid·əl ,här'män·iks }

ellipsoidal wave functions See Lamé wave functions. { ə|lip,sȯid·əl 'wāv ,fəŋk·shənz }

ellipsoid of revolution An ellipsoid generated by rotation of an ellipse about one of its axes. Also known as spheroid. { ə'lip,sȯid əv ,rev·ə'lü·shən }

elliptic cone A cone whose base is an ellipse. { ə|lip·tik 'kōn }

elliptic conical surface A conical surface whose directrix is an ellipse. { ə|lip·tik |kän·ə·kəl 'sər·fəs }

elliptic coordinates The coordinates of a point in the plane determined by confocal ellipses and hyperbolas. { ə'lip·tik kō'ȯrd·ən·əts }

elliptic cylinder A cylinder whose directrix is an ellipse. { ə|lip·tik 'sil·ən·dər }

elliptic differential equation A general type of second-order partial differential equation which includes Laplace's equation and has the form $\sum_{i,j=1}^{n} A_{ij} \left(\partial^2 u/\partial x_i\,\partial x_j\right) +$ $\sum_{i=1}^{n} B_i \left(\partial u/\partial x_i\right) + Cu + F = 0$ where A_{ij}, B_i, C, and F are suitably differentiable real functions of $x_1, x_2, \ldots, x_n$, and there exists at each point $(x_1, x_2, \ldots, x_n)$ a real linear transformation on the x_i which reduces the quadratic form $\sum_{i,j=1}^{n} A_{ij}\, x_i\, x_j$ to a sum of n squares, all of the same sign. Also known as elliptic partial differential equation. { ə'lip·tik dif·ə|ren·chəl i'kwā·zhən }

elliptic function An inverse function of an elliptic integral; alternatively, a doubly periodic, meromorphic function of a complex variable. { ə'lip·tik 'fəŋk·shən }

elliptic geometry The geometry obtained from Euclidean geometry by replacing the parallel line postulate with the postulate that no line may be drawn through a given point, parallel to a given line. Also known as Riemannian geometry. { ə'lip·tik jē'äm·ə·trē }

elliptic integral An integral over x whose integrand is a rational function of x and the square root of $p(x)$, where $p(x)$ is a third- or fourth-degree polynomial without multiple roots. { ə'lip·tik 'int·ə·grəl }

elliptic integral of the first kind Any elliptic integral which is finite for all values of the limits of integration and which approaches a finite limit when one of the limits of integration approaches infinity. { ə|lip·tik |int·ə·grəl əv thə |fərst ,kīnd }

elliptic integral of the second kind Any elliptic integral which approaches infinity as one of the limits of integration y approaches infinity, or which is infinite for some value of y, but which has no logarithmic singularities in y. { ə|lip·tik |int·ə·grəl əv thə |sek·ənd ,kīnd }

elliptic integral of the third kind Any elliptic integral which has logarithmic singularities when considered as a function of one of its limits of integration. { ə|lip·tik |int·ə·grəl əv thə |thərd ,kīnd }

ellipticity Also known as oblateness. **1.** For an ellipse, the difference between the semimajor and semiminor axes of the ellipse, divided by the semimajor axis. **2.** For an oblate spheroid, the difference between the equatorial diameter and the axis of revolution, divided by the equatorial diameter. { ē,lip'tis·əd·ē }

elliptic paraboloid A surface which can be so situated that sections parallel to one coordinate plane are parabolas while those parallel to the other plane are ellipses. { ə|lip·tik pə'rab·ə,lȯid }

elliptic partial differential equation See elliptic differential equation. { ə'lip·tik |pär·shəl dif·ə|ren·chəl i'kwā·zhən }

elliptic point A point on a surface at which the total curvature is strictly positive. { ə'lip·tik 'pȯint }

elliptic Riemann surface See elliptic type. { i¦lip·tik 'rē,män ,sər·fəs }

elliptic type A type of simply connected Riemann surface that can be mapped conformally on the closed complex plane, including the point at infinity. Also known as elliptic Riemann surface. { ə¦lip·tik 'tīp }

elliptic wedge The surface generated by a moving straight line that remains parallel to a given plane and intersects both a given straight line and an ellipse whose plane is parallel to the given line but does not contain it. { ə¦lip·tik 'wej }

embedding An injective homomorphism between two algebraic systems of the same type. { em'bed·iŋ }

empirical curve A smooth curve drawn through or close to points representing measured values of two variables on a graph. { em'pir·ə·kəl 'kərv }

empirical probability The ratio of the number of times an event has occurred to the total number of trials performed. Also known as a posteriori probability. { em'pir·ə·kəl ,präb·ə'bil·əd·ē }

empty set The set with no elements. { 'em·tē 'set }

Encke roots For any two numbers a_1 and a_2, the numbers $-x_1$ and $-x_2$, where x_1 and x_2 are the roots of the equation $x^2 + a_1x + a_2 = 0$, with $|x_1| < |x_2|$. { 'eŋ·kə ,rüts }

endogenous variables In a mathematical model, the dependent variables; their values are to be determined by the solution of the model equations. { en'däj·ə·nəs 'ver·ē·ə·bəlz }

endomorphism A function from a set with some structure (such as a group, ring, vector space, or topological space) to itself which preserves this structure. { 'en·də'mȯr,fiz·əm }

end point Either of two values or points that mark the ends of an interval or line segment. { 'end ,pȯint }

end-vertex A vertex of a graph that has exactly one edge incident to it. { 'end ,vər,teks }

enneagon See nonagon. { 'en·ē·ə,gän }

entire function A function of a complex variable which is analytic throughout the entire complex plane. Also known as integral function. { en¦tīr ¦faŋk·shən }

entire ring See integral domain. { en¦tīr 'riŋ }

entire series A power series which converges for all values of its variable; a power series with an infinite radius of convergence. { en¦tīr 'sir·ēz }

entire surd A surd that does not contain a rational factor or term. { en¦tīr 'sərd }

entropy In a mathematical context, this concept is attached to dynamical systems, transformations between measure spaces, or systems of events with probabilities; it expresses the amount of disorder inherent or produced. { 'en·trə·pē }

entropy of a partition If ξ is a finite partition of a probability space, the entropy of ξ is the negative of the sum of the products of the probabilities of elements in ξ with the logarithm of the probability of the element. { 'en·trə·pē əv ə pär'tish·ən }

entropy of a transformation See Kolmogorov-Sinai invariant. { 'en·trə·pē əv ə tranz·fər'mā·shən }

entropy of a transformation given a partition If T is a measure preserving transformation on a probability space and ξ is a finite partition of the space, the entropy of T given ξ is the limit as $n \rightarrow \infty$ of $1/n$ times the entropy of the partition which is the common refinement of ξ, $T^{-1}\xi$, ..., $T^{-n+1}\xi$. { 'en·trə·pē əv ə tranz·fər'mā·shən 'giv·ən ə pär'tish·ən }

enumerable See countable. { ē'nüm·rə·bəl }

envelope 1. The envelope of a one-parameter family of curves is a curve which has a common tangent with each member of the family. 2. The envelope of a one-parameter family of surfaces is the surface swept out by the characteristic curves of the family. { 'en·və,lōp }

epicenter The center of a circle that generates an epicycloid or hypocycloid. { 'ep·ə,sen·tər }

epicycle The circle which generates an epicycloid or hypocycloid. { 'ep·ə,sī·kəl }

epicycloid The curve traced by a point on a circle as it rolls along the outside of a fixed circle. { ‚ep·ə'sī‚klȯid }

epi spiral A plane curve whose equation in polar coordinates (r, θ) is $r \cos n\theta = a$, where a is a constant and n is an integer. { 'ep·ē ‚spī·rəl }

epitrochoid A curve traced by a point rigidly attached to a circle at a point other than the center when the circle rolls without slipping on the outside of a fixed circle. { ¦ep·ə'trō‚kȯid }

epsilon chain A finite sequence of points such that the distance between any two successive points is less than the positive real number epsilon (ϵ). { 'ep·sə‚län ‚chān }

epsilon neighborhood The set of all points in a metric space whose distance from a given point is less than some number; this number is designated ϵ. { 'ep·sə‚län 'nā·bər‚hůd }

epsilon symbols The symbols $\epsilon^{i_1 i_2 \cdots i_n}$ and $\epsilon_{i_1 i_2 \cdots i_n}$ which are $+1$ if $i_1, i_2, \ldots, i_n$ is an even permutation of $1, 2, \ldots, n$; -1 if it is an odd permutation; and zero otherwise. { 'ep·sə‚län ‚sim·bəlz }

equal Being the same in some sense determined by context. { 'ē·kwəl }

equality The state of being equal. { ē'kwal·əd·ē }

equally likely cases All simple events in a trial have the same probability. { ¦ē·kwə·lē ¦līk·lē 'kās·əs }

equal ripple property For any continuous function $f(x)$ on the interval $-1,1$, and for any positive integer n, a property of the polynomial of degree n, which is the best possible approximation to $f(x)$ in the sense that the maximum absolute value of $e_n(x) = f(x) - p_n(x)$ is as small as possible; namely, that $e_n(x)$ assumes its extreme values at least $n + 2$ times, with the consecutive extrema having opposite signs. { ¦ē·kwəl 'rip·əl ‚präp·ərd·ē }

equal sets Sets with precisely the same elements. { ¦ē·kwəl 'sets }

equals relation *See* equivalence relation. { 'ē·kwəlz ri‚lā·shən }

equal tails test A technique for choosing two critical values for use in a two-sided test; it consists of selecting critical values c and d so that the probability of acceptance of the null hypothesis if the test statistic does not exceed c is the same as the probability of acceptance of the null hypothesis if the test statistic is not smaller than d. { ¦ē·kwəl 'tālz ‚test }

equate To state algebraically that two expressions are equal to one another. { ē'kwāt }

equation A statement that each of two expressions is equal to the other. { i'kwā·zhən }

equation of mixed type A partial differential equation which is of hyperbolic, parabolic, or elliptic type in different parts of a region. { i¦kwā·zhən əv ¦mikst 'tīp }

equiangular polygon A polygon all of whose interior angles are equal. { ¦ē·kwē¦aŋ·gyə·lər 'päl·ə‚gän }

equiangular spiral *See* logarithmic spiral. { ¦ē·kwē¦aŋ·gyə·lər 'spī·rəl }

equicontinuous at a point A family of functions is equicontinuous at a point x if for any $\epsilon > 0$ there is a $\delta > 0$ such that, whenever $|x - y| < \delta$, $|f(x) - f(y)| < \epsilon$ for every function $f(x)$ in the family. { ¦ē·kwē·kən'tin·yə·wəs at ə 'pȯint }

equicontinuous family of functions A family of functions with the property that given any $\epsilon > 0$ there is a $\delta > 0$ such that whenever $|x - y| < \delta$, $|f(x) - f(y)| < \epsilon$ for every function $f(x)$ in the family. Also known as uniformly equicontinuous family of functions. { ¦ē·kwē·kən'tin·yə·wəs 'fam·lē əv 'fəŋk·shənz }

equidecomposable The property of two plane or space regions, either of which can be disassembled into finite number of pieces and reassembled to form the other one. { ¦ek·wē‚dē·kəm'pōz·ə·bəl }

equidistant Being the same distance from some given object. { ¦ē·kwə¦dis·tənt }

equidistant system A system of parametric curves on a surface obtained by setting surface coordinates u and v equal to various constants, where the coordinates are chosen so that an element of length ds on the surface is given by $ds^2 = du^2 + F \, du \, dv + dv^2$, where F is a function of u and v. { ¦ē·kwə¦dis·tənt 'sis·təm }

equilateral polygon A polygon all of whose sides are the same length. { ¦ē·kwə¦lad·ə·rəl 'päl·ə‚gän }

75

equilateral polyhedron A polyhedron all of whose faces are identical. { ¦ē·kwə¦lad· ə·rəl ˌpäl·ə'hē·drən }

equinumerable sets *See* equivalent sets. { ˌek·wə¦nüm·rə·bəl 'sets }

equipotent sets *See* equivalent sets. { ˌek·wə¦pōt·ənt 'sets }

equitangential curve *See* tractrix. { ˌē·kwə·tan'jen·chəl 'kərv }

equivalence A logic operator having the property that if P, Q, R, etc., are statements, then the equivalence of P, Q, R, etc., is true if and only if all statements are true or all statements are false. { i'kwiv·ə·ləns }

equivalence classes The collection of pairwise disjoint subsets determined by an equivalence relation on a set; two elements are in the same equivalence class if and only if they are equivalent under the given relation. { i'kwiv·ə·ləns ˌklas·əs }

equivalence law of ordered sampling If a random ordered sample of size s is drawn from a population of size N, then on any particular one of the s draws each of the N items has the same probability, $1/N$, of appearing. { i'kwiv·ə·ləns ¦lò əv ¦òr·dərd 'sam·pliŋ }

equivalence relation A relation which is reflexive, symmetric, and transitive. Also known as equals functions. { i'kwiv·ə·ləns ri'lā·shən }

equivalence transformation A mapping which associates with each square matrix A the matrix $B = SAT$, where S and T are nonsingular matrices. Also known as equivalent transformation. { i'kwiv·ə·ləns ˌtranz·fər,mā·shən }

equivalent angles Two rotation angles that have the same measure. { i,kwiv·ə·lənt 'aŋ·gəls }

equivalent continued fractions Continued fractions whose values to n terms are the same for $n = 1, 2, 3, \ldots$. { i¦kwiv·ə·lənt kən¦tin·yüd 'frak·shənz }

equivalent elements *See* associates. { ə¦kwiv·ə·lənt 'el·ə·məns }

equivalent equations Equations that have the same set of solutions. { i¦kwiv·ə·lənt i'kwā·zhənz }

equivalent inequalities Inequalities that have the same set of solutions. { i¦kwiv·ə· lənt ˌin·i'kwäl·əd·ēz }

equivalent propositional functions Propositional functions that have the same truth sets. { i,kwiv·ə·lənt ˌpräp·ə¦zish·ən·əl 'fəŋk·shənz }

equivalent propositions Two propositions, either of which is true if and only if the other is true. { i,kwiv·ə·lənt ˌpräp·ə'zish·ənz }

equivalent sets Sets which have the same cardinal number; sets whose elements can be put into one-to-one correspondence with each other. Also known as equinumerable sets; equipotent sets. { i¦kwiv·ə·lənt 'sets }

ergodic 1. Property of a system or process in which averages computed from a data sample over time converge, in a probabilistic sense, to ensemble or special averages. **2.** Pertaining to such a system or process. { ər'gäd·ik }

ergodic theory The study of measure-preserving transformations. { ər'gäd·ik 'thē· ə·rē }

ergodic transformation A measure-preserving transformation on X with the property that whenever X is written as a union of two disjoint invariant subsets, one of these must have measure zero. { ər'gäd·ik tranz·fər'mā·shən }

Erlang distribution *See* gamma distribution. { 'er,läŋ ˌdis·trə,byü·shən }

error equation The equation of a normal distribution. { 'er·ər i,kwā·zhən }

error function The real function defined as the integral from 0 to x of $e^{-t^2} dt$ or $e^{t^2} dt$, or the integral from x to ∞ of $e^{-t^2} dt$. { 'er·ər ˌfəŋk·shən }

error of the first kind *See* type I error. { ¦er·ər əv thə ¦fərst ˌkīnd }

error of the second kind *See* type II error. { ¦er·ər əv thə ¦sekənd ˌkīnd }

error range The difference between the highest and lowest error values; a measure of the uncertainty associated with a number. { 'er·ər ˌrānj }

error sum of squares In analysis of variance, the sum of squares of the estimates of the contribution from the stochastic component. Also known as residual sum of squares. { ¦er·ər ¦səm əv ¦skwerz }

escribed circle For a triangle, a circle that lies outside of the triangle and is tangent to one side of the triangle and to the extensions of the other two sides. Also known as excircle. { ə¦skrībd 'sər·kəl }

essential bound For a function f, a number A such that the set of points x for which the absolute value of $f(x)$ is greater than A is of measure zero. { i¦sen·chəl 'baùnd }

essential constants A set of constants in an equation that cannot be replaced by a smaller number of constants in another equation that has the same solutions. { i¦sen·chəl 'kän·stəns }

essentially bounded function A function that has an essential bound. { i¦sen·chə·lē ¦baùnd·əd 'fəŋk·shən }

essential mapping A mapping between topological spaces that is not homotopic to a mapping whose range is a single point. { i¦sen·chəl 'map·iŋ }

essential singularity An isolated singularity of a complex function which is neither removable nor a pole. { i'sen·chəl siŋ·gyə'lar·əd·ē }

essential supremum For an essentially bounded function, the greatest lower bound of the essential bounds. { i¦sen·chəl sə'prēm·əm }

estimation theory A branch of probability and statistics concerned with deriving information about properties of random variables, stochastic processes, and systems based on observed samples. { ˌes·tə'mā·shən ˌthē·ə·rē }

estimator A random variable or a function of it used to estimate population parameters. { 'es·tə,mäd·ər }

Euclidean algorithm A method of finding the greatest common divisor of a pair of integers. { yü'klid·ē·ən 'al·gə,rith·əm }

Euclidean geometry The study of the properties preserved by isometries of two- and three-dimensional Euclidean space. { yü'klid·ē·ən jē'äm·ə·trē }

Euclidean ring A commutative ring, together with a function, f, from the nonzero elements of the ring to the nonnegative integers, such that (1) $f(xy) \geq f(x)$ if $xy \neq 0$, and (2) for any members of the ring, x and y, with $x \neq 0$, there are members q and r such that $y = qx + r$ and either $r = 0$ or $f(r) < f(x)$. { yü,klid·ē·ən 'riŋ }

Euclidean space A space consisting of all ordered sets $(x_1, \ldots, x_n)$ of n numbers with the distance between $(x_1, \ldots, x_n)$ and $(y_1, \ldots, y_n)$ being given by $\left[\sum_{i=1}^{n} (x_i - y_i)^2\right]^{1/2}$; the number n is called the dimension of the space. { yü'klid·ē·ən 'spās }

Euler characteristic of a topological space X The number $\chi(X) = \Sigma(-1)^q \beta_q$, where β_q is the qth Betti number of X. { 'òi·lər ,kar·ik·tə'ris·tik əv ə ,täp·ə¦läj·i·kəl ¦spās 'eks }

Euler diagram A diagram consisting of closed curves, used to represent relations between logical propositions or sets; similar to a Venn diagram. { 'òi·lər ,dī·ə,gram }

Eulerian graph A graph that has an Eulerian path. { òi¦ler·ē·ən 'graf }

Eulerian path A path that traverses each of the lines in a graph exactly once. { òi'ler·ē·ən 'path }

Euler-Lagrange equation A partial differential equation arising in the calculus of variations, which provides a necessary condition that $y(x)$ minimize the integral over some finite interval of $f(x,y,y')dx$, where $y' = dy/dx$; the equation is $[\delta f(x,y,y')/\delta y] - (d/dx)[\delta f(x,y,y')/\delta y'] = 0$. Also known as Euler's equation. { ¦òi·lər lə'gränj i,kwā·zhən }

Euler-Maclaurin formula A formula used in the numerical evaluation of integrals, which states that the value of an integral is equal to the sum of the value given by the trapezoidal rule and a series of terms involving the odd-numbered derivatives of the function at the end points of the interval over which the integral is evaluated. { ¦òi·lər mə'klòr·ən ,fòr·myə·lə }

Euler method A method of obtaining an approximate solution of an ordinary differential equation of the form $dy/dx = f(x,y)$, where f is a specified function of x and y. Also known as Eulerian description. { 'oi·lər ,meth·əd }

Euler's constant The limit as n approaches infinity, of $1 + 1/2 + 1/3 + \cdots + 1/n - \ln n$, equal to approximately 0.5772. Denoted γ. Also known as Mascheroni's constant. { 'òi·lərz ¦kän·stənt }

Euler's criterion

Euler's criterion A criterion for the congruence $x^n \equiv a \pmod{m}$ to have a solution, namely that $a^{\phi/d} \equiv 1 \pmod{m}$, where $\phi = \phi(m)$ is Euler's phi function evaluated at m, and d is the greatest common divisor of ϕ and n. { 'ȯi·lərz krī'tir·ē·ən }

Euler's equation *See* Euler-Lagrange equation. { 'ȯi·lərz i¦kwā·zhən }

Euler's formula The formula $e^{ix} = \cos x + i \sin x$, where $i = \sqrt{-1}$. { 'ȯi·lərz ˌför·myə·lə }

Euler's numbers The numbers E_{2n} defined by the equation

$$\frac{1}{\cos z} = \sum_{n=0}^{\infty} (-1)^n \frac{E_{2n}}{(2n)!} z^{2n}$$

{ 'ȯi·lərz ˌnəm·bərz }

Euler's phi function A function ϕ, defined on the positive integers, whose value $\phi(n)$ is the number of integers equal to or less than n and relatively prime to n. Also known as indicator; phi function; totient. { 'ȯi·lərz 'fī ˌfəŋk·shən }

Euler's spiral *See* Cornu's spiral. { 'ȯi·lərz ¦spī·rəl }

Euler's theorem For any polyhedron, $V - E + F = 2$, where V, E, F represent the number of vertices, edges, and faces respectively. { 'ȯi·lərz ˌthir·əm }

Euler transformation A method of obtaining from a given convergent series a new series which converges faster to the same limit, and for defining sums of certain divergent series; the transformation carries the series $a_0 - a_1 + a_2 - a_3 + \cdots$ into a series whose nth term is $\sum_{r=0}^{n-1} (-1)^r \binom{n-1}{r} a_r/2^n$. { 'ȯi·lər ˌtranz·fər'mā·shən }

even function A function with the property that $f(x) = f(-x)$ for each number x. { 'ē·vən ˌfəŋk·shən }

even number An integer which is a multiple of 2. { 'ē·vən ˌnəm·bər }

even permutation A permutation which may be represented as a result of an even number of transpositions. { ¦ē·vən pər·myə'tā·shən }

event A mathematical model of the result of a conceptual experiment; this model is a measurable subset of a probability space. { i'vent }

eventually in A net is eventually in a set if there is an element a of the directed system that indexes the net such that, if b is also an element of this directed system and $b \geq a$, then x_b (the element indexed by b) is in this set. { i'ven·chəl·ē ˌin }

even vertex A vertex whose degree is an even number. { ¦ēv·ən 'vər,teks }

Everett's interpolation formula A formula for estimating the value of a function at an intermediate value of the independent variable, when its value is known at a series of equally spaced points (such as those that appear in a table), in terms of the central differences of the function of even order only and coefficients which are polynomial functions of the independent variable. { ¦ev·rəts ˌin·tər·pə'lā·shən ˌför·myə·lə }

evolute 1. The locus of the centers of curvature of a curve. **2.** The two surfaces of center of a given surface. { 'ev·əˌlüt }

exact differential equation A differential equation obtained by setting the total differential of some function equal to zero. { ig'zakt dif·ə'ren·chəl i,kwā·zhən }

exact differential form A differential form which is the differential of some other form. { ig'zakt dif·ə'ren·chəl ˌförm }

exact division Division wherein the remainder is zero. { ig'zakt di'vizh·ən }

exact divisor A divisor that leaves a remainder of zero. { ig'zakt di'vī·zər }

exact sequence A sequence of homomorphisms with the property that the kernel of each homomorphism is precisely the image of the previous homomorphism. { ig'zakt 'sē·kwəns }

excenter The center of the escribed circle of a given triangle. { ¦ek'sen·tər }

except A logical operator which has the property that if P and Q are two statements, then the statement "P except Q" is true only when P alone is true; it is false for the other three combinations (P false Q false, P false Q true, and P true Q true). { ek'sept }

exceptional group One of five Lie groups which leave invariant certain forms constructed out of the Cayley numbers; they are Lie groups with maximum symmetry in the sense that, compared with other simple groups with the same rank (number of independent invariant operators), they have maximum dimension (number of generators). { ek¦sep·shən·əl ¦grüp }

exceptional Jordan algebra A Jordan algebra that cannot be written as a symmetrized product over a matrix algebra; used in formulating a generalization of quantum mechanics. { ek'sep·shən·əl ¦jȯrd·ən 'al·jə·brə }

excircle *See* escribed circle. { ¦ek'sər·kəl }

exclusive or A logic operator which has the property that if P is a statement and Q is a statement, then P exclusive or Q is true if either but not both statements are true, false if both are true or both are false. { ik¦sklü·siv 'ȯr }

existence proof An argument that establishes the truth of an existence theorem. { ig'zis·təns ‚prüf }

existence theorem The theorem that at least one object of a specified type exists. { ig'zis·təns ‚thir·əm }

existential quantifier A logical relation, often symbolized $\exists$, that may be expressed by the phrase "there is a" or "there exists"; if P is a predicate, the statement $(\exists x)P(x)$ is true if there exists at least one value of x in the domain of P for which $P(x)$ is true, and is false otherwise. { ‚eg·zə¦sten·chəl 'kwän·tə‚fī·ər }

exogenous variables In a mathematical model, the independent variables, which are predetermined and given outside the model. { ‚ek'säj·ə·nəs 'ver·ē·ə·bəlz }

exotic four-space A four-dimensional manifold that is homeomorphic, but not diffeomorphic, to four-dimensional Euclidean space. { ig¦zäd·ik 'fȯr‚spās }

exotic sphere A smooth manifold that is homeomorphic, but not diffeomorphic, to a sphere. { ig¦zäd·ik 'sfir }

expanded notation The representation of a number as the sum of a series of terms, each of which is written explicitly as the product of a digit and the base of the number system raised to some power. { ik¦spand·əd nō'tā·shən }

expanded numeral A number expressed in expanded notation. { ik¦spand·əd 'nüm·rəl }

expansion The expression of a quantity as the sum of a finite or infinite series of terms, as a finite or infinite product of factors, or, in general, in any extended form. { ik'span·shən }

expectation *See* expected value. { ‚ek‚spek'tā·shən }

expected value 1. For a random variable x with probability density function $f(x)$, this is the integral from $-\infty$ to ∞ of $xf(x)dx$. Also known as expectation. **2.** For a random variable x on a probability space (Ω, P), the integral of x with respect to the probability measure P. { ek'spek·təd 'val·yü }

experimental design A pattern for setting up experiments and making observations about the relationship between several variables in which one attempts to obtain as much information as possible for a fixed expenditure level. { ik‚sper·ə'ment·əl di'zīn }

explementary angles *See* conjugate angles. { ‚ek·splə¦men·tə·rē 'aŋ·gəlz }

exponent A number or symbol placed to the right and above some given mathematical expression. { ik'spō·nənt }

exponential For a bounded linear operator A on a Banach space, the sum of a series which is formally the exponential series in A. { ‚ek·spə'nen·chəl }

exponential curve A graph of the function $y = a^x$, where a is a positive constant. { ‚ek·spə'nen·chəl 'kərv }

exponential density function A probability density function obtained by integrating a function of the form $\exp(-|x - m|/\sigma)$, where m is the mean and σ the standard deviation. { ‚ek·spə'nen·chəl den·səd·ē ‚fəŋk·shən }

exponential distribution A continuous probability distribution whose density function is given by $f(x) = ae^{-ax}$, where $a > 0$, for $x > 0$, and $f(x) = 0$ for $x \le 0$; the mean and standard deviation are both $1/a$. { ‚ek·spə'nen·chəl dis·trə'byü·shən }

79

exponential equation An equation containing e^x (the Naperian base raised to a power) as a term. { ,ek·spə'nen·chəl i'kwā·zhən }

exponential function The function $f(x) = e^x$, written $f(x) = \exp(x)$. { ,ek·spə'nen·chəl 'fəŋk·shən }

exponential generating function A function, $G(x)$, corresponding to a sequence, a_0, $a_1, \ldots$, where $G(x) = a_0 + (a_1x/1!) + (a_2x^2/2!) + \cdots$. { ,eks·pə¦nen·chəl ¦jen·ə,rād·iŋ 'fəŋk·shən }

exponential integral The function defined to be the integral from x to ∞ of $(e^{-t}/t)\,dt$ for x positive. { ,ek·spə'nen·chəl 'int·ə·grəl }

exponential law See law of exponents. { ,ek·spə'nen·chəl 'lò }

exponential series The Maclaurin series expansion of e^x, namely, $e^x = 1 + \sum\limits_{n=1}^{\infty} \dfrac{x^n}{n!}$.
{ ,ek·spə'nen·chəl 'sir·ēz }

exradius The radius of an escribed circle of a triangle. { ,eks'rād·ē·əs }

exsecant The trigonometric function defined by subtracting unity from the secant, that is exsec $\theta = \sec \theta - 1$. { ,ek'sē·kant }

extended mean-value theorem See second mean-value theorem. { ik¦sten·dəd ,mēn ,val·yü 'thir·əm }

extension See extension field. { ik'sten·chən }

extension field An extension field of a given field E is a field F such that E is a subfield of F. Also known as extension. { ik'sten·chən ,fēld }

extension map An extension map of a map f from a set A to a set L is a map g from a set B to L such that A is a subset of B and the restriction of g to A equals f. { ik'sten·chən ,map }

exterior 1. For a set A in a topological space, the largest open set contained in the complement of A. 2. For a plane figure, the set of all points that are neither on the figure nor inside it. 3. For an angle, the set of points that lie in the plane of the angle but not between the rays defining the angle. 4. For a simple closed plane curve, one of the two regions into which the curve divides the plane according to the Jordan curve theorem, namely, the region that is not bounded. { ek'stir·ē·ər }

exterior algebra An algebra whose structure is analogous to that of the collection of differential forms on a Riemannian manifold. Also known as Grassmann algebra. { ek'stir·ē·ər 'al·jə·brə }

exterior angle 1. An angle between one side of a polygon and the prolongation of an adjacent side. 2. An angle made by a line (the transversal) that intersects two other lines, and either of the latter on the outside. { ek'stir·ē·ər 'aŋ·gəl }

exterior content See exterior Jordan content. { ek'stir·ē·ər 'kän,tent }

exterior Jordan content Also known as exterior content. 1. For a set of points on a line, the largest number C such that the sum of the lengths of a finite number of closed intervals that includes every point in the set is always equal to or greater than C. 2. The exterior Jordan content of a set of points, X, in n-dimensional Euclidean space (where n is a positive integer) is the greatest lower bound on the hypervolume of the union of a finite set of hypercubes that contains X. { ek¦stir·ē·ər ¦jórd·ən 'kän,tent }

exterior measure See Lebesgue exterior measure. { ek¦stir·ē·ər 'mezh·ər }

external angle The angle defined by an arc around the boundaries of an internal angle or included angle. { ek¦stərn·əl 'aŋ·gəl }

external dominating set See dominating vertex set. { ek¦stərn·əl ,däm·ə,nād·iŋ 'set }

externally tangent circles Two circles, neither of which is inside the other, that have a single point in common. { ek¦stərn·əl·ē ¦tan·jənt 'sər·kəlz }

external operation For a set S, a function of one or more independent variables such that at least one of the independent variables has values in S but either one or more of the independent variables or the dependent variable fails to have values in S. { ek¦stərn·əl ,äp·ə'rā·shən }

external stability number See vertex domination number. { ek,stərn·əl stə'bil·əd·ē ,nəm·bər }

external tangent For two circles, each exterior to the other, a line that is tangent to both circles such that both circles are on the same side of this line. { ek¦stərn·əl 'tan·jənt }

extract a root To determine a root of a given number, usually a positive real root, or a negative real odd root of a negative number. { ik'strakt ə 'rüt }

extraneous root A root that is introduced into an equation in the process of solving another equation, but is not a solution of the equation to be solved. { ik¦strān·ē·əs 'rüt }

extrapolation Estimating a function at a point which is larger than (or smaller than) all the points at which the value of the function is known. { ik‚strap·ə'lā·shən }

extremals For a variational problem in the calculus of variaitons entailing use of the Euler-Lagrange equation, the extremals are the solutions of this equation. { ek'strem·əlz }

extreme *See* extremum. { ek'strēm }

extreme and mean ratio *See* golden section. { ek'strēm ən 'mēn ‚rā·shō }

extreme point 1. A maximum or minimum value of a function. **2.** A point in a convex subset K of a vector space is called extreme if it does not lie on the interior of any line segment contained in K. { ek'strēm 'póint }

extreme terms The first and last terms in a proportion. { ek¦strēm 'tərmz }

extreme value problem A set of mathematical conditions which may be met by values that are less than or greater than an upper or a lower bound, that is, an extreme value. { ek¦strēm 'val·yü präb·ləm }

extremum A maximum or minimum value of a function. Also known as extreme. { ek'strēm·əm }

F

face 1. One of the plane polygons bounding a polyhedron. **2.** A face of a simplex is the subset obtained by setting one or more of the coordinates a_i, defining the simplex, equal to 0; for example, the faces of a triangle are its sides and vertices. **3.** The face of a half space is the plane that bounds it. **4.** One of the regions bounded by edges of a planar graph. { fās }

face angle An angle between two successive edges of a polyhedral angle. { 'fās ,aŋ·gəl }

facet A proper face of a convex polytope that is not contained in any larger face. { 'fas·ət }

factor 1. For an integer n, any integer which gives n when multiplied by another integer. **2.** For a polynomial p, any polynomial which gives p when multiplied by another polynomial. **3.** For a graph G, a spanning subgraph of G with at least one edge. **4.** A quantity or a variable being studied in an experiment as a possible cause of variation. { 'fak·tər }

factorable integer An integer that has factors other than unity and itself. { 'fak·trə· bəl 'int·ə·jər }

factorable polynomial A polynomial which has polynomial factors other than itself. { 'fak·tə·rə·bəl ,päl·ə'nō·mē· əl }

factor analysis Given sets of variables which are related linearly, factor analysis studies techniques of approximating each set relative to the others; usually the variables denote numbers. { 'fak·tər ə,nal·ə·səs }

factor group *See* quotient group. { 'fak·tər ,grüp }

factorial The product of all positive integers less than or equal to n; written $n!$; by convention $0! = 1$. { fak'tór·ē·əl }

factorial design A design for an experiment that allows the experimenter to find out the effect levels of each factor on levels of all the other factors. { fak'tór·ē· əl di,zīn }

factorial moment The nth factorial moment of a random variable X is the expected value of $X(X-1)(X-2)\cdots(X-n+1)$. { fak,tór·ē·əl 'mō·mənt }

factorial ring *See* unique-factorization domain. { fak'tór·ē·əl ,riŋ }

factorial series The series $1 + (1/1!) + (1/2!) + (1/3!) + \cdots$, whose $(n+1)$st term is $1/n!$ for $n = 1, 2, \cdots$; its sum is the number e. { fak'tór·ē·əl 'sir,ēz }

factoring Finding the factors of an integer or polynomial. { 'fak·tə·riŋ }

factoring of the secular equation Factoring the polynomial that results from expanding the secular determinant of a matrix, in order to find the roots of this polynomial, which are the eigenvalues of the matrix. { 'fak·tə·riŋ əv thə ¦sek·yə· lər i'kwā· zhən }

factor model Any one of the probability models which goes into the construction of a product model. { 'fak· tər ,mäd·əl }

factor module The factor module of a module M over a ring R by a submodule N is the quotient group M/N, where the product of a coset $x + N$ by an element a in R is defined to be the coset $ax + N$. { 'fak·tər ,mä·jül }

factor of proportionality Two quantities A and B are related by a factor of proportionality μ if either $A = \mu B$ or $B = \mu A$. { 'fak·tər əv prə,pórsh·ən'al·əd·ē }

factor-reversal test A test for index numbers in which an index number of quantity, obtained if symbols for price and quantity are interchanged in an index number of price, is multiplied by the original price index to give an index of changes in total value.   { ¦fak·tər ri'vər·səl ˌtest }

factor ring See quotient ring.   { 'fak·tər ˌriŋ }

factor space See quotient space.   { 'fak·tər ˌspās }

factor theorem of algebra A polynomial $f(x)$ has $(x - a)$ as a factor if and only if $f(a) = 0$.   { 'fak·tər ˌthir·əm əv 'al·jə·brə }

fair game A game in which all of the participants have equal expectation of gain.   { ¦fer 'gām }

faithful module A module M over a commutative ring R such that if a is an element in R for which $am = 0$ for all m in M, then $a = 0$.   { ¦fāth,fúl 'mä·jül }

faithful representation A homomorphism h of a group onto some group of matrices or linear operators such that h is an injection.   { ¦fāth,fúl ˌrep·rə·zen'tā·shən }

falling factorial polynomials The polynomials $[x]_n = x(x - 1)(x - 2) \cdots (x - n + 1)$.   { ˌfȯl·iŋ fak,tȯr·ē·əl ˌpäl·ə'nō·mē·əlz }

false acceptance Accepting on the basis of a statistical test a hypothesis which is wrong.   { ¦fȯls ak'sep·təns }

false rejection Rejecting on the basis of a statistical test a hypothesis which is correct.   { ¦fȯls ri'jek·shən }

faltung A family of functions where the convolution of any two members of the family is also a member of the family.   Also known as convolution family.   { 'fäl,túŋ }

family of curves A set of curves whose equations can be obtained by varying a finite number of parameters in a particular general equation.   { ˌfam·lē əv 'kərvz }

Fano plane A projective plane in which the points of intersection of the three possible pairs of opposite sides of a quadrilateral are collinear.   { 'fä·nō ˌplān }

Fano's axiom The postulate that the points of intersection of the three possible pairs of opposite sides of any quadrilateral in a given projective plane are noncollinear; thus a projective plane satisfying Fano's axiom is not a Fano plane, and a Fano plane does not satisfy Fano's axiom.   { ¦fä·nōz 'ak·sē·əm }

Farey sequence The Farey sequence of order n is the increasing sequence, from 0 to 1, of fractions whose denominator is equal to or less that n, with each fraction expressed in lowest terms.   { 'far·ē ˌsē·kwəns }

fast Fourier transform A Fourier transform employing the Cooley-Tukey algorithm to reduce the number of operations.   Abbreviated FFT.   { ¦fast ˌfúr·ē,ā 'tranz,fȯrm }

Fatou-Lebesgue lemma Given a sequence f_n of positive measurable functions on a measure space (X, μ), then

$$\int_X (\lim_{n \to \infty} \inf f_n) d\mu \le \lim_{n \to \infty} \int_X f_n d\mu$$

{ ˌfä'tü lə'beg ¦lem·ə }

F distribution The ratio of two independent chi-square variables each divided by its degree of freedom; used to test hypotheses in the analysis of variance and hypotheses about whether or not two normal populations have the same variance.   { 'ef ˌdis·trə,byü·shən }

feasible flow A flow on a directed network such that the net flow at every intermediate vertex is zero.   { ¦fē·zə·bəl 'flō }

Feit-Thompson theorem The proposition that every group of odd order is solvable.   { ¦fīt ¦täm·sən ˌthir·əm }

Fermat numbers The numbers of the form $F_n = 2^{(2^n)} + 1$ for $n = 0, 1, 2, \ldots$.   { 'fer·mä ˌnəm·bərz }

Fermat's last theorem The proposition, proven in 1995, that there are no positive integer solutions of the equation $x^n + y^n = z^n$ for $n \ge 3$.   { fer'mäz ¦last 'thir·əm }

Fermat's spiral A plane curve whose equation in polar coordinates (r,θ) is $r^2 = a^2\theta$, where a is a constant.   { fer'mäz ˌspī·rəl }

Fermat's theorem The proposition that, if p is a prime number and a is a positive integer which is not divisible by p, then $a^{p-1} - 1$ is divisible by p .   { 'fer,mäz ˌthir·əm }

Ferrers diagram An array of dots associated with an integer partition $n = a_1 + \cdots + a_k$, whose ith row contains a_i dots. { 'fer·ərz ‚di·ə‚gram }

FFT *See* fast Fourier transform.

fiber The set of points in the total space of a bundle which are sent into the same element of the base of the bundle by the projection map. { 'fī·bər }

fiber bundle A bundle whose total space is a G-space X, whose base is the homomorphic image of the orbit space of X, and whose fibers are isomorphic to the orbits of points in the base space under the action of G. { 'fī·bər ‚bən·dəl }

Fibonacci number A number in the Fibonacci sequence whose first two terms are $f_1 = f_2 = 1$. { ¦fib·ə¦nä·chē 'nəm·bər }

Fibonacci sequence The sequence 1, 1, 2, 3, 5, 8, 13, 21, . . . , or any sequence where each entry is the sum of the two previous entries. { ‚fē·bə'näch·ē ‚sē·kwəns }

fiducial inference A type of inference whose purpose is to make probabilistic statements about values of unknown parameters; based on the distribution of population values about which the inference is to be made. { fə¦dü·shəl ‚in·tər'fir·əns }

fiducial limits The boundaries within which a parameter is considered to be located; a concept in fiducial inference. { fə¦dü·shəl 'lim·əts }

field An algebraic system possessing two operations which have all the properties that addition and multiplication of real numbers have. { fēld }

field of planes on a manifold A continuous assignment of a vector subspace of tangent vectors to each point in the manifold. Also known as plane field. { 'fēld əv 'plänz òn ə 'man·ə‚fōld }

field of vectors on a manifold A continuous assignment of a tangent vector to each point in the manifold. Also known as vector field. { 'fēld əv 'vek·tərz òn ə 'man·ə‚fōld }

field theory The study of fields and their extensions. { 'fēld ‚thē·ə·rē }

filter A family of subsets of a set S: it does not include the empty set, the intersection of any two members of the family is also a member, and any subset of S containing a member is also a member. { 'fil·tər }

filter base A family of subsets of a given set with the property that it does not include the empty set, and the intersection of any finite number of members of the family includes another member. { 'fil·tər ‚bās }

final-value theorem The theorem that if $f(t)$ is a function which has a Laplace transform $F(s)$, and if the derivative of $f(t)$ with respect to t is also Laplace transformable, and if the limit of $f(t)$ as t approaches infinity exists, then this limit is equal to the limit of $sF(s)$ as s approaches zero. { ¦fīn·əl ¦val·yü ¦thir·əm }

fineness 1. For a partition of a metric space, the least upper bound on distances between points in the same member of the partition. **2.** For a partition of an interval into subintervals, the length of the longest subinterval. Also known as mesh; norm. { 'fīn·nəs }

finer A partition P of a set is finer than another partition Q of the same set if each member of P is a subset of a member of Q. { 'fīn·ər }

finite character 1. A property of a family C of sets such that any finite subset of a member of C belongs to C, and C includes any set all of whose finite subsets belong to C. **2.** A characteristic of a property of subsets of a set such that a subset S has the property if and only if all the nonempty finite subsets of S have the property. { 'fī‚nīt 'kar·ik·tər }

finite decimal *See* terminating decimal. { ¦fī‚nīt 'des·məl }

finite difference The difference between the values of a function at two discrete points, used to approximate the derivative of the function. { ¦fī‚nīt 'dif·rəns }

finite-difference equations Equations arising from differential equations by substituting difference quotients for derivatives, and then using these equations to approximate a solution. { ¦fī‚nīt ¦dif·rəns i‚kwā·zhənz }

finite discontinuity A discontinuity of a function that lies at the center of an interval on which the function is bounded. { 'fī‚nīt ‚dis·kän·tə'nü·əd·ē }

finite extension An extension field F of a given field E such that F, viewed as a vector space over E, has finite dimension. { ¦fī‚nīt ik'sten·chən }

finite group A group which contains a finite number of distinct elements. { ¦fī,nīt 'grüp }

finite intersection property of a family of sets If the intersection of any finite number of them is nonempty, then the intersection of all the members of the family is nonempty. { ¦fī,nīt,in·tər'sek·shən ,präp·ərd·ē əv ə 'fam·lē əv 'sets }

finitely additive set function *See* additive set function. { ¦fī,nīt·lē ¦ad·ə·div 'set ,fəŋk·shən }

finitely generated extension A finitely generated extension of a field k is the smallest field which contains k and some finite set of elements. { ¦fī,nīt·lē ¦gen·ə,rād·əd ik'sten·chən }

finitely generated left module A left module over a ring A that has a finite subset, x_1, $x_2, \ldots, x_n$, such that any member of the module has the form $a_1x_1 + \cdots + a_nx_n$, where $a_1, \ldots, a_n$ are members of A. { ¦fī,nīt·lē ¦jen·ə,rād·əd ¦left 'mäj·əl }

finitely representable A Banach space A is said to be finitely representable in a Banach space B if every finite-dimensional subspace of A is nearly isometric to a subspace of B. { ¦fī,nīt·lē ,rep·rə'zen·tə·bəl }

finite mathematics 1. Those parts of mathematics which deal with finite sets. 2. Those fields of mathematics which make no use of the concept of limit. Also known as discrete mathematics. { ¦fī,nīt ,math·ə'mad·iks }

finite matrix A matrix with a finite number of rows and columns. { ¦fī,nīt 'mā·triks }

finite measure space A measure space in which the measure of the entire space is a finite number. { 'fī,nīt ¦mezh·ər ,späs }

finite moment theorem The theorem that if $f(x)$ is a continuous function, and if the integral of $f(x) x^n$ over a finite interval is zero for all positive integers n, then $f(x)$ is identically zero in that interval. { ¦fī,nīt 'mō·mənt ,thir·əm }

finite plane In projective geometry, a plane with a finite number of points and lines. { ¦fī,nīt 'plān }

finite population A population of finite individuals or elements. { ¦fī,nīt ,päp·yə'lā·shən }

finite quantity Any bounded quantity. { ¦fī,nīt 'kwän·əd·ē }

finite sequence 1. A listing of some finite number, n, of mathematical entities that is indexed by the first n positive integers, 1, 2, . . . , n. 2. More precisely, a function whose domain is the first n positive integers. { ¦fī,nīt 'sē·kwəns }

finite series A series that has a limited number of terms. { 'fī,nīt 'sir,ēz }

finite set A set whose elements can be indexed by integers 1, 2, 3, . . . , n inclusive. { ¦fī,nīt 'set }

Finsler geometry The study of the geometry of a manifold in terms of the various possible metrics on it by means of Finsler structures. { 'fin·slər jē'äm·ə·trē }

Finsler structure on a manifold A family of metrics varying continuously from point to point. { 'fin·slər ,strək·chər ón ə 'man·ə,fōld }

first category 1. A set is of first category if it is a countable union of nowhere dense sets. 2. A set S is of first category at a point x if there is a neighborhood of x whose intersection with S is of first category. { 'fərst 'kad·ə,gòr·ē }

first countable topological space A topological space in which every point has a countable number of open neighborhoods so that any neighborhood of this point contains one of these. { ¦fərst 'kaùnt·ə·bəl ,täp·ə¦läj·ə·kəl 'späs }

first derivative The derivative of a function, considered as a function of the independent variable just as was the original function from which the derivative was taken. { ¦fərst də'riv·əd·iv }

first derived curve *See* derived curve. { ¦fərst də¦rīvd 'kərv }

first law of the mean *See* mean value theorem. { 'fərst ,lò əv thə 'mēn }

first law of the mean for integrals The proposition that the definite integral of a continuous function over an interval equals the length of the interval multiplied by the value of the function at some point in the interval. { ¦fərst ,lò əv thə ¦mēn fòr 'int·ə·grəlz }

first negative pedal *See* negative pedal. { 'fərst 'neg·əd·iv 'ped·əl }

first-order difference A member of a sequence that is formed from a given sequence

by subtracting each term of the original sequence from the next succeeding term.
{ ¦fərst ¦ȯrd·ər 'dif·rəns }

first-order theory A logical theory in which predicates are not allowed to have other functions or predicates as arguments and in which predicate quantifiers and function quantifiers are not permitted. { ¦fərst ˌȯrd·ər 'thē·ə·rē }

first pedal curve See pedal curve. { ¦fərst 'ped·əl ˌkərv }

first positive pedal curve See pedal curve. { 'fərst 'päz·əd·iv ˌped·əl ˌkərv }

first quadrant 1. The range of angles from 0 to 90°. **2.** In a plane with a system of cartesian coordinates, the region in which the x and y coordinates are both positive.
{ ¦fərst 'kwäd·rənt }

first species The class of sets G_0 such that one of the sets G_n is the null set, where, in general, G_n is the derived set of G_{n-1}. { 'fərst 'spē,shēz }

Fischer's distribution Given data from a normal population with S_1^2 and S_2^2 two independent estimates of variance, the distribution $1/2$ log (S_1^2/S_2^2). { 'fish·ərz ˌdis·trə'byü·shən }

Fischer-Yates test A test of independence of data arranged in a 2×2 contingency table. { ¦fish·ər ¦yāts ˌtest }

Fisher-Irwin test A method for testing the null hypothesis in an experiment with quantal response. { ¦fish·ər ¦ər·wən ˌtest }

Fisher's ideal index The geometric mean of Laspeyres and Paasche index numbers. Also known as ideal index number. { ¦fish·ərz ¦ī,dēl 'in,deks }

Fisher's inequality The inequality whereby the number b of blocks in a balanced incomplete block design is equal to or greater than the number v of elements arranged among the blocks. { ¦fish·ərz ˌin·i'kwäl·əd·ē }

five-dimensional space A vector space whose basis has five vectors. { 'fīv də,men·chən·əl 'spās }

fixed-base index In a time series, an index number whose base period for computing the index number is constant throughout the lifetime of the index. { ¦fikst ¦bās 'in,deks }

fixed point For a function f mapping a set S to itself, any element of S which f sends to itself. { ¦fikst 'pȯint }

fixed-point theorem Any theorem, such as the Brouwer theorem or Schauder's fixed-point theorem, which states that a certain type of mapping of a set into itself has at least one fixed point. { ¦fikst 'pȯint ˌthir·əm }

fixed radix notation A form of positional notation in which successive digits are interpreted as coefficients of successive powers of an integer called the base or radix. { ¦fikst 'rā,diks nō,tā·shən }

flat space A Riemannian space for which a coordinate system exists such that the components of the metric tensor are constants throughout the space; equivalently, a space in which the Riemann-Christoffel tensor vanishes throughout the space. { ¦flat ˌspās }

flecnode A node that is also a point of inflection of one of the two branches of the curve that cross at the node. { 'flek,nōd }

floating arithmetic See floating-point arithmetic. { ¦flōd·iŋ ə'rith·mə·tik }

floating-decimal arithmetic See floating-point arithmetic. { ¦flōd·iŋ ¦des·məl ə'rith·mə·tik }

floating-point arithmetic A method of performing arithmetical operations, used especially by automatic computers, in which numbers are expressed as integers multiplied by the radix raised to an integral power, as 87×10^{-4} instead of 0.0087. Also known as floating arithmetic; floating-decimal arithmetic. { ¦flōd·iŋ ¦pȯint ə'rith·mə·tik }

Floquet theorem A second-order linear differential equation whose coefficients are periodic single-valued functions of an independent variable x has a solution of the form $e^{\mu x}P(x)$ where μ is a constant and $P(x)$ a periodic function. { flō'kā ˌthir·əm }

flow A function from the set of arcs in an s-t network to the nonnegative integers whose value at each arc is equal to or less than the weight of the arc. { flō }

flow value For a feasible flow on an s-t network, the outflow from the source. { 'flō ˌval·yü }

fluctuation noise See random noise. { ˌflək·chə'wā·shən ˌnȯiz }

F martingale A stochastic process $\{X_t, t > 0\}$ such that the conditional expectation of X_t given F_s equals X_s whenever $s < t$, where $F = \{F_t, t \geq 0\}$ is an increasing family of sigma algebras that represents the amount of information increasing with time. { ¦ef 'mart·ən,gāl }

focal chord For a conic, a chord that passes through a focus of the conic. { 'fō· kəl ¦kȯrd }

focal property 1. The property of an ellipse or hyperbola whereby lines drawn from the foci to any point on the conic make equal angles with the tangent to the conic at that point. **2.** The property of a parabola whereby a line from the focus to any point on the parabola, and a line through this point parallel to the axis of the parabola, make equal angles with the tangent to the parabola at this point. { 'fō· kəl 'präp·ər·dē }

focal radius For a conic, a line segment from a focus to any point on the conic. { 'fō· kəl 'rād·ē·əs }

focus A point in the plane which together with a line (directrix) defines a conic section. { 'fō·kəs }

folium A plane curve that is a pedal curve (first positive pedal) of the deltoid. { 'fō· lē·əm }

folium of Descartes A plane cubic curve whose equation in cartesian coordinates x and y is $x^3 + y^3 = 3axy$, where a is some constant. Also known as leaf of Descartes. { 'fō·lē·əm əv dā'kärt }

Ford-Fulkerson theorem The theorem that in any s-t network there exists a feasible flow and an s-t cut such that (1) the flow equals the weight of the cut, (2) on any arc belonging to the cut, this flow equals the weight of the arc, and (3) on any arc, that would belong to the cut if its orientation were reversed, the flow equals zero. Also known as max-flow min-cut theorem. { ¦fȯrd 'fül·kər·sən ˌthir·əm }

forecast To assess the magnitude that a quantity will have at a specified time in the future. Also known as predict. { 'fȯr,kast }

forest See acyclic graph. { 'fär·əst }

formal derivative For a polynomial, $a_n x^n + a_{n-1} x^{n-1} + \cdots + a_1 x + a_0$, where the coefficients $a_0, a_1, \ldots, a_n$ are elements of a ring, the formal derivative is the polynomial $na_n x^{n-1} + (n-1) a_{n-1} x^{n-2} + \cdots + a_1$. { ˌfȯrm·əl də'riv·əd·iv }

formal logic The study of the permissible relationships between propositions, a study that concerns the form rather than the content. { ¦fȯr·məl 'läj·ik }

formal power series A power series whose convergence is disregarded, but which is subject to the operations of addition and multiplication with other such series. { ¦fȯr·məl 'paů·ər ˌsir·ēz }

formula An equation or rule relating mathematical objects or quantities. { 'fȯr·myə·lə }

forward difference One of a series of quantities obtained from a function whose values are known at a series of equally spaced points by repeatedly applying the forward difference operator to these values; used in interpolation or numerical calculation and integration of functions. { ¦fȯr·wərd 'dif·rəns }

forward difference operator A difference operator, denoted Δ, defined by the equation $\Delta f(x) = f(x + h) - f(x)$, where h is a constant indicating the difference between successive points of interpolation or calculation. { ¦fȯr·wərd ¦dif·rəns 'äp·ə ˌrād·ər }

forward shift operator See displacement operator. { ¦fȯr·wərd ¦shift 'äp·ə,rād·ər }

four-color problem The problem of proving the statement that, given any map in the plane, it is possible to color the regions with four colors so that any two regions with a common boundary have different colors. { ¦fȯr 'kəl·ər ˌpräb·ləm }

four-group The only group of order 4 other than the cyclic group. { 'fȯr ˌgrüp }

Fourier analysis The study of convergence of Fourier series and when and how a function is approximated by its Fourier series or transform. { ˌfur·ē,ā ə,nal·ə·səs }

Fourier-Bessel integrals Given a function $F(r,\theta)$ independent of θ where r,θ are the polar coordinates in the plane, these integrals have the form

$$\int_0^\infty u\,du \int_0^\infty F(r)J_m(ur)r\,dr$$

where J_m is a Bessel function order m.   { ˌfùr·ē͵ā ˈbes·əl ˈint·ə·grəlz }

Fourier-Bessel series For a function $f(x)$, the series whose mth term is $a_m J_0(j_m x)$, where $j_1, j_2, \ldots$ are positive zeros of the Bessel function J_0 arranged in ascending order, and a_m is the product of $2/J_1^2$ (j_m) and the integral over t from 0 to 1 of $tf(t)J_0(j_m t)$; J_1 is a Bessel function.   { ˌfùr·ē͵ā ˈbes·əl ˌsir·ēz }

Fourier-Bessel transform *See* Hankel transform.   { ˌfùr·ē͵ā ˈbes·əl ˈtranz͵fòrm }

Fourier expansion *See* Fourier series.   { ˌfùr·ē͵ā ik'span·chən }

Fourier integrals For a function $f(x)$ the Fourier integrals are

$$\frac{1}{\pi}\int_0^\infty du \int_{-\infty}^\infty f(t)\cos u(x-t)dt$$

$$\frac{1}{\pi}\int_0^\infty du \int_{-\infty}^\infty f(t)\sin u(x-t)dt$$

{ ˌfùr·ē͵ā ˈint·ə·grəlz }

Fourier kernel Any kernel $K(x,y)$ of an integral transform which may be written in the form $K(x,y) = k(xy)$ and which is identical with the kernel of the inverse transform.   { ˈfòr·ē·ā ͵kər·nəl }

Fourier-Legendre series Given a function $f(x)$, the series from $n = 0$ to infinity of $a_n P_n$ (x), where (x), $n = 0, 1, 2, \ldots$, are the Legendre polynomials, and a_n is the product of $(2n + 1)/2$ and the integral over x from -1 to 1 of $f(x)P_n(x)$.   { ˌfùr·ē͵ā lə'zhän·drə ˌsir·ēz }

Fourier series The Fourier series of a function $f(x)$ is

$$\frac{1}{2}a_0 + \sum_{n=1}^\infty (a_n \cos nx + b_n \sin nx)$$

with

$$a_n = \frac{1}{\pi}\int_{-\pi}^\pi f(x)\cos nx\,dx$$

$$b_n = \frac{1}{\pi}\int_{-\pi}^\pi f(x)\sin nx\,dx$$

Also known as Fourier expansion.   { ˌfùr·ē͵ā ˌsir·ēz }

Fourier's half-range series A Fourier series that either contains only terms that are even in the independent variable (the cosine series) or contains only terms that are odd (the sine series).   { ˈfòr·ē͵āz ˈhaf ˈrānj ˌsir͵ēz }

Fourier space The space in which the Fourier transform of a function is defined.   { ˌfùr·ē͵ā ͵spās }

Fourier's theorem If $f(x)$ satisfies the Dirichlet conditions on the interval $-\pi < x < \pi$, then its Fourier series converges to $f(x$ for all values of x in this interval at which $f(x)$ is continuous, and approaches $\frac{1}{2}[f(x + 0) + f(x - 0)]$ at points at which $f(x)$ is discontinuous, where $f(x - 0)$ is the limit on the left of f at x and $f(x + 0)$ is the limit on the right of f at x.   { ˌfùr·ē͵āz ͵thir·əm }

Fourier-Stieltjes series For a function $f(x)$ of bounded variation on the interval $[0,2\pi]$, the series from $n = 0$ to infinity of $c_n \exp\,(inx)$, where c_n is $\frac{1}{2}\pi$ times the integral from $x = 0$ to $x = 2\pi$ of $\exp\,(-inx)df(x)$.   { ˌfùr·ē͵ā 'stēl·yes ˌsir·ēz }

Fourier-Stieltjes transform For a function $f(y)$ of bounded variation on the interval $(-\infty, \infty)$, the function $F(x)$ equal to $1/\sqrt{2\pi}$ times the integral from $y = -\infty$ to $y = \infty$ of $\exp\,(-ixy)df(y)$.   { ˌfùr·ē͵ā 'stēl·yes ͵tranz͵fòrm }

Fourier synthesis The determination of a periodic function from its Fourier components.   { ˌfùr·ē͵ā 'sin·thə·səs }

Fourier transform

Fourier transform For a function $f(t)$, the function $F(x)$ equal to $1/\sqrt{2\pi}$ times the integral over t from $-\infty$ to ∞ of $f(t)$ exp (itx). { ˌfu̇r·ē͵ā 'tranzˌfȯrm }

four-point A set of four points in a plane, no three of which are collinear. Also known as complete four-point. { 'fȯr ˌpȯint }

fourth proportional For numbers a, b, and c, a number x such that $a/b = c/x$. { 'fȯrth prə'pȯr·shən·əl }

fourth quadrant 1. The range of angles from 270 to 360°. 2. In a plane with a system of Cartesian coordinates, the region in which the x coordinate is positive and the y coordinate is negative. { ¦fȯrth 'kwäd·rənt }

F process A stochastic process $\{X_t, t > 0\}$ whose value at time t is determined by the information up to time t; more precisely, the events $\{X_t \leq a\}$ belong to F_t for every t and a, where $F = \{F_t, t \geq 0\}$ is an increasing family of sigma algebras that represents the amount of information increasing with time. { 'ef ˌpräs·əs }

fractal A geometrical shape whose structure is such that magnification by a given factor reproduces the original object. { 'frakt·əl }

fractal dimensionality A number D associated with a fractal which satisfies the equation $N = b^D$, where b is the factor by which the length scale changes under a magnification in each step of a recursive procedure defining the object, and N is the factor by which the number of basic units increases in each such step. Also known as Mandelbrot dimensionality. { 'frak·təl diˌmen·shə'nal·əd·ē }

fraction An expression which is the product of a real number or complex number with the multiplicative inverse of a real or complex number. { 'frak·shən }

fractional equation 1. Any equation that contains fractions. 2. An equation in which the unknown variable appears in the denominator of one or more terms. { ¦frak·shən·əl i'kwā·zhən }

fractional factorial experiment An experiment in which certain properly chosen levels of factors are left out. Also known as fractional replicate. { ¦frak·shən·əl fak¦tȯr·ē·əl ik'sper·ə·mənt }

fractional ideal A submodule of the quotient field of an integral domain. { ¦frak·shən·əl i'dēl }

fractional replicate See fractional factorial experiment. { ¦frak·shən·əl 'rep·lə·kət }

fraction in lowest terms A fraction from which all common factors have been divided out of the numerator and denominator. { 'frak·shən in ¦lō·əst 'tərmz }

Fréchet space 1. A topological vector space that is locally convex, metrizable, and complete. 2. A topological vector space that is metrizable and complete. 3. See T1 space. { frā'shā ˌspās }

Fredholm determinant A power series obtained from the function $K(x,y)$ of the Fredholm equation which provides solutions to the equation under certain conditions. { 'fredˌhōm di¦tər·mə·nənt }

Fredholm integral equations Given functions $f(x)$ and $K(x,y)$, the Fredholm integral equations with unknown function y are

$$\text{type 1:} f(x) = \int_a^b K(x,t)y(t)dt$$

$$\text{type 2:} y(x) = f(x) + \lambda \int_a^b K(x,t)y(t)dt$$

{ 'fredˌhōm ¦int·ə·grəl i'kwā·zhənz }

Fredholm operator A linear operator between Banach spaces which has closed range, and both the Fredholm operator and its adjoint have finite dimensional null space. { 'fredˌhōm ˌäp·ə͵rād·ər }

Fredholm theorem A Fredholm equation of type 2 with continuous $f(x)$ has a unique continuous solution, or else the corresponding equation of type 1 has a positive number of linearly independent solutions. { 'fredˌhōm ˌthir·əm }

Fredholm theory The study of the solutions of the Fredholm equations. { 'fredˌhōm ˌthē·ə·rē }

free group A group whose generators satisfy the equation $x \cdot y = e$ (e is the identity element in the group) only when $x = y^{-1}$ or $y = x^{-1}$. { 'frē ˌgrüp }

free module A module which is a free group with respect to its additive group. { ¦frē ¦mäj·yül }

Freeth's nephroid The strophoid of a circle with respect to a pole located at the center and a fixed point located on the circumference. Also known as nephroid of Freeth. { 'frāths 'nef₂rȯid }

free tree A tree graph in which there is no node which is distinguished as the root. { 'frē ˌtrē }

free ultrafilter An ultrafilter of a set, S, that contains any subset of S whose complement is finite. { ¦frē 'əl·trəˌfil·tər }

free variable In logic, a variable that has an occurrence which is not within the scope of a quantifier and thus can be replaced by a constant. { ¦frē 'ver·ē·ə·bəl }

Frenet-Serret formulas Formulas in the theory of space curves, which give the directional derivatives of the unit vectors along the tangent, principal normal and binormal of a space curve in the direction tangent to the curve. Also known as Serret-Frenet formulas. { fre'nā sə'rā ˌfȯr·myə·ləz }

frequency The number of times an event or item falls into or is expected to fall into a certain class or category. { 'frē·kwən·sē }

frequency curve A graphical representation of a continuous frequency distribution; the value of the variable is the abscissa and the frequency is the ordinate. { 'frē·kwən·sē ˌkərv }

frequency distribution A function which measures the relative frequency or probability that a variable can take on a set of values. { ¦frē·kwən·sē ˌdis·trə'byü·shən }

frequency function See probability density function. { 'frē·kwən·sē ˌfəŋk·shən }

frequency polygon A graph obtained from a frequency distribution by joining with straight lines points whose abscissae are the midpoints of successive class intervals and whose ordinates are the corresponding class frequencies. { 'frē·kwən·sē 'päl·əˌgän }

frequency probabilities See objective probabilities. { 'frē·kwən·sē ˌpräb·ə₂bil·əd·ēz }

frequency table A tabular arrangement of the distribution of an event or item according to some specified category or class intervals. { 'frē·kwən·sē ˌtā·bəl }

frequently in A net is frequently in a set if, for each element a of the directed system that indexes the set, there is an element b of the directed system such that $b \geq a$ and x_b (the element indexed by b) is in this set. { 'frē·kwənt·lē ˌin }

Fresnel integrals Given a parameter x, the integrals over t from 0 to x of sin t^2 and of cos t^2 or from x to ∞ of (cos $t)/t^{1/2}$ and of (sin $t)/t^{1/2}$. { frā'nel 'int·ə·grəlz }

friendship theorem The proposition that, among a finite set of people, if every pair of people has exactly one common friend, then there is someone who knows everyone else. { 'fren₂ship ˌpräb·ləm }

Frobenius method A method of finding a series solution near a point for a linear homogeneous ordinary differential equation. { frō'ben·yüs ˌmeth·əd }

frontier For a set in a topological space, all points in the closure of the set but not in its interior. Also known as boundary. { frən'tir əv ə 'set }

frustum The part of a solid between two cutting parallel planes. { 'frəs·təm }

F test See variance ratio test. { 'ef ˌtest }

Fubini's theorem The theorem stating conditions under which

$$\int \int f(u,v)dudv = \int du \int f(u,v)dv = \int dv \int f(u,v)du$$

{ fü'bē·nēz ˌthir·əm }

Fuchsian differential equation A homogeneous, linear differential equation whose coefficients are analytic functions whose only singularities, if any, are poles of order one. { ¦fyük·sē·ən ˌdif·ə¦ren·chəl i'kwā·zhən }

Fuchsian group A Kleinian group G for which there is a region D in the complex plane, consisting of either the interior of a circle or the portion of the plane on one side of a straight line, such that D is mapped onto itself by every element of G. { 'fyük·sē·ən ˌgrüp }

full linear group The group of all nonsingular linear transformations of a complex vector space whose group operation is composition. { ¦fül 'lin·ē·ər ‚grüp }

fully parenthesized notation A method of writing arithmetic expressions in which parentheses are placed around each pair of operands and its associated operator. { 'fül·ē pə‚ren·thə‚sīzd nō'tā·shən }

function A mathematical rule between two sets which assigns to each member of the first, exactly one member of the second. { 'fəŋk·shən }

functional Any function from a vector space into its scalar field. { 'fəŋk·shən·əl }

functional analysis A branch of analysis which studies the properties of mappings of classes of functions from one topological vector space to another. { ¦'fəŋk·shən·əl ə'nal·ə·səs }

functional constraint A mathematical equation which must be satisfied by the independent parameters in an optimization problem, representing some physical principle which governs the relationship among these parameters. { 'fəŋk·shən·əl kən'stränt }

function space A metric space whose elements are functions. { 'fəŋk·shən ‚spās }

function table A table that lists the values of a function for various values of the variable. { 'fəŋk·shən ‚tā·bəl }

functor A function between categories which associates objects with objects and morphisms with morphisms. { 'fəŋk·tər }

fundamental affine connection An affine connection whose coefficients arise from the covariant and contravariant metric tensors of a space. { ¦fən·də¦ment·əl ə'fīn kə'nek·shən }

fundamental forms of a surface Differential forms which express the area and curvature of the surface. { ¦fən·də¦ment·əl 'förmz əv ə 'sər·fəs }

fundamental group For a topological space, the group of homotopy classes of all closed paths about a point in the space; this group yields information about the number and type of "holes" in a surface. { ¦fən·də¦ment·əl 'grüp }

fundamental region Any region in the complex plane that can be mapped conformally onto all of the complex plane. { ¦fən·də¦ment·əl 'rē·jən }

fundamental sequence *See* Cauchy sequence. { ¦fən·də¦ment·əl 'sē·kwəns }

fundamental tensor *See* metric tensor. { ¦fən·də¦ment·əl 'ten·sər }

fundamental theorem of algebra Every polynomial of degree n with complex coefficients has exactly n roots counted according to multiplicity. { ¦fən·də¦ment·əl ¦thir·əm əv 'al·jə·brə }

fundamental theorem of arithmetic Every positive integer greater than 1 can be factored uniquely into the form $P_1{}^{n_1} \ldots P_i{}^{n_i} \ldots P_k{}^{n_k}$, where the P_i are primes, the n_i positive integers. { ¦fən·də¦ment·əl ¦thir·əm əv ə'rith·mə·tik }

fundamental theorem of calculus Given a continuous function $f(x)$ on the closed interval $[a,b]$ the functional

$$F(x) = \int_a^x f(t)\, dt$$

is differentiable on $[a,b]$ and $F'(x) = f(x)$ for every x in $[a,b]$, and if G is any function on $[a,b]$ such that $G'(x) = f(x)$ for all x in $[a,b]$, then

$$\int_a^b f(t)\, dt = G(b) - G(a)$$

{ ¦fən·də¦ment·əl ¦thir·əm əv 'kal·kyə·ləs }

fuzzy logic The logic of approximate reasoning, bearing the same relation to approximate reasoning that two-valued logic does to precise reasoning. { ¦fəz·ē 'läj·ik }

fuzzy mathematics A methodology for systematically handling concepts that embody imprecision and vagueness. { ¦fəz·ē ‚math·ə'mad·iks }

fuzzy model A finite set of fuzzy relations that form an algorithm for determining the outputs of a process from some finite number of past inputs and outputs. { ¦fəz·ē 'mäd·əl }

fuzzy relation A fuzzy subset of the Cartesian product $X \times Y$, denoted as a relation from a set X to a set Y. { ¦fəz·ē ri'lā·shən }

fuzzy relational equation An equation of the form $A \cdot R = B$, where A and B are fuzzy sets, R is a fuzzy relation, and $A \cdot R$ stands for the composition of A with R. { ¦fəz· ē ri¦lā·shən·əl i'kwā·zhən }

fuzzy set An extension of the concept of a set, in which the characteristic function which determines membership of an object in the set is not limited to the two values 1 (for membership in the set) and 0 (for nonmembership), but can take on any value between 0 and 1 as well. { 'fəz·ē 'set }

fuzzy value A membership function of a fuzzy set that serves as the value assigned to a variable. { ¦fəz·ē 'val,yü }

G

Galois field A type of field extension obtained from considering the coefficients and roots of a given polynomial. Also known as root field; splitting field. { 'gal,wä ,fēld }

Galois group A group of isomorphisms of a particular field extension associated with a polynomial's roots. { 'gal,wä ,grüp }

Galois theory The study of the Galois field and Galois group corresponding to a polynomial. { 'gal,wä ,thē·ə·rē }

Galtonian curve A graph showing the variation of any quantity from its normal value. { gȯl'tō·nē·ən 'kərv }

gambler's ruin A game of chance which can be considered to be a series of Bernoulli trials at which each player wins a specified sum of money for every success and loses another sum for every failure; play goes on until the initial capital is lost and the player is ruined. { ¦gam·blərz 'rü·ən }

game A mathematical model expressing a contest between two or more players under specified rules. { gām }

game theory The mathematical study of games or abstract models of conflict situations from the viewpoint of determining an optimal policy or strategy. Also known as theory of games. { 'gām ,thē·ə·rē }

game tree A tree graph used in the analysis of strategies for a game, in which the vertices of the graph represent positions in the game, and a given vertex has as its successors all vertices that can be reached in one move from the given position. Also known as lookahead tree. { 'gām ,trē }

gamma distribution A normal distribution whose frequency function involves a gamma function. Also known as Erlang distribution. { ¦gam·ə ,dis·trə'byü·shən }

gamma function The complex function given by the integral with respect to t from 0 to ∞ of $e^{-t}t^{z-1}$; this function helps determine the general solution of Gauss' hypergeometric equation. { 'gam·ə ,fəŋk·shən }

gamma random variable A random variable that has a gamma distribution. { ¦gam·ə ¦ran·dəm 'ver·ē·ə·bəl }

Gaskin's theorem A theorem in projective geometry which states that if a circle circumscribes a triangle which is identical with its conjugate triangle with respect to a given conic, then the tangent to the circle at either of its intersections with the director circle of the conic is perpendicular to the tangent to the director circle at the same intersection. { 'gas·kinz ,thir·əm }

Gauss-Bonnet theorem The theorem that the Euler characteristic of a compact Riemannian surface is $1/(2\pi)$ times the integral over the surface of the Gaussian curvature. { ¦gaus bə'nā ,thir·əm }

Gauss-Codazzi equations Equations dealing with the components of the fundamental tensor and Riemann-Christoffel tensor of a surface. { ¦gaus kō'dat·sē i,kwā·zhənz }

Gauss' error curve *See* normal distribution. { 'gaus 'er·ər ,kərv }

Gauss formulas Formulas dealing with the sine and cosine of angles in a spherical triangle. Also known as Delambre analogies. { 'gaus ,fȯr·myə·ləz }

Gauss' hypergeometric equation The differential equation, arising in many physical contexts, $x(1 - x)y'' + [c - (a + b + 1)x]y' - aby = 0$. { 'gaus ,hī·pər,jē·ə'me·trik i'kwā·zhən }

Gaussian complex integers

Gaussian complex integers Complex numbers whose real and imaginary parts are both integers. { ¦gaù·sē·ən ¦käm,pleks 'int·ə·jərz }

Gaussian curvature The invariant of a surface specified by Gauss' theorem. Also known as total curvature. { ¦gaù·sē·ən 'kər·və·chər }

Gaussian curve The bell-shaped curve corresponding to a population which has a normal distribution. Also known as normal curve. { ¦gaù·sē·ən 'kərv }

Gaussian distribution *See* normal distribution. { ¦gaù·sē·ən ,dis·trə'byü·shən }

Gaussian elimination A method of solving a system of n linear equations in n unknowns, in which there are first $n - 1$ steps, the mth step of which consists of subtracting a multiple of the mth equation from each of the following ones so as to eliminate one variable, resulting in a triangular set of equations which can be solved by back substitution, computing the nth variable from the nth equation, the $(n - 1)$st variable from the $(n - 1)$st equation, and so forth. { ¦gaù·sē·ən ə,lim·ə'nā·shən }

Gaussian integer A complex number whose real and imaginary parts are both ordinary (real) integers. Also known as complex integer. { ¦gäùs·ē·ən 'int·ə·jər }

Gaussian noise *See* Wiener process. { ¦gaù·sē·ən 'nòiz }

Gaussian reduction A procedure of simplification of the rows of a matrix which is based upon the notion of solving a system of simultaneous equations. Also known as Gauss-Jordan elimination. { ¦gaù·sē·ən ri'dək·shən }

Gaussian representation *See* spherical image. { ¦gaùs·ē·ən ,rep·rə·zen'tā·shən }

Gauss-Jordan elimination *See* Gaussian reduction. { ¦gaùs ¦jórd·ən ə,lim·ə'nā·shən }

Gauss' law of the arithmetic mean The law that a harmonic function can attain its maximum value only on the boundary of its domain of definition, unless it is a constant. { 'gaùs ,lò əv thə ,a·rith¦med·ik 'mēn }

Gauss-Legendre rule An approximation technique of definite integrals by a finite series which uses the zeros and derivatives of the Legendre polynomials. { ¦gaùs lə'zhän·drə ,rül }

Gauss' mean value theorem The value of a harmonic function at a point in a planar region is equal to its integral about a circle centered at the point. { 'gaùs 'mēn ,val·yü ,thir·əm }

Gauss-Seidel method *See* Seidel method. { ¦gaùs 'zīd·əl ,meth·əd }

Gauss test In an infinite series with general term a_n, if $a_{n+1}/a_n = 1 - (x/n) - [f(n)/n^\lambda]$ where x and λ are greater than 1, and $f(n)$ is a bounded integer function, then the series converges. { 'gaùs ,test }

Gauss' theorem 1. The assertion, under certain light restrictions, that the volume integral through a volume V of the divergence of a vector function is equal to the surface integral of the exterior normal component of the vector function over the boundary surface of V. Also known as divergence theorem; Green's theorem in space; Ostrogradski's theorem. 2. At a point on a surface the product of the principal curvatures is an invariant of the surface, called the Gaussian curvature. { 'gaùs ,thir·əm }

gcd *See* greatest common divisor.

Gegenbauer polynomials A family of polynomials solving a special case of the Gauss hypergeometric equation. Also known as ultraspherical polynomials. { 'gāg·ən,baùr ,päl·i'nō·mē·əlz }

Gelfond-Schneider theorem The theorem that if a and b are algebraic numbers, where a is not equal to 0 or 1, and b is not a rational number, then a^b is a transcendental number. { ¦gel,fänd 'shnīd·ər ,thir·əm }

general continuum hypothesis A generalization of the continuum hypothesis which asserts that the smallest cardinal number greater than the cardinal number of an infinite set, S, is the cardinal number of the set of subsets of S. { ¦jen·rəl kən'tin·yə·wəm hī,päth·ə·səs }

general integral *See* general solution. { ¦gen·rəl 'int·ə·grəl }

generalized binomial trials model A product model in which the nth factor model has two simple events with probabilities p_n and $q_n = 1 - p_n$. Also known as Poisson binomial trials model. { 'jen·rə,līzd bī¦nō·mē·əl 'trīlz ,mäd·əl }

96

generalized Euclidean space *See* inner-product space. { ˌjen·rəˌlīzd yü̇ˌklid·ē·ən ′spās }

generalized feasible flow A feasible flow in a generalized *s-t* network such that the outflow at any intermediate vertex does not exceed the weight of that vertex. { ˌjen·rəˌlīzd ˌfēz·ə·bəl ′flō }

generalized function *See* distribution. { ′jen·rəˌlīzd ′fəŋk·shən }

generalized max-flow min-cut theorem The theorem that in a generalized *s-t* network the maximum possible flow value of a generalized feasible flow equals the minimum possible weight of a generalized *s-t* cut. { ˌjen·rəˌlīzd ˈmaksˌflō ˌminˈkət ˌthir·əm }

generalized mean-value theorem *See* second mean-value theorem. { ˈjen·rəˌlīzd ′mēn ˈvalˈyü ˌthir·əm }

generalized permutation Any ordering of a finite set of elements that are not necessarily distinct. { ˌjen·rəˌlīzd ˌpər·myəˈtā·shən }

generalized Poincaré conjecture The question as to whether every closed *n*-manifold which has the homotopy type of the *n*-sphere is homeomorphic to the *n*-sphere. { ′jen·rəˌlīzd ˈpwän·kaˌrä kənˈjekˈchər }

generalized power For a positive number a and an irrational number x, the number a^x defined by the equation $a^x = e^x \log{}^a$, where e is the base of the natural logarithms and log a is taken to that base. { ′jen·rəˌlīzd ′pau·ər }

generalized ratio test *See* d'Alembert's test for convergence. { ˌjen·rəˌlīzd ′rā·shō ˌtest }

generalized s-t cut A set of arcs and vertices in a generalized *s-t* network such that any directed path from the source to the terminal includes at least one element of this set. { ˌjen·rəˌlīzd ′esˈtē ′kət }

generalized s-t network An *s-t* network on which is defined a weight function from the vertices of the network to the nonnegative integers. { ˌjen·rəˌlīzd ′esˈtē ′netˌwərk }

general solution For an *n*th-order differential equation, a function of the independent variables of the equation and of *n* parameters such that assignment of any numerical values to the parameters yields a solution to the equation. Also known as general integral. { ˈjen·rəl səˈlü·shən }

general term The general term of a sequence or series is an expression subscripted by an integer which determines any desired entry. { ˈjen·rəl ′tərm }

general topology The branch of topology that studies the relationships between the basic topological properties that spaces may possess. Also known as point-set topology. { ˈjen·rəl təˈpäl·ə·jē }

generating function 1. A function $g(x,y)$ corresponding to a family of orthogonal polynomials $f_0(x), f_1(x), \ldots$, where a Taylor series expansion of $g(x,y)$ in powers of y will have the polynomial $f_n(x)$ as the coefficient for the term y^n. 2. A function, $g(y)$, corresponding to a sequence $a_0, a_1, \ldots$, where $g(y) = a_0 + a_1 y + a_2 y^2 + \cdots$. Also known as ordinary generating function. { ′jen·əˌrād·iŋ ˌfəŋk·shən }

generator 1. One of the set of elements of an algebraic system such as a group, ring, or module which determine all other elements when all admissible operations are performed upon them. 2. *See* generatrix. { ′jen·əˌrād·ər }

generatrix The straight line generating a ruled surface. Also known as generator. { ˈjen·əˈrā·triks }

Genocchi number An integer of the form $G_n = 2(2^{2n} - 1)B_n$, where B_n is the *n*th Bernoulli number. { gəˈnäk·ēˌnəm·bər }

genus An integer associated to a surface which measures the number of holes in the surface. { ′jē·nəs }

geodesic A curve joining two points in a Riemannian manifold which has minimum length. { ˈjē·əˈdes·ik }

geodesic circle The locus of all points on a given surface whose geodesic distance from a given point on the surface (called the center of the circle) is a given constant. { ˈjē·əˈdes·ik ′sər·kəl }

geodesic curvature For a point on a curve lying on a surface, the curvature of the orthogonal projection of the curve onto the tangent plane to the surface at the point; it measures the departure of the curve from a geodesic. Also known as tangential curvature. { ¦jē·ə¦des·ik 'kərv·ə·chər }

geodesic distance For two points in a Riemannian manifold, the length of a geodesic connecting them. { ¦jē·ə¦des·ik 'di·stəns }

geodesic ellipse The locus of all points on a given surface at which the sum of geodesic distances from a fixed pair of points is a constant. { ¦jē·ə¦des·ik i'lips }

geodesic hyperbola The locus of all points on a given surface at which the difference between the geodesic distances to two fixed points is a constant. { ¦jē·ə¦des·ik hī'pər·bə·lə }

geodesic line The shortest line between two points on a mathematically derived surface. { ¦jē·ə¦des·ik 'līn }

geodesic parallels Two curves on a given surface such that the lengths of geodesics between the curves that intersect both curves orthogonally is a constant. { ¦jē·ə¦des·ik 'par·ə,lelz }

geodesic parameters Coordinates u and v of a surface such that the curves obtained by setting u equal to various constants form a family of geodesic parallels, while the curves obtained by setting v equal to various constants form the corresponding orthogonal family, of length $u_2 - u_1$ between the points (u_1,v) and (u_2,v). { ¦jē·ə¦des·ik pə'ram·əd·ərz }

geodesic polar coordinates Coordinates u and v of a surface such that the curves obtained by setting u equal to various constants are geodesic circles with a common center P and geodesic radius u, and the curves obtained by setting v equal to various constants are geodesics passing through P such that v_0 is the angle between the tangents at P to the lines $v = 0$ and $v = v_0$. { ¦jē·ə¦des·ik ¦pōl·ər kō'órd·ə·nəts }

geodesic radius For a geodesic circle on a surface, the geodesic distance from the center of a circle to the points on the circle. { ¦jē·ə¦des·ik 'rād·ē·əs }

geodesic torsion 1. For a given point on a surface and a given direction, the torsion of the geodesic on the surface through the point and in the given direction. **2.** For a given curve on a surface at a given point, the torsion of the geodesic through the point in the same direction as the given curve. { ¦jē·ə¦des·ik 'tór·shən }

geodesic triangle The figure formed by three geodesics joining three points on a given surface. { ¦jē·ə¦des·ik 'trī,aŋ·gəl }

geodetic triangle *See* spheroidal triangle. { ¦jē·ə¦ded·ik 'trī,aŋ·gəl }

geometric average *See* geometric mean. { ¦jē·ə¦me·trik 'av·rij }

geometric complex *See* simplicial complex. { ¦jē·ə,me·trik 'käm,pleks }

geometric distribution A discrete probability distribution whose probability function is given by the equation $P(x) = p(1 - p)^{x-1}$ for x any positive integer, $p(x) = 0$ otherwise, when $0 \le p \le 1$; the mean is $1/p$. { ¦jē·ə¦me·trik ,dis·trə'byü·shən }

geometric duals Two polyhedra such that the vertices of one are in unique correspondence with the faces of the other. { ¦jē·ə,me·trik 'dülz }

geometric mean The geometric mean of n given quantities is the nth root of their product. Also known as geometric average. { ¦jē·ə¦me·trik 'mēn }

geometric moment of inertia The geometric moment of inertia of a plane figure about an axis in or perpendicular to the plane is the integral over the area of the figure of the square of the distance from the axis. Also known as second moment of area. { ¦jē·ə¦me·trik ¦mō·mənt əv i'nər·shə }

geometric number theory The branch of number theory studying relationships among numbers by examining the geometric properties of ordered pair sets of such numbers. { ¦jē·ə¦me·trik 'nəm·bər ,thē·ə·rē }

geometric progression A sequence which has the form a, ar, ar^2, ar^3, ... { ¦jē·ə¦me·trik prə'gresh·ən }

geometric sequence A sequence in which the ratio of a term to its predecessor is the same for one term as for any other. { ¦jē·ə,me·trik 'sē·kwəns }

geometric series An infinite series of the form $a + ar + ar^2 + ar^3 + \cdots$. { ¦jē·ə¦me·trik 'sir·ēz }

geometry The qualitative study of shape and size. { jē'äm·ə·trē }

Gershgorin's method A method of obtaining bounds on the eigenvalue of a matrix, based on the fact that the absolute value of any eigenvalue is equal to or less than the maximum over the rows of the matrix of the sum of the absolute values of the entries in a row, and is also equal to or less than the maximum over the columns of the matrix of the sum of the absolute values of the entries in a column. { gərsh'gór·ənz ,meth·əd }

gibbous Bounded by convex curves. { 'jib·əs }

Gibbs' phenomenon A convergence phenomenon occurring when a function with a discontinuity is approximated by a finite number of terms from a Fourier series. { 'gibz fə,näm·ə,nän }

Gibrat's distribution The distribution of a variable whose logarithm has a normal distribution. { zhē'bräz di·strə'byü·shən }

give-and-take lines Straight lines which are used to approximate the boundary of an irregular, curvilinear figure for the purpose of approximating its area; they are placed so that small portions excluded from the area under consideration are balanced by other small portions outside the boundary. { ¦giv ən 'tāk ,līnz }

Givens's method A transformation method for finding the eigenvalues of a matrix, in which each of the orthogonal transformations that reduce the original matrix to a triple-diagonal matrix makes one pair of elements, a_{ij} and a_{ji}, lying off the principal diagonal and the diagonals immediately above and below it, equal to zero, without affecting zeros obtained earlier. { 'giv·ən·zəz ,meth·əd }

given-year method *See* Paasche's index. { ¦giv·ən 'yir ,meth·əd }

glb *See* greatest lower bound.

glisette A curve, such as Watt's curve, traced out by a point attached to a curve which moves so that it always touches two fixed curves, or the envelope of any line or curve attached to the moving curve. { gli'set }

Glivenko-Cantelli lemma The empirical distribution functions of a random variable converge uniformly in probability to the distribution function of the random variable. { gli'veŋ·kō kan'tel·ē 'lem·ə }

global property A property of an object (such as a space, function, curve, or surface) whose specification requires consideration of the entire object, rather than merely the neighborhoods of certain points. { ¦glō·bəl 'präp·ərd·ē }

gnomon A geometric figure formed by removing from a parallelogram a similar parallelogram that contains one of its corners. { 'nō·mən }

Gödel numbering *See* arithmetization. { 'gərd·əl ,nəm·bə·riŋ }

Gödel's proof Any formal arithmetical system is incomplete in the sense that, given any consistent set of arithmetical axioms, there are true statements in the resulting arithmetical system that cannot be derived from these axioms. { 'gərd·əlz 'prüf }

Gödel's second theorem The theorem that any formal arithmetical system is incomplete in the sense that, if it is consistent, it cannot prove its own consistency. { ¦gərd·əlz ,sek·ənd 'thir·əm }

Goldbach conjecture The unestablished conjecture that every even number except the number 2 is the sum of two primes. { 'gōl,bäk̠ kəŋ,jek·chər }

golden mean *See* golden section. { ,gōld·ən 'mēn }

golden ratio *See* golden section. { ,gōld·ən 'rā·shō }

golden rectangle A rectangle that can be divided into a square and another rectangle similar to itself; its sides have the ratio $(1+\sqrt{5})/2$. { ,göl·dən 'rek,taŋ·gəl }

golden section The division of a line so that the ratio of the whole line to the larger interval equals the ratio of the larger interval to the smaller. Also known as divine proportion; extreme and mean ratio; golden mean; golden ratio. { 'gōl·dən 'sek·shən }

Gompertz curve A curve similar to the exponential curve except that the constant a is raised to the b^x power instead of the x power; used in fitting a trend line to a nonlinear time series. { 'gäm,pərts ,kərv }

gon *See* grade. { gän }

goodness of fit The degree to which the observed frequencies of occurrence of events in an experiment correspond to the probabilities in a model of the experiment. Also known as best fit. { ¦gùd·nəs əv 'fit }

googol

googol A name for 10 to the power 100. { 'gü,gȯl }

googolplex A name for 10 to the power googol. { 'gü,gȯl,pleks }

grade A unit of plane angle, equal to 0.01 right angle, or $\pi/200$ radians, or $0.9°$. Also known as gon. { grād }

graded Lie algebra A generalization of a Lie algebra in which both commutators and anticommutators occur. { ¦grād·əd ¦lē 'al·jə·brə }

gradient A vector obtained from a real function $f(x_1, x_2, \ldots, x_n)$ whose components are the partial derivatives of f; this measures the maximum rate of change of f in a given direction. { 'grād·ē·ənt }

gradient method A finite iterative procedure for solving a system of n equations in n unknowns. { 'grād·ē·ənt ,meth·əd }

gradient projection method Computational method used in nonlinear programming when constraint functions are linear. { 'grād·ē·ənt prə'jek·shən ,meth·əd }

Graeffe's method A method of solving algebraic equations by means of squaring the exponents and making appropriate substitutions. { 'gref·əz ,meth·əd }

Gram determinant The Gram determinant of vectors $v_1, \ldots, v_n$ from an inner product space is the determinant of the $n \times n$ matrix with the inner product of v_i and v_j as entry in the ith column and jth row; its vanishing is a necessary and sufficient condition for linear dependence. { 'gram di'tərm·ə·nənt }

Gram-Schmidt orthogonalization process A process by which an orthogonal set of vectors is obtained from a linearly independent set of vectors in an inner product space. { ¦gram 'shmit ,ȯr¦thäg·ən·əl·ə'zā·shən ,präs·əs }

Gram's theorem A set of vectors are linearly dependent if and only if their Gram determinant vanishes. { 'gramz ,thir·əm }

graph **1.** The planar object, formed from points and line segments between them, used in the study of circuits and networks. **2.** The graph of a function f is the set of all ordered pairs $[x,f(x)]$, where x is in the domain of f. **3.** The set of all points that satisfy a particular equation, inequality, or system of equations or inequalities. **4.** *See* graphical representation. { graf }

graph component A particular type of maximal connected subgraph of a graph. { 'graf kəm'pō·nənt }

graphical analysis The study of interdependent phenomena by analyzing graphical representations. { ¦graf·ə·kəl ə'nal·ə·səs }

graphical representation The plot of the points in the plane which constitute the graph of a given real function or a pictorial diagram depicting interdependence of variables. Also known as graph. { ¦graf·ə·kəl ,rep·rə·zen'tā·shən }

graphical vector A finite, nonincreasing sequence of nonnegative integers that is the degree vector of some simple graph. { ¦graf·ə·kəl 'vek·tər }

graph theory **1.** The mathematical study of the structure of graphs and networks. **2.** The body of techniques used in graphing functions in the plane. { 'graf ,thē·ə·rē }

Grassmann algebra *See* exterior algebra. { 'gräs·mən ,al·jə·brə }

Grassmannian *See* Grassmann manifold. { ¦gräs¦man·ē·ən }

Grassmann manifold The differentiable manifold whose points are all k-dimensional planes passing through the origin in n-dimensional Euclidean space. Also known as Grassmannian. { 'gräs·mən 'man·ə,fōld }

great circle The circle on the two-sphere produced by a plane passing through the center of the sphere. { 'grāt ¦sər·kəl }

greatest common divisor The greatest common divisor of integers $n_1, n_2, \ldots, n_k$ is the largest of all integers that divide each n_i. Abbreviated gcd. Also known as highest common factor (hcf). { 'grād·əst ¦käm·ən di'vīz·ər }

greatest lower bound The greatest lower bound of a set of numbers S is the largest number among the lower bounds of S. Abbreviated glb. Also known as infimum (inf). { 'grād·əst ¦lō·ər 'baünd }

greatest-lower-bound axiom The statement that any set of real numbers that has a lower bound also has a greatest lower bound. { ¦grād·əst ,lō·ər,baünd 'ak·sē·əm }

Greco-Latin square An arrangement of combinations of two sets of letters (one set Greek, the other Roman) in a square array, in such a way that no letter occurs more than once in the array. Also known as orthogonal Latin square. { ¦grek·ō¦lat·ən 'skwer }

Green's dyadic A vector operator which plays a role analogous to a Green's function in a partial differential equation expressed in terms of vectors. { 'grēnz dī'ad·ik }

Green's function A function, associated with a given boundary value problem, which appears as an integrand for an integral representation of the solution to the problem. { 'grēnz ,fəŋk·shən }

Green's identities Formulas, obtained from Green's theorem, which relate the volume integral of a function and its gradient to a surface integral of the function and its partial derivatives. { 'grēnz i'den·ə,dēz }

Green's theorem Under certain general conditions, an integral along a closed curve C involving the sum of functions $P(x,y)$ and $Q(x,y)$ is equal to a surface integral, over the region D enclosed by C, of the partial derivatives of P and Q; namely, $\int_C P\,dx + Q\,dy = \iint_D \left(\frac{\partial Q}{\partial x} - \frac{\partial P}{\partial y}\right) dx\,dy.$ { 'grēnz ,thir·əm }

Green's theorem in space See Gauss' theorem. { ¦grēnz ,thir·əm in 'spās }

Gregory formula A formula used in the numerical evaluation of integrals derived from the Newton formula. { 'greg·ə·rē ,för·myə·lə }

gross errors Errors that occur when a measurement process is subject occasionally to large inaccuracies. { ¦grōs 'er·ərz }

group A set G with an associative binary operation where $g_1 \cdot g_2$ always exists and is an element of G, each g has an inverse element g^{-1}, and G contains an identity element. { grüp }

groupoid A set having a binary relation everywhere defined. { 'grü,pȯid }

group theory The study of the structure of groups which especially deals with the classification of finite groups. { 'grüp ,thē·ə·rē }

group without small subgroups A topological group in which there is a neighborhood of the identity element that contains no subgroup other than the subgroup consisting of the identity element alone. { ¦grüp with,aůt ,smȯl 'səb,grüps }

growth index For a function of bounded growth f, the smallest real number a such that for some positive real constant M the quantity Me^{ax} is greater than the absolute value of $f(x)$ for all positive x; for a function that is not of bounded growth, the quantity $+\infty$. { 'grōth ,in,deks }

G space A topological space X together with a topological group G and a continuous function on the Cartesian product of X and G to X such that if the values of this function at (x,g) are denoted by xg, then $x(g_1g_2) = (xg_1)g_2$ and $xe = x$ where e is the identity in G and g_1, g_2 are elements in G. { 'jē ,spās }

Gudermannian The function y of the variable x satisfying $\tan y = \sinh x$ or $\sin y = \tanh x$; written gdx. { 'güd·ər,män·ē·ən }

Gutschoven's curve See kappa curve. { 'güt,shō·fənz ,kərv }

Haar measure A measure on the Borel subsets of a locally compact topological group whose value on a Borel subset U is unchanged if every member of U is multiplied by a fixed element of the group. { 'här ˌmezh·ər }

Hadamard's conjecture The conjecture that any partial differential equation that is essentially different from the wave equation fails to satisfy Huygens' principle. { 'had·əˌmärdz kən'jek·chər }

Hadamard's inequality An inequality that gives an upper bound for the square of the absolute value of the determinant of a matrix in terms of the squares of the matrix entries; the upper bound is the product, over the rows of the matrix, of the sum of the squares of the absolute values of the entries in a row. { 'had·əˌmärdz ˌin·əˌkwäl·əd·ē }

Hadamard's three-circle theorem The theorem that if the complex function $f(z)$ is analytic in the ring $a < |z| < b$, and if $m(r)$ denotes the maximum value of $|f(z)|$ on the circle $|z| = r$ with $a < r < b$, then log $m(r)$ is a convex function of log r. { 'had·əˌmärdz ˌthrē ˌsər·kəl 'thir·əm }

Hahn-Banach extension theorem The theorem that every continuous linear functional defined on a subspace or linear manifold in a normed linear space X may be extended to a continuous linear functional defined on all of X. { ˌhän ˌbän·äk ek'sten·chən ˌthir·əm }

Hahn decomposition The Hahn decomposition of a measurable space X with signed measure m consists of two disjoint subsets A and B of X such that the union of A and B equals X, A is positive with respect to m, and B is negative with respect to m. { ˌhän dēˌkäm·pə'zish·ən }

half-angle formulas In trigonometry, formulas that express the trigonometric functions of an angle in terms of trigonometric functions of the angle. { 'haf ˌaŋ·gəl ˌfȯr·myə·ləz }

half line See ray. { 'haf ˌlīn }

half plane The portion of a plane lying on one side of some line in the plane; in particular, all points of the complex plane either above or below the real axis. { 'haf ˌplān }

half-side formulas In trigonometry, formulas that express the tangents of one-half of each of the sides of a spherical triangle in terms of its angles. { 'haf ˌsīd ˌfȯr·myə·ləz }

half space A space bounded only by an infinite plane. { 'haf ˌspās }

half-width For a function which has a maximum and falls off rapidly on either side of the maximum, the difference between the two values of the independent variable for which the dependent variable has one-half its maximum value. { 'haf ˌwidth }

Hall's theorem See marriage theorem. { 'hȯlz ˌthir·əm }

Hamel basis For a normed space, a collection of vectors with every finite subset linearly independent, while any vector of the space is a linear combination of at most countably many vectors from this subset. { 'ham·əl ˌbā·səs }

Hamilton-Cayley theorem See Cayley-Hamilton theorem. { 'ham·əl·tən 'kā·lē ˌthir·əm }

Hamiltonian circuit See Hamiltonian path. { ˌham·əlˌtō·nē·ən 'sər·kət }

Hamiltonian cycle See Hamiltonian path. { ˌham·əl'tō·nē·ən ˌsī·kəl }

Hamiltonian graph

Hamiltonian graph A graph which has a Hamiltonian path. { ,ham·əl'tō·nē·ən ,graf }

Hamiltonian path A path along the edges of a graph that traverses every vertex exactly once and terminates at its starting point. Also known as Hamiltonian circuit; Hamiltonian cycle. { ,ham·əl'tō·nē·ən ,path }

Hamilton-Jacobi equation A particular partial differential equation useful in studying certain systems of ordinary equations arising in the calculus of variations, dynamics, and optics: $H(q_1, \ldots, q_n, \partial\phi/\partial q_1, \ldots, \partial\phi/\partial q_n, t) + \partial\phi/\partial t = 0$, where $q_1, \ldots, q_n$ are generalized coordinates, t is the time coordinate, H is the Hamiltonian function, and ϕ is a function that generates a transformation by means of which the generalized coordinates and momenta may be expressed in terms of new generalized coordinates and momenta which are constants of motion. { 'ham·əl·tən jə'kō·bē i,kwä·zhən }

Hamilton-Jacobi theory The study of the solutions of the Hamilton-Jacobi equation and the information they provide concerning solutions of the related systems of ordinary differential equations. { 'ham·əl·tən jə'kō·bē ,thē·ə·rē }

ham sandwich 1. The theorem that if the functions f and h have the same limit L, and if the value of a third function g is greater to or equal than that of f and less than or equal to than that of h for all values of the independent variable, then g also has the limit L. **2.** The theorem that there is a plane that cuts each of three bounded, connected, open sets in space into two sets of equal volume. { ,ham 'san,wich }

handshaking lemma The result that the sum of the degrees of a graph is twice the number of its edges. { 'han,shāk·iŋ ,lem·ə }

Hankel functions The Bessel functions of the third kind, occurring frequently in physical studies. { 'häŋk·əl ,fəŋk·shənz }

Hankel transform The Hankel transform of order m of a real function $f(t)$ is the function $F(s)$ given by the integral from 0 to ∞ of $f(t)tJ_m(st)dt$, where J_m denotes the mth-order Bessel function. Also known as Bessel transform; Fourier-Bessel transform. { 'häŋk·əl ,tranz,fōrm }

harmonic A solution of Laplace's equation which is separable in a specified coordinate system. { här'män·ik }

harmonic analysis A study of functions by attempting to represent them as infinite series or integrals which involve functions from some particular well-understood family; it subsumes studying a function via its Fourier series. { här'man·ik ə'nal·ə·səs }

harmonic average *See* harmonic mean. { här,män·ik 'av·rij }

harmonic conjugates 1. Two points, P_3 and P_4, that are collinear with two given points, P_1 and P_2, such that P_3 lies in the line segment P_1P_2 while P_4 lies outside it, and, if x_1, x_2, x_3, and x_4 are the abscissas of the points, $(x_3 - x_1)/(x_3 - x_2) = -(x_4 - x_1)/(x_4 - x_2)$. **2.** A pair of harmonic functions, u and v, such that $u + iv$ is an analytic function, or, equivalently, u and v satisfy the Cauchy-Riemann equations. { här'män·ik 'kän·jə·gəts }

harmonic division The division of a line segment externally and internally in the same ratio; that is, the division of a line segment by the harmonic conjugates of its end points. { här'män·ik di'vizh·ən }

harmonic function 1. A function of two real variables which is a solution of Laplace's equation in two variables. **2.** A function of three real variables which is a solution of Laplace's equation in three variables. { här'män·ik 'fəŋk·shən }

harmonic mean For n positive numbers x_1, x_2, $\ldots$, x_n their harmonic mean is the number $n/(1/x_1 + 1/x_2 + \cdots + 1/x_n)$. Also known as harmonic average. { här'män·ik 'mēn }

harmonic measure Let D be a domain in the complex plane bounded by a finite number of Jordan curves Γ, and let Γ be the disjoint union of α and β, where α and β are Jordan arcs; the harmonic measure of α with respect to D is the harmonic function on D which assumes the value 1 on α and the value 0 on β. { här,män·ik 'mezh·ər }

harmonic pencil The configuration of four lines, passing through a single point, such that any line that is not parallel to one of the four cuts the four lines at points which are harmonic conjugates. { här¦män·ik 'pen·səl }

harmonic progression A sequence of numbers whose reciprocals form an arithmetic progression. Also known as harmonic sequence. { här'män·ik prə'gresh·ən }

harmonic range The configuration of four collinear points which are harmonic conjugates. { här¦män·ik 'rānj }

harmonic ratio A cross ratio that is equal to −1. { här¦män·ik 'rā·sho }

harmonic sequence *See* harmonic progression. { här‚män·ik 'sē·kwəns }

harmonic series A series whose terms form a harmonic progression. { här¦män· ik 'sir‚ēz }

Harnack's first convergence theorem The theorem that if a sequence of functions harmonic in a common domain of three-dimensional space and continuous on the boundary of the domain converges uniformly on the boundary, then it converges uniformly in the domain to a function which is itself harmonic; the sequence of any partial derivative of the functions in the original sequence converges uniformly to the corresponding partial derivative of the limit function in every closed subregion of the domain. { 'här·naks ¦fərst kən'vər·jəns ‚thir·əm }

Harnack's second convergence theorem The theorem that if a sequence of functions is harmonic in a common domain of three-dimensional space and their values are monotonically decreasing at any point in the domain, then convergence of the sequence at any point in the domain implies uniform convergence of the sequence in every closed subregion of the domain to a function which is itself harmonic. { 'här·naks ¦sek·ənd kən'vər·jəns ‚thir·əm }

Hartley transform An analog of the Fourier transform for finite, real-valued data sets; for a function f defined at N data values, $0, 1, 2, \ldots, N - 1$, the Hartley transform is a function, F, also defined on the set $(0, 1, 2, \ldots, N - 1)$, whose value at n is the sum over the variable r, from 0 through $N - 1$, of the quantity $N^{-1}f(r)$ cas $(2\pi nr/N)$, where cas $\theta = \cos \theta + \sin \theta$. { }

Hasse diagram A representation of a partially ordered set as a directed graph, in which elements of the set are represented by vertices of the graph, and there is a directed arc from x to y if and only if y covers x. { 'häs·ə ‚dī·ə‚gram }

Hausdorff maximal principle The principle that every partially ordered set has a linearly ordered subset S which is maximal in the sense that S is not a proper subset of another linearly ordered subset. { 'haûs·dôrf 'mak·sə·məl ‚prin·sə·pəl }

Hausdorff paradox The theorem that a sphere can be represented as the union of four disjoint sets, A, B, C, and D, where D is a countable set, and A is congruent to each of the three sets B, C, and the union of B and C. { 'haûs‚dôrf ‚par·ə‚däks }

Hausdorff space A topological space where each pair of distinct points can be enclosed in disjoint open neighborhoods. Also known as T₂ space. { 'haûs·dôrf ‚spās }

hav *See* haversine.

haversine The haversine of an angle A is half of the versine of A, or is $\frac{1}{2}(1 - \cos A)$. Abbreviated hav. { 'ha·vər‚sīn }

hcf *See* greatest common divisor.

Heaviside calculus A type of operational calculus that is used to completely analyze a linear dynamical system which represents some vibrating physical system. { 'hev· ē‚sīd ‚kal·kyə·ləs }

Heaviside's expansion theorem A theorem providing an infinite series representation for the inverse Laplace transforms of functions of a particular type. { 'hev·ē‚sīdz ik'span·chən ‚thir·əm }

Heaviside unit function The real function $f(x)$ whose value is 0 if x is negative and whose value is 1 otherwise. { 'hev·ē‚sīd 'yü·nət ¦fəŋk·shən }

hei function A function that is expressed in terms of Hankel functions in a manner similar to that in which the bei function is expressed in terms of Bessel functions. { 'hī ‚fəŋk·shən }

height 1. The perpendicular distance between horizontal lines or planes passing through the top and bottom of an object. **2.** The height of a rational number q is the maximum of $|m|$ and $|n|$, where m and n are relatively prime integers such that $q = m/n$. { hīt }

Heine-Borel theorem The theorem that the only compact subsets of the real line are those which are closed and bounded. { 'hī·nə bò'rel ˌthir·əm }

helical Pertaining to a cylindrical spiral, for example, a screw thread. { 'hel·ə·kəl }

helicoid A surface generated by a curve which is rotated about a straight line and also is translated in the direction of the line at a rate that is a constant multiple of its rate of rotation. { 'hel·əˌkòid }

helix A curve traced on a cylindrical or conical surface where all points of the surface are cut at the same angle. { 'hē,liks }

helix angle The constant angle between the tangent to a helix and a generator of the cylinder upon which the helix lies. { 'hē,liks ˌaŋ·gəl }

Helly's theorem The theorem that there is a point that belongs to each member of a collection of bounded closed convex sets in an n-dimensional Euclidean space if the collection has at least $n + 1$ members and any $n + 1$ members of the collection have a common point. { 'hel·ēz ˌthir·əm }

Helmholtz equation A partial differential equation obtained by setting the Laplacian of a function equal to the function multiplied by a negative constant. { 'helmˌhōlts iˌkwā·zhən }

Helmholtz's theorem The theorem determining a general class of vector fields as being everywhere expressible as the sum of an irrotational vector with a divergence-free vector. { 'helmˌhōlt·səz ˌthir·əm }

hemicycle A curve in the form of a semicircle. { 'he·mēˌsī·kəl }

hemisphere One of the two pieces of a sphere divided by a great circle. { 'he·mēˌsfir }

hemispheroid One of the halves into which a spheroid is divided by a plane of symmetry. { ˌhe·mē'sfirˌòid }

heptahedron A polyhedron with seven faces. { ˌhep·tə'hē·drən }

heptagon A seven-sided polygon. { 'hep·təˌgän }

heptakaidecagon A polygon with 17 sides. { ˌhep·təˌkī'dek·əˌgän }

heptomino One of the 108 plane figures that can be formed by joining seven unit squares along their sides. { hep'täm·əˌnō }

her function A function that is expressed in terms of Hankel functions in a manner similar to that in which the ber function is expressed in terms of Bessel functions. { 'her ˌfəŋk·shən }

Hermite polynomials A family of orthogonal polynomials which arise as solutions to Hermite's differential equation, a particular case of the hypergeometric differential equation. { er'mēt ˌpäl·ə'nō·mē·əlz }

Hermite's differential equation A particular case of the hypergeometric equation; it has the form $w'' - 2zw' + 2nw = 0$, where n is an integer. { er'mēts dif·əˌren·chəl i'kwä·zhən }

Hermitian conjugate For a matrix A, the transpose of the complex conjugate of A. Also known as adjoint; associate matrix. { er'mish·ən 'kän·jə·gət }

Hermitian conjugate operator See adjoint operator. { er'mish·ən 'kän·jə·gət 'äp·əˌrād·ər }

Hermitian form 1. A polynomial in n real or complex variables where the matrix constructed from its coefficients is Hermitian. 2. More generally, a sesquilinear form g such that $g(x,y) = \overline{g(y,x)}$ for all values of the independent variables x and y, where $\overline{g(x,y)}$ is the image of $g(x,y)$ under the automorphism of the underlying ring. { er'mish·ən 'fòrm }

Hermitian inner product See inner product. { er'mish·ən 'in·ər ˌpräd·əkt }

Hermitian kernel A kernel $K(x,t)$ of an integral transformation or integral equation is Hermitian if $K(x,t)$ equals its adjoint kernel, $K^*(t,x)$. { er'mish·ən 'kər·nəl }

Hermitian matrix A matrix which equals its conjugate transpose matrix, that is, is self-adjoint. { er'mish·ən 'mā·triks }

Hermitian operator A linear operator A on vectors in a Hilbert space, such that if x and y are in the range of A then the inner products (Ax,y) and (x,Ay) are equal. { er'mish·ən 'äp·əˌrād·ər }

Hermitian scalar product *See* inner product. { er'mish·ən 'skāl·ər ‚präd·əkt }

Hermitian space *See* inner product space. { er'mish·ən 'spās }

hermit point *See* isolated point. { 'hər·mit ‚pòint }

Heron's formula *See* Hero's formula. { 'her·ənz ‚fòr·myə·lə }

Hero's formula A formula expressing the area of a triangle in terms of the sides a, b, and c as $\sqrt{s(s-a)(s-b)(s-c)}$ where $s = (1/2)(a + b + c)$ Also known as Heron's formula. { 'hir·ōz ‚fòr·myə·lə }

Hesse's theorem A theorem in projective geometry which states that, from the three pairs of lines containing the two pairs of opposite sides and the diagonals of a quadrilateral, if any two pairs are conjugate lines with respect to a given conic, then so is the third. { 'hes·əz ‚thir·əm }

Hessian For a function $f(x_1, \ldots, x_n)$ of n real variables, the real-valued function of $(x_1, \ldots, x_n)$ given by the determinant of the matrix with entry $\partial^2 f / \partial x_i \partial x_j$ in the ith row and jth column; used for analyzing critical points. { 'hesh·ən }

heterogeneous Pertaining to quantities having different degrees or dimensions. { ‚hed·ə'räj·ə·nəs }

heuristic method A method of solving a problem in which one tries each of several approaches or methods and evaluates progress toward a solution after each attempt. { hyü'ris·tik 'meth·əd }

hexadecimal Pertaining to a number system using the base 16. Also known as sexadecimal. { ‚hek·sə'des·məl }

hexadecimal number system A digital system based on powers of 16, as compared with the use of powers of 10 in the decimal number system. Also known as sexadecimal number system. { ‚hek·sə'des·məl 'nəm·bər ‚sis·təm }

hexafoil A multifoil consisting of six congruent arcs of a circle arranged around a regular hexagon. { 'hek·sə‚fòil }

hexagon A six-sided polygon. { 'hek·sə‚gän }

hexahedron A polyhedron with six faces. { ‚hek·sə'hē·drən }

hexomino One of the 35 plane figures that can be formed by joining six unit squares along their sides. { hek'säm·ə‚nō }

hidden Markov model A finite-state machine that is also a doubly stochastic process involving at least two levels of uncertainty: a random process associated with each state, and a Markov chain, which characterizes the probabilistic relationship among the states in terms of how likely one state is to follow another. { ‚hid·ən 'mär·kəf ‚mäd·əl }

higher plane curve Any algebraic curve whose degree exceeds 2. { 'hī·ər ‚plān ‚kərv }

highest common factor *See* greatest common divisor. { 'hī·əst 'käm·ən 'fak·tər }

Hilbert cube The topological space which is the Cartesian product of a countable number of copies of I, the unit interval. { 'hil·bərt ‚kyüb }

Hilbert parallelotope 1. A subset of an infinite-dimensional Hilbert space with coordinates $x_1, x_2, \ldots$, for which the absolute value of x_n is equal to or less that $(1/2)^n$ for each n. 2. The subset of this space for which the absolute value of x_n is equal to or less that $1/n$ for each n. { 'hil·bərt ‚par·ə'lel·ə‚tōp }

Hilbert-Schmidt theory A body of theorems which investigates the kernel of an integral equation via its eigenfunctions, and then applies these functions to help determine solutions of the equation. { ‚hil·bərt 'shmit ‚thē·ə·rē }

Hilbert space A Banach space which also is an inner-product space with the inner product of a vector with itself being the same as the square of the norm of the vector. { 'hil·bərt ‚spās }

Hilbert's theorem The proposition that the ring of polynomials with coefficients in a commutative Noetherian ring is itself a Noetherian ring. { 'hil‚bərts ‚thir·əm }

Hilbert transform The transform $F(y)$ of a function $f(x)$ realized by taking the Cauchy principal value of the integral over the real numbers of $(1/\pi) f(x)[1/(x-y)] \, dx$. { 'hil·bərt ‚tranz‚fòrm }

hill-climbing Any numerical procedure for finding the maximum or maxima of a function. { 'hil ‚klim·iŋ }

Hindu-Arabic numerals *See* arabic numerals. { ‚hin·dü ‚ar·ə·bik 'nüm·rəlz }

hippopede

hippopede A plane curve whose equation in polar coordinates r and θ is $r^2 = 4b(a - b\sin^2\theta)$, where a and b are positive constants. Also known as horse fetter. { 'hip·ə,pēd }

histogram A graphical representation of a distribution function by means of rectangles whose widths represent intervals into which the range of observed values is divided and whose heights represent the number of observations occurring in each interval. { 'his·tə,gram }

Hitchcock transportation problem The problem in linear programming of minimizing the cost of moving ships between two configurations in both of which there is a specified number of ships in each of a finite number of ports, when the costs of moving one ship from each of the ports in the first configuration to each of the ports in the second are specified. { ¦hich,käk ,tranz·pər'tā·shən ,präb·ləm }

Hjelmslev plane *See* affine Hjelmslev plane. { 'hyelm,slev ,plān }

Hodge conjecture The $2p$-dimensional rational cohomology classes in an n-dimensional algebraic manifold M which are carried by algebraic cycles are those with dual cohomology classes representable by differential forms of bidegree $(n - p, n - p)$ on M. { 'häj kən,jek·chər }

Hölder condition 1. A function $f(x)$ satisfies the Hölder condition in a neighborhood of a point x_0 if $|f(x) - f(x_0)| \le c|(x - x_0)|^n$, where c and n are constants. **2.** A function $f(x)$ satisfies a Hölder condition in an interval or in a region of the plane if $|f(x) - f(y)| \le c|x - y|^n$ for all x and y in the interval or region, where c and n are constants. { 'hel·dər kən,dish·ən }

Hölder's inequality Generalization of the Schwarz inequality: for real functions $|\int f(x)g(x)dx| \le (\int|f(x)|^p dx)^{1/p} \cdot (\int|g(x)|^q dx)^{1/q}$ where $1/p + 1/q = 1$. { 'hel·dərz ,in·i'kwäl·əd·ē }

Hölder summation A method of attributing a sum to certain divergent series in which a new series is formed, each of whose partial sums is the average of the first n partial sums of the original series, and this process is repeated until a stage is reached where the limit of this average exists. { 'hel·dər sə,mā·shən }

holomorphic function *See* analytic function. { ¦häl·ō¦mȯr·fik 'fəŋk·shən }

homeomorphic spaces Two topological spaces with a homeomorphism existing between them; intuitively one can be obtained from the other by stretching, twisting, or shrinking. { ¦hō·mē·ə¦mȯr·fik 'spās·əz }

homeomorphism A continuous map between topological spaces which is one-to-one, onto, and its inverse function is continuous. Also known as bicontinuous function; topological mapping. { ¦hō·mē·ə¦mȯr,fiz·əm }

homogeneity Equality of the distribution functions of several populations. { ,hō·mə·jə'nē·əd·ē }

homogeneous Pertaining to a group of mathematical symbols of uniform dimensions or degree. { ,hä·mə'jē·nē·əs }

homogeneous coordinates To a point in the plane with Cartesian coordinates (x,y) there corresponds the homogeneous coordinates (x_1,x_2,x_3), where $x_1/x_3 = x$, $x_2/x_3 = y$; any polynomial equation in Cartesian coordinates becomes homogeneous if a change into these coordinates is made. { ,hä·mə'jē·nē·əs kō'ȯrd·ən·əts }

homogeneous differential equation A differential equation where every scalar multiple of a solution is also a solution. { ,hä·mə'jē·nē·əs ,dif·ə'ren·chəl i,kwā·zhən }

homogeneous equation An equation that can be rewritten into the form having zero on one side of the equal sign and a homogeneous function of all the variables on the other side. { ,hä·mə'jē·nē·əs i'kwā·zhən }

homogeneous function A real function $f(x_1, x_2, \ldots, x_n)$ is homogeneous of degree r if $f(ax_1, ax_2, \ldots, ax_n) = a^r f(x_1, x_2, \ldots, x_n)$ for every real number a. { ,hä·mə'jē·nē·əs 'fəŋk·shən }

homogeneous integral equation An integral equation where every scalar multiple of a solution is also a solution. { ,hä·mə'jē·nē·əs 'int·ə·grəl i,kwā·zhən }

homogeneous polynomial A polynomial all of whose terms have the same total degree; equivalently it is a homogenous function of the variables involved. { ,hä·mə'jē·nē·əs ,päl·ə'nō·mē·əl }

homogeneous space A topological space having a group of transformations acting upon it, that is, a transformation group, where for any two points x and y some transformation from the group will send x to y. { ¸hä·mə'jē·nē·əs 'spās }

homogeneous transformation *See* linear transformation. { ¸hä·mə'jē·nē·əs ¸tranz·fər'mā·shən }

homographic transformations *See* Möbius transformations. { ¦hä·mə¦graf·ik ¸tranz·fər'mā·shənz }

homological algebra The study of the structure of modules, particularly by means of exact sequences; it has application to the study of a topological space via its homology groups. { ¦hä·mə¦läj·ə·kəl 'al·jə·brə }

homology group Associated to a topological space X, one of a sequence of Abelian groups $H_n(X)$ that reflect how n-dimensional simplicial complexes can be used to fill up X and also help determine the presence of n-dimensional holes appearing in X. Also known as Betti group. { hə'mäl·ə·jē ¸grüp }

homology theory Theory attempting to compare topological spaces and investigate their structures by determining the algebraic nature and interrelationships appearing in the various homology groups. { hə'mäl·ə·jē ¸thē·ə·rē }

homomorphism A function between two algebraic systems of the same type which preserves the algebraic operations. { ¸hä·mə'mȯr¸fiz·əm }

homoscedastic 1. Pertaining to two or more distributions whose variances are equal. 2. Pertaining to a variate in a bivariate distribution whose variance is the same for all values of the other variate. { ¸hä·mō·skə¦das·tik }

homothetic center The fixed point through which pass lines joining corresponding points of homothetic figures. Also known as center of similitude; ray center. { ¸häm·ə¸thed·ik 'sen·tər }

homothetic curves For a given point, a set of curves such that any straight line through the point intersects all the curves in the set at the same angle. { ¦häm·ə¸thed·ik 'kərvz }

homothetic figures Similar figures which are placed so that lines joining corresponding points pass through a common point and are divided in a constant ratio by this point. Also known as radially related figures. { ¦häm·ə¸thed·ik 'fig·yərz }

homothetic ratio *See* ratio of similitude. { ¦hō·mə¦thed·ik 'rā·shō }

homothetic transformation A transformation that leaves the origin of coordinates fixed and multiplies the distance between any two points by the same fixed constant. Also known as transformation of similitude. { ¦häm·ə¸thed·ik ¸tranz·fər'mā·shən }

homotopy Between two mappings of the same topological spaces, a continuous function representing how, in a step-by-step fashion, the image of one mapping can be continuously deformed onto the image of the other. { hō'mäd·ə·pē }

homotopy groups Associated to a topological space X, the groups appearing for each positive integer n, which reflect the number of different ways (up to homotopy) than an n-dimensional sphere may be mapped to X. { hō'mäd·ə·pē ¸grüps }

homotopy theory The study of the topological structure of a space by examining the algebraic properties of its various homotopy groups. { hō'mäd·ə·pē ¸thē·ə·rē }

horn angle A geometric figure formed by two tangent plane curves that lie on the same side of their mutual tangent line in the neighborhood of the point of tangency. { 'hȯrn ¸aŋ·gəl }

Horner's method A technique for approximating the real roots of an algebraic equation; a root is located between consecutive integers, then a successive search is performed. { 'hȯrn·ərz ¸meth·əd }

horse fetter *See* hippopede. { 'hȯrs ¸fed·ər }

Householder's method A transformation method for finding the eigenvalues of a symmetric matrix, in which each of the orthogonal transformations that reduce the original matrix to a triple-diagonal matrix reduces one complete row to the required form. { 'haủs¸hōl·dərz ¸meth·əd }

Hughes plane A finite projective plane with nine points on each line that can be represented by a nonlinear ternary ring generated by a four-point in the plane. { 'hyüz ¸plān }

hull

hull *See* span. { həl }

Hurwitz polynomial A polynomial whose zeros all have negative real parts. { 'hər·vitz ,päl·ə'nō·mē·əl }

Hurwitz's criterion A criterion that determines whether a polynomial is a Hurwitz polynomial, based on the signs of a set of determinants formed from the polynomial's coefficients. { 'hər,wit·səz krī,tir·ē·ən }

Huygens' approximation The length of a small circular arc is approximately $\frac{1}{3}(8c' - c)$, where c is the chord of the arc and c' is the chord of half the arc. { 'hī·gənz ə,präk·sə,mā·shən }

hyperbola The plane curve obtained by intersecting a circular cone of two nappes with a plane parallel to the axis of the cone. { hī·pər·bə·lə }

hyperbolic cosecant A function whose value is equal to the reciprocal of the value of the hyperbolic sine. Abbreviated csch. { ¦hī·pər¦bäl·ik kō'sē,kant }

hyperbolic cosine A function whose value for the complex number z is one-half the sum of the exponential of z and the exponential of $-z$. Abbreviated cosh. { ¦hī·pər¦bäl·ik 'kō,sīn }

hyperbolic cotangent A function whose value is equal to the value of the hyperbolic cosine divided by the value of the hyperbolic sine. Abbreviated coth. { ¦hī·pər¦bäl·ik kō'tan·jənt }

hyperbolic cylinder A cylinder whose directrix is a hyperbola. { ¦hī·pər¦bäl·ik 'sil·ən·dər }

hyperbolic differential equation A general type of second-order partial differential equation which includes the wave equation and has the form

$$\sum_{i,j=1}^{n} A_{ij}(\partial^2 u/\partial x_i \partial x_j) + \sum_{i=1}^{n} B_i(\partial u/\partial x_i) + Cu + F = 0$$

where the A_{ij}, B_i, C, and F are suitably differentiable real functions of $x_1, x_2, \ldots, x_n$, and there exists at each point $(x_1, x_2, \ldots, x_n)$ a real linear transformation on the x_i which reduces the quadratic form $\sum_{i,j=1}^{n}$ to a sum of n squares not all of the same sign. { ¦hī·pər¦bäl·ik ,dif·ə¦ren·chəl i'kwā·zhən }

hyperbolic form A nondegenerate, symmetric or alternating form on a vector space E such that E is a hyperbolic space under this form. { ¦hī·pər,bäl·ik 'fȯrm }

hyperbolic functions The real or complex functions sinh (x), cosh (x), tanh (x), coth (x), sech (x), csch (x); they are related to the hyperbola in somewhat the same fashion as the trigonometric functions are related to the circle, and have properties analogous to those of the trigonometric functions. { ¦hī·pər¦bäl·ik 'fəŋk·shənz }

hyperbolic geometry *See* Lobachevski geometry. { ¦hī·pər¦bäl·ik jē'äm·ə·trē }

hyperbolic logarithm *See* logarithm. { ¦hī·pər¦bäl·ik 'läg·ə,rith·əm }

hyperbolic paraboloid A surface which can be so situated that sections parallel to one coordinate plane are parabolas while those parallel to the other plane are hyperbolas. { ¦hī·pər¦bäl·ik pə'rab·ə,lȯid }

hyperbolic plane A two-dimensional vector space E on which there is a nondegenerate, symmetric or alternating form $f(x,y)$ such that there exists a nonzero element w in E for which $f(w,w) = 0$. { ¦hī·pər,bäl·ik 'plān }

hyperbolic point A point on a surface where the Gaussian curvature is strictly negative. { ¦hī·pər¦bäl·ik 'pȯint }

hyperbolic Riemann surface *See* hyperbolic type. { ,hī·pər¦bäl·ik 'rē,män ,sər·fəs }

hyperbolic secant A function whose value is equal to the reciprocal of the value of the hyperbolic cosine. Abbreviated sech. { ¦hī·pər¦bäl·ik 'sē,kant }

hyperbolic sine A function whose value for the complex number z is one-half the difference between the exponential of z and the exponential of $-z$. Abbreviated sinh. { ¦hī·pər¦bäl·ik 'sīn }

hyperbolic space A space described by hyperbolic rather than Cartesian coordinates. { ¦hī·pər¦bäl·ik 'spās }

hyperbolic spiral A plane curve for which the radius vector is inversely proportional to the polar angle. Also known as reciprocal spiral. { ¦hī·pər¦bäl·ik 'spi·rəl }

hyperbolic tangent A function whose value is equal to the value of the hyperbolic sine divided by the value of the hyperbolic cosine. Abbreviated tanh. { ¦hī·pər¦bäl·ik 'tan·jənt }

hyperbolic type A type of simply connected Riemann surface that can be mapped conformally on the interior of the unit circle. Also known as hyperbolic Riemann surface. { ¦hī·pər¦bäl·ik 'tīp }

hyperboloid A quadric surface given by an equation of the form $(x^2/a^2) \pm (y^2/b^2) - (z^2/c^2) = 1$; in certain cases it is a hyperboloid of revolution, which can be realized by rotating the pieces of a hyperbola about an appropriate axis. { hī'pər·bə,lȯid }

hyperboloid of one sheet A surface whose equation in stardard form is $(x^2/a^2) + (y^2/b^2) - (z^2/c^2) = 1$, so that it is in one piece, and cuts planes perpendicular to the x or y axes in hyperbolas and planes perpendicular to the z axis in ellipses. { hī'pər·bə,lȯid əv 'wən ,shēt }

hyperboloid of revolution A surface generated by rotating a hyperbola about one of its axes. { hī'pər·bə,lȯid əv ,rev·ə'lü·shən }

hyperboloid of two sheets A surface whose equation in standard form is $(x^2/a^2) - (y^2/b^2) - (z^2/c^2) = 1$, so that it is in two pieces, and cuts planes perpendicular to the y and z axes in hyperbolas and planes perpendicular to the x axis in ellipses, except for the interval $-a < x < a$, where there is no intersection. { hī'pər·bə,lȯid əv 'tü ,shēts }

hypercircle method A geometric method of obtaining approximate solutions of linear boundary value problems of mathematical physics that cannot be solved exactly, in which a correspondence is made between physical variables and vectors in a function space. { ¦hī·pər¦sər·kəl ,meth·əd }

hypercomplex number 1. An element of a division algebra. **2.** *See* quaternion. { ¦hī·pər¦käm,pleks 'nəm·bər }

hypercomplex system *See* algebra. { ¦hī·pər¦käm,pleks 'sis·təm }

hypercube The analog of a cube in n dimensions ($n = 2, 3, \ldots$), with 2^n vertices, $n2^{n-1}$ edges, and $2n$ cells; for an object with edges of length $2a$, the coordinates of the vertices are $(\pm a, \pm a, \ldots, \pm a)$. { 'hī·pər ,kyüb }

hypergeometric differential equation *See* Gauss' hypergeometric equation. { ,hī·pər,jē·ə'me·trik ,dif·ə¦ren·chəl i'kwā·zhən }

hypergeometric distribution The distribution of the number D of special items in a random sample of size s drawn from a population of size N that contains r of the special items:

$$P(D = d) = \binom{r}{d}\binom{N-r}{s-d}\Big/\binom{N}{s}$$

{ ,hī·pər,jē·ə'me·trik ,dis·trə'byü·shən }

hypergeometric function A function which is a solution to the hypergeometric equation and obtained as an infinite series expansion. { ,hī·pər,jē·ə'me·trik 'fəŋk·shən }

hypergeometric series A particular infinite series which in certain cases is a solution to the hypergeometric equation, and having the form:

$$1 + \frac{ab}{c}z + \frac{1}{2!}\frac{a(a+1)b(b+1)}{c(c+1)}z^2 + \cdots$$

{ ,hī·pər,jē·ə'me·trik 'sir·ēz }

hyperplane A hyperplane is an $(n-1)$-dimensional subspace of an n-dimensional vector space. { ¦hī·pər,plān }

hyperplane of support Relative to a convex body in a normed vector space, a hyperplane whose distance from the body is zero, and which separates the normed vector space into two halves, one of which contains no points of the convex body. { 'hī·pər,plān əv sə'pȯrt }

hyperreal numbers *See* nonstandard numbers. { ¦hī·pər,rēl 'nəm·bərz }

hypersurface The analog of a surface in n-dimensional Euclidean space, where n is a positive integer; the set of points, $(x_1, x_2, \ldots, x_n)$, satisfying an equation of the form $f(x_1, \ldots, x_n) = 0$.

hypervolume 1. The hypervolume of the direct product of open or closed intervals in each of the coordinates of a Euclidean n-space (where n is a positive integer) is the product of the lengths of the intervals. **2.** The Jordan content of any set in Euclidean n-space whose exterior Jordan content equals its interior Jordan content. { 'hī·pər‚väl·yəm }

hypocycloid The curve which is traced in the plane as a given point fixed on a circle moves while this circle rolls along the inside of another circle. { ‚hī·pō'sī‚klȯid }

hypotenuse On a right triangle, the side opposite the right angle. { hī'pät·ən‚üs }

hypothesis A statement which specifies a population or distribution, and whose truth can be tested by sample evidence. { hī'päth·ə·səs }

hypothesis testing The branch of statistics which considers the problem of choosing between two actions on the basis of the observed value of a random variable whose distribution depends on a parameter, the value of which would indicate the correct action. { hī'päth·ə·səs ‚test·iŋ }

hypotrochoid A curve traced by a point rigidly attached to a circle at a point other than the center when the circle rolls without slipping on the inside of a fixed circle. { ‚hī·pō'trō‚kȯid }

icosahedral group The group of motions of three-dimensional space that transform a regular icosahedron into itself. { ī̩käs·ə¦hē·drəl 'grüp }

icosahedron A 20-sided polyhedron. { ī¦kä·sə¦hē·drən }

ideal A subset I of a ring R where $x - y$ is in I for every x,y in I and either rx is in I for every r in R and x in I or xr is in I for every r in R and x in I; in the first case I is called a left ideal, and in the second a right ideal; an ideal is two-sided if it is both a left and a right ideal. { ī'dēl }

ideal index number *See* Fisher's ideal index. { ī'dēl 'in̩deks ̩nəm·bər }

ideal line The collection of all ideal points, each corresponding to a given family of parallel lines. Also known as line at infinity. { ī'dēl 'līn }

ideal point In projective geometry, all lines parallel to a given line are hypothesized to meet at a point at infinity, called an ideal point. Also known as point at infinity. { ī'dēl 'point }

ideal theory The branch of algebra studying the properties of ideals. { ī'dēl 'thē·ə·rē }

idem factor The dyadic $I = ii + jj + kk$ such that scalar multiplication of I by any vector yields that vector. { 'i̩dem ̩fak·tər }

idempotent **1.** An element x of an algebraic system satisfying the equation $x^2 = x$. **2.** An algebraic system in which every element x satisfies $x^2 = x$. { ¦i̩dem¦pōt·ənt }

idempotent law A law which states that an element x of an algebraic system satisfies $x^2 = x$. { ¦i̩dem¦pōt·ənt 'lò }

idempotent matrix A matrix E satisfying the equation $E^2 = E$. { ¦i̩dem¦pōt·ənt 'mā·triks }

identity **1.** An equation satisfied for all possible choices of values for the variables involved. **2.** *See* identity element. { ī'den·ə̩dē }

identity element The unique element e of a group where $g · e = e · g = g$ for every element g of the group. Also known as identity. { ī'den·ə̩dē ̩el·ə·mənt }

identity function The function of a set to itself which assigns to each element the same element. Also known as identity operator. { ī'den·ə̩dē ̩fəŋk·shən }

identity matrix The square matrix all of whose entries are zero except along the principal diagonal where they all are 1. { ī'den·ə̩dē ̩mā·triks }

identity operator *See* identity function. { ī'den·ə̩dē ̩äp·ə̩rād·ər }

if and only if operation *See* biconditional operation. { ¦if ən 'ōn·lē ¦if ̩äp·ə̩rā·shən }

if-then operation *See* implication.

ill-posed problem A problem which may have more than one solution, or in which the solutions depend discontinuously upon the initial data. Also known as improperly posed problem. { 'il ¦pōzd 'präb·ləm }

illusory correlation *See* nonsense correlation. { i̩lü·zə·rē ̩kä·rə'lā·shən }

image **1.** For a point x in the domain of a function f, the point $f(x)$. **2.** For a subset A of the domain of a function f, the set of all points that are equal to $f(x)$ for some point x in A. { 'im·ij }

imaginary axis All complex numbers $x + iy$ where $x = 0$; the vertical coordinate axis for the complex plane. { ə'maj·ə̩ner·ē 'ak·səs }

imaginary circle The set of points in the x-y plane that satisfy the equation $x^2 + y^2 = -r^2$, or $(x - h)^2 + (y - k)^2 = -r^2$, where r is greater than zero, and x, y, h, and k are allowed to be complex numbers. { i¦maj·ə̩ner·ē 'sər·kəl }

imaginary number A complex number of the form $a + bi$, with b not equal to zero, where a and b are real numbers, and $i = \sqrt{-1}$; some mathematicians require also that $a = 0$. Also known as imaginary quantity. { ə'maj·ə,ner·ē 'nəm·bər }

imaginary part For a complex number $x + iy$ the imaginary part is the real number y. { ə'maj·ə,ner·ē ,pärt }

imaginary quantity See imaginary number. { ə'maj·ə,ner·ē 'kwän·əd·ē }

imbedding A homeomorphism of one topological space to a subspace of another topological space. { im'bed·iŋ }

immersion A mapping f of a topological space X into a topological space Y such that for every x ;nl X there exists a neighborhood N of x, such that f is a homeomorphism of N onto $f(N)$. { ə'mər·zhən }

implication 1. The logical relation between two statements p and q, usually expressed as "if p then q." 2. A logic operator having the characteristic that if p and q are statements, the implication of p and q is false if p is true and q is false, and is true otherwise. Also known as conditional implication; if-then operation; material implication. { ,im·plə'kā·shən }

implicit differentiation The process of finding the derivative of one of two variables with respect to the other by differentiating all the terms of a given equation in the two variables and solving the resulting equation for this derivative. { im'plis·it ,dif·ə,ren·chē'ā·shən }

implicit enumeration A method of solving integer programming problems, in which tests that follow conceptually from using implied upper and lower bounds on variables are used to eliminate all but a tiny fraction of the possible values, with implicit treatment of all other possibilities. { im'plis·ət i,nü·mə'rā·shən }

implicit function A function defined by an equation $f(x,y) = 0$, when x is considered as an independent variable and y, called an implicit function of x, as a dependent variable. { im'plis·ət 'fəŋk·shən }

implicit function theorem A theorem that gives conditions under which an equation in variables x and y may be solved so as to express y directly as a function of x; it states that if $F(x,y)$ and $\partial F(x,y)/\partial y$ are continuous in a neighborhood of the point (x_0,y_0) and if $F(x,y) = 0$ and $\partial F(x,y)/\partial y \neq 0$, then there is a number $\epsilon > 0$ such that there is one and only one function $f(x)$ that is continuous and satisfies $F[x,f(x)] = 0$ for $|x - x_0| < \epsilon$, and satisfies $f(x_0) = y_0$. { im'plis·ət ¦fəŋk·shən ,thir·əm }

improper divisor An improper divisor of an element x in a commutative ring with identity is any unit of the ring or any associate of x. { im'präp·ər di'vī·zər }

improper face For a convex polytope, either the empty set or the polytope itself. { ,im¦präp·ər 'fās }

improper fraction 1. In arithmetic, the quotient of two integers in which the numerator is greater than or equal to the denominator. 2. In algebra, the quotient of two polynomials in which the degree of the numerator is greater than or equal to that of the denominator. { im'präp·ər 'frak·shən }

improper integral Any integral in which either the integrand becomes unbounded on the domain of integration, or the domain of integration is itself unbounded. { im'präp·ər 'int·ə·grəl }

improperly posed problem See ill-posed problem. { im'präp·ər·lē ¦pōzd 'präb·ləm }

improper orthogonal transformation An orthogonal transformation such that the determinant of its matrix is -1. { im¦präp·ər ȯr¦thäg·ə·nəl ,tranz·fər'mā·shən }

impulse function An idealized or generalized function defined not by its values but by its behavior under integration, such as the (Dirac) delta function. { 'im,pəls ,fəŋk·shən }

incenter The center of the inscribed circle of a given triangle. { ¦in¦sen·tər }

incidence function The function that assigns a pair of vertices to each edge of a graph. { 'in·səd·əns ,fəŋk·shən }

incidence matrix In a graph, the $p \times q$ matrix (b_{ij}) for which $b_{ij} = 1$ if the ith vertex is an end point of the jth edge, and $b_{ij} = 0$ otherwise. { 'in·səd·əns ,mā·triks }

incircle See inscribed circle. { ¦in¦sər·kəl }

114

inclination 1. The inclination of a line in a plane is the angle made with the positive x axis. 2. The inclination of a line in space with respect to a plane is the smaller angle the line makes with its orthogonal projection in the plane. 3. The inclination of a plane with respect to a given plane is the smaller of the dihedral angles which it makes with the given plane. { ‚iŋ·klə'nā·shən }

inclusion-exclusion principle The principle that, if A and B are finite sets, the number of elements in the union of A and B can be obtained by adding the number of elements in A to the number of elements in B, and then subtracting from this sum the number of elements in the intersection of A and B. { ¦in‚klü·zhən 'eks‚klü·zhən ‚prin·sə·pəl }

inclusion relation 1. A set theoretic relation, usually denoted by the symbol $\subset$, such that, if A and B are two sets, A $\subset$ B if and only if every element of A is an element of B. 2. Any relation on a Boolean algebra which is reflexive, antisymmetric, and transitive. { iŋ'klü·zhən ri‚lā·shən }

incommensurable line segments Two line segments the ratio of whose lengths is irrational. { ‚in·kə¦mens·ə·rə·bəl 'līn ‚seg·məns }

incommensurable numbers Two numbers whose ratio is irrational. { ‚in·kə'mens·ə·rə·bəl 'nəm·bərz }

incompatible equations Two or more equations that are not satisfied by any set of values for the variables appearing. Also known as inconsistent equations. { ‚in·kəm'pad·ə·bəl i¦kwā·zhənz }

incompatible inequalities Two or more inequalities that are not satisfied by any set of values of the variables involved. Also known as inconsistent inequalities. { ‚in·kəm'pad·ə·bəl in·ə'kwäl·əd·ēz }

incomplete beta function The function $\beta_x(p,q)$ defined by

$$\beta_x(p,q) = \int_0^x t^{p-1}(1-t)^{q-1}\,dt$$

where $0 \le x \le 1$, $p > 0$, and $q > 0$. { ‚in·kəm'plēt 'bād·ə ‚fəŋk·shən }

incomplete gamma function Either of the functions $\gamma(a,x)$ and $\Gamma(a,x)$ defined by

$$\gamma(a,x) = \int_0^x t^{a-1} e^{-t}\,dt$$

$$\Gamma(a,x) = \int_x^\infty t^{a-1} e^{-t}\,dt$$

where $0 \le x \le \infty$ and $a > 0$. { ‚in·kəm'plēt 'gam·ə ‚fəŋk·shən }

incomplete Latin square *See* Yonden square. { ¦iŋ·kəm‚plēt ¦lat·ən 'skwer }

inconsistent axioms A set of axioms from which both a proposition and its negation can be deduced. { ‚in·kən‚sis·tənt 'ak·sē·əmz }

inconsistent equations *See* incompatible equations. { ‚in·kən'sis·tənt i'kwā·zhənz }

inconsistent inequalities *See* incompatible inequalities. { ‚in·kən¦sis·tənt ‚in·ə'kwäl·əd·ēz }

increasing function A function, f, of a real variable, x, whose value gets larger as x gets larger; that is, if $x < y$, then $f(x) < f(y)$. Also known as strictly increasing function. { in'krēs·iŋ ‚fəŋk·shən }

increasing sequence A sequence of real numbers in which each term is greater than the preceding term. { in¦krēs·iŋ 'sē·kwəns }

increment A change in the argument or values of a function, usually restricted to being a small positive or negative quantity. { 'iŋ·krə·mənt }

indefinite integral An indefinite integral of a function $f(x)$ is a function $F(x)$ whose derivative equals $f(x)$. Also known as antiderivative; integral. { in'def·ə·nət 'int·ə·grəl }

indegree For a vertex, v, in a directed graph, the number of arcs directed from other vertices to v. { 'in·di‚grē }

independence number For a graph, the largest possible number of vertices in an

independent set. Also known as internal stability number. { ,in·di'pen·dəns ,nəm·bər }

independent axiom A member of a set of axioms that cannot be deduced as a consequence of the other axioms in the set. { ,in·di,pen·dənt 'ak·sē·əm }

independent edge set See matching. { ,in·di,pen·dənt 'eg ,set }

independent equations A system of equations such that no one of them is necessarily satisfied by a solution to the rest. { ,in·də'pen·dənt i'kwā·zhənz }

independent events Two events in probability such that the occurrence of one of them does not affect the probability of the occurrence of the other. { ,in·də'pen·dənt i'vens }

independent functions A set of functions such that knowledge of the values obtained by all but one of them at a point is insufficient to determine the value of the remaining function. { ,in·də'pen·dənt 'fəŋk·shənz }

independent random variables The discrete random variables $X_1, X_2, \ldots, X_n$ are independent if for arbitrary values $x_1, x_2, \ldots, x_n$ of the variables the probability that $X_1 = x_1$ and $X_2 = x_2$, etc., is equal to the product of the probabilities that $X_i = x_i$ for $i = 1, 2, \ldots, n$; random variables which are unrelated. { ,in·də'pen·dənt ¦ran·dəm ,ver·ē·ə·bəls }

independent set A set of vertices in a simple graph such that no two vertices in this set are adjacent. Also known as internally stable set. { ,in·di,pen·dənt 'set }

independent variable In an equation $y = f(x)$, the input variable x. Also known as argument. { ,in·də'pen·dənt 'ver·ē·ə·bəl }

indeterminate equations A set of equations possessing an infinite number of solutions. { ,in·də'tərm·ə·nət i'kwā·zhənz }

indeterminate forms Products, quotients, differences, or powers of functions which are undefined when the argument of the function has a certain value, because one or both of the functions are zero or infinite; however, the limit of the product, quotient, and so on as the argument approaches this value is well defined. { ,in·də'tərm·ə·nət 'fòrmz }

index 1. Unity of a logarithmic scale, as the C scale of a slide rule. **2.** A subscript or superscript used to indicate a specific element of a set or sequence. **3.** The number above and to the left of a radical sign, indicating the root to be extracted. **4.** For a subgroup of a finite group, the order of the group divided by the order of the subgroup. **5.** For a continuous complex-valued function defined on a closed plane curve, the change in the amplitude of the function when traversing the curve in a counterclockwise direction, divided by 2π. **6.** For a quadratic or Hermitian form, the number of terms with positive coefficients when the form is reduced by a linear transformation to a sum of squares or a sum of squares of absolute values. **7.** For a symmetric or Hermitian matrix, the number of positive entries when the matrix is transformed to diagonal form. **8.** See winding number. { 'in,deks }

index line See isopleth. { 'in,deks ,līn }

index number A number indicating change in magnitude, as of cost or of volume of production, as compared with the magnitude at a specified time, usually taken as 100; for example, if production volume in 1970 was two times as much as the volume in 1950 (taken as 100), its index number is 200. { 'in,deks ,nəm·bər }

index of precision The constant h in the normal curve $y = K \exp [-h^2(x - u)^2]$; a large value of h indicates a high precision, or small standard deviation. { 'in,deks əv prə'sizh·ən }

indicator See Euler's phi function. { 'in·də,kād·ər }

indicator function See characteristic function. { 'in·də,kād·ər ,fəŋk·shən }

indirect proof 1. A proof of a proposition in which another theorem is first proven from which the given theorem follows. **2.** See reductio ad absurdum. { ,in·də,rekt 'prüf }

indiscreet topology See trivial topology. { ,in·də,skrēt tə'päl·ə·jē }

induced orientation An orientation of a face of a simplex S opposite a vertex p_i

obtained by deleting p_i from the ordering defining the orientation of S. { in¦düsd ‚ȯr·ē·ən'tā·shən }

induced subgraph *See* vertex-induced subgraph. { in‚düst 'səb‚graf }

inequality A statement that one quantity is less than, less than or equal to, greater than, or greater than or equal to another quantity. { ‚in·i'kwäl·əd·ē }

inessential mapping A mapping between topological spaces that is homotopic to a mapping whose range is a single point. { ‚in·ə¦sen·chəl 'map·iŋ }

inf *See* greatest lower bound.

infimum *See* greatest lower bound. { 'in·fə·məm }

infinite Larger than any fixed number. { 'in·fə·nət }

infinite discontinuity A discontinuity of a function for which the absolute value of the function can have arbitrarily large values arbitrarily close to the discontinuity. { 'in·fə·nit ‚dis‚känt·ən'ü·əd·ē }

infinite extension An extension field F of a given field E such that F, viewed as a vector space over E, has infinite dimension. { 'in·fi·nit ik'sten·chən }

infinite group A group that contains an infinite number of distinct elements. { 'in·fə·nit 'grüp }

infinite integral An integral at least one of whose limits of integration is infinite. { 'in·fə·nət 'int·ə·grəl }

infinite population A universe which contains an infinite number of elements; it can be continuous or discrete. { 'in·fi·nit ‚päp·yə'lā·shən }

infinite root An equation $f(x) = 0$ is said to have an infinite root if the equation $f(1/y) = 0$ has a root at $y = 0$. { ‚in‚fə·nət 'rüt }

infinite series An indicated sum of an infinite sequence of quantities, written $a_1 + a_2 + a_3 + \cdots$, or $\sum_{k=1}^{\infty} a_k$. { 'in·fə·nət 'sir·ēz }

infinite set A set with more elements than any fixed integer; such a set can be put into a one to one correspondence with a proper subset of itself. { 'in·fə·nət 'set }

infinitesimal A function whose value approaches 0 as its argument approaches some specified limit. { ¦in‚fin·ə¦tes·ə·məl }

infinitesimal generator A closed linear operator defined relative to some semigroup of operators and which uniquely determines that semigroup. { ¦in‚fin·ə¦tes·ə·məl 'jen·ə‚rād·ər }

infinity The concept of a value larger than any finite value. { in'fin·əd·ē }

infix notation A method of forming mathematical or logical expressions in which operators are written between the operands on which they act. { 'in‚fiks nō‚tā·shən }

inflectional tangent A tangent to a curve at a point of inflection. { in¦flek·shə·nəl 'tan·jənt }

inflection point *See* point of inflection. { in'flek·shən ‚pȯint }

inflow The inflow to a vertex in an s-t network is the sum of the flows of all the arcs that terminate at that vertex. { 'in‚flō }

information function of a partition If ξ is a finite partition of a probability space, the information function of ξ is a step function whose sets of constancy are the elements of ξ and whose value on an element of ξ is the negative of the logarithm of the probability of this element. { ‚in·fər'mā·shən ¦fəŋk·shən əv ə pär'tish·ən }

information theory The branch of probability theory concerned with the likelihood of the transmission of messages, accurate to within specified limits, when the bits of information composing the message are subject to possible distortion. { ‚in·fər'mā·shən ‚thē·ə·rē }

initial line One of the two rays that form an angle and that may be regarded as remaining stationary while the other ray (the terminal line) is rotated about a fixed point on it to form the angle. { i¦nish·əl 'līn }

initial-value problem An nth-order ordinary or partial differential equation in which the solution and its first $(n-1)$ derivatives are required to take on specified values at a particular value of a given independent variable. { i'nish·əl ¦val·yü ‚präb·ləm }

initial-value theorem The theorem that, if a function $f(t)$ and its first derivative have

Laplace transforms, and if $g(s)$ is the Laplace transform of $f(t)$, and if the limit of $sg(s)$ as s approaches infinity exists, then this limit equals the limit of $f(t)$ as t approaches zero. { i'nish·əl ¦val·yü ¸thir·əm }

injection A mapping f from a set A into a set B which has the property that for any element b of B there is at most one element a of A for which $f(a) = b$. Also known as injective mapping; one-to-one mapping; univalent function. { in'jek· shən }

inner automorphism An automorphism h of a group where $h(g) = g_0^{-1} \cdot g \cdot g_0$, for every g in the group with g_0 some fixed group element. { ¦in·ər ¸òd·ō'mór·fiz·əm }

inner function A continuous open mapping of a topological space X into a topological space Y where the inverse image of each point in Y is zero dimensional. { ¦in·ər 'fəŋk·shən }

inner measure *See* Lebesgue interior measure. { ¸in·ər 'mezh·ər }

inner product **1.** A scalar valued function of pairs of vectors from a vector space, denoted by (x,y) where x and y are vectors, and with the properties that (x,x) is always positive and is zero only if $x = 0$, that $(ax + by,z) = a(x,z) + b(y,z)$ for any scalars a and b, and that $(x,y) = (y,x)$ if the scalars are real numbers, $(x,y) = (y,x)$ if the scalars are complex numbers. Also known as Hermitian inner product; Hermitian scalar product. **2.** The inner product of vectors $(x_1, \ldots, x_n)$ and $(y_1, \ldots, y_n)$ from n-dimensional Euclidean space is the sum of $x_i y_i$ as i ranges from 1 to n. Also known as dot product; scalar product. **3.** The inner product of two functions f and g of a real or complex variable is the integral of $f(x)g(x)dx$, where $g(x)$ denotes the conjugate of $g(x)$. **4.** The inner product of two tensors is the contracted tensor obtained from their product by means of pairing contravariant indices of one with covariant indices of the other. { ¦in·ər 'präd·əkt }

inner product space A vector space that has an inner product defined on it. Also known as generalized Euclidean space; Hermitian space; pre-Hilbert space. { ¦in· ər 'präd·əkt ¸spās }

inradius The radius of the inscribed circle of a triangle. { 'in¸rād·ē·əs }

inscribed circle A circle that lies within a given triangle and is tangent to each of its sides. Also known as incircle. { in¦skrībd 'sər·kəl }

inscribed polygon A polygon that lies within a given circle or curve and whose vertices all lie on the circle or curve. { in¦skrībd 'päl·ə¸gän }

inseparable degree Let E be a finite extension of a field F; the inseparable degree of E over F is the dimension of E viewed as a vector space over F divided by the separable degree of E over F. { in¦sep·rə·bəl di'grē }

integer Any positive or negative counting number or zero. { 'int·ə·jər }

integer partition For a positive integer n, a nonincreasing sequence of positive integers whose sum equals n. { 'int·ə·jər pär'tish·ən }

integrable differential equation A differential equation that either is exact or can be transformed into an exact differential equation by multiplying each equation term by a common factor. { ¸int·i·grə·bəl ¸dif·ə¸ren·chəl i'kwā·zhən }

integrable function A function whose integral, defined in a specific manner, exists and is finite. { ¦int·i·grə·bəl 'fəŋk·shən }

integral **1.** A solution of a differential equation is sometimes called an integral of the equation. **2.** An element a of a ring B is said to be integral over a ring A contained in B if it is the root of a polynomial with coefficients in A and with leading coefficient 1. **3.** *See* definite Riemann integral; indefinite integral. { 'int·ə·grəl }

integral calculus The study of integration and its applications to finding areas, volumes, or solutions of differential equations. { 'int·ə·grəl 'kal·kyəl·ləs }

integral closure The integral closure of a subring A of a ring B is the set of all elements in B that are integral over A. { 'int·ə·grəl 'klō·zhər }

integral curvature For a given region on a surface, the integral of the Gaussian curvature over the region. { 'int·ə·grəl 'kərv·ə·chər }

integral curves A family of curves that satisfy a particular differential equation. { 'int· ə·grəl 'kərvz }

integral domain A commutative ring with identity where the product of nonzero elements is never zero. Also known as entire ring. { 'int·ə·grəl dō'mān }

integral equation An equation where the unknown function occurs under an integral sign. { 'int·ə·grəl i'kwā·zhən }

integral extension An integral extension of a commutative ring A is a commutative ring B containing A such that every element of B is integral over A. { 'int·ə·grəl ik'sten·chən }

integral function 1. A function taking on integer values. **2.** *See* entire function. { 'int·ə·grəl ,fəŋk·shən }

integrally closed ring An integral domain which is equal to its integral closure in its quotient field. { in¦teg·rə·lē ¦klōzd 'riŋ }

integral map A homomorphism from a commutative ring A into a commutative ring B such that B is an integral extension of $f(A)$. { 'int·ə·grəl ,map }

integral operator A rule for transforming one function into another function by means of an integral; this often is in context a linear transformation on some vector space of functions. { 'int·ə·grəl 'äp·ə,rād·ər }

integral test If $f(x)$ is a function that is positive and decreasing for positive x, then the infinite series with nth term $f(n)$ and the integral of $f(x)$ from 1 to ∞ are either both convergent (finite) or both infinite. { 'int·ə·grəl ,test }

integral transform *See* integral transformation. { 'int·ə·grəl 'tranz,fórm }

integral transformation A transform of a function $F(x)$ given by the function

$$f(y) = \int_a^b K(x,y)F(x)\ dx$$

where $K(x,y)$ is some function. Also known as integral transform. { 'int·ə·grəl ,tranz·fər'mā·shən }

integrand The function which is being integrated in a given integral. { 'int·ə,grand }

integrating factor A factor which when multiplied into a differential equation makes the portion involving derivatives an exact differential. { 'int·ə,grād·iŋ 'fak·tər }

integration The act of taking a definite or indefinite integral. { ,int·ə'grā·shən }

integration by parts A technique used to find the integral of the product of two functions by means of an identity involving another simpler integral; for functions of one variable the identity is

$$\int_a^b fg'\ dx + \int_a^b gf'\ dx = f(b)g(b) - f(a)g(a);$$

for functions of several variables the technique is tantamount to using Stokes' theorem or the divergence theorem. { ,int·ə'grā·shən bī 'parts }

integration constant *See* constant of integration. { ,int·ə'grā·shən ¦kän·stənt }

integrodifferential equation An equation relating a function, its derivatives, and its integrals. { in¦teg·rō,dif·ə¦ren·chəl i'kwā·zhən }

intensification An operation that increases the value of the membership function of a fuzzy set if the value is equal to or greater than 0.5, and decreases it if it is less than 0.5. { in,tens·ə·fə'kā·shən }

interaction The phenomenon which causes the response to applying two treatments not to be the simple sum of the responses to each treatment. { ¦in·tə¦rak·shən }

intercept The point where a straight line crosses one of the axes of a Cartesian coordinate system. { ¦in·tər¦sept }

interior 1. For a set A in a topological space, the set of all interior points of A. **2.** For a plane figure, the set of all points inside the figure. **3.** For an angle, the set of points that lie in the plane of the angle and between the rays defining the angle. **4.** For a simple closed plane curve, one of the two regions into which the curve divides the plane according to the Jordan curve theorem, namely, the region that is bounded. { in'tir·ē·ər }

interior angle 1. An angle between two adjacent sides of a polygon that lies within the polygon. **2.** For a line (called the transversal) that intersects two other lines, an angle between the transversal and one of the two lines that lies within the space between the two lines. { in'tir·ē·ər 'aŋ·gəl }

interior content *See* interior Jordan content. { in¦tir·ē·ər 'kän,tent }

interior Jordan content Also known as interior content. **1.** For a set a points on a line, the smallest number C such that the sum of the lengths of a finite number of open, nonoverlapping intervals that are completely contained in the set is always equal to or less than C. **2.** The interior Jordan content of a set of points, X, in n-dimensional Euclidean space (where n is a positive integer) is the least upper bound on the hypervolume of the union of a finite set of hypercubes that is contained in X. { in¦tir·ē·ər 'jord·ən ¦kän,tent }

interior measure *See* Lebesgue interior measure. { in¦tir·ē·ər 'mezh·ər }

interior point A point p in a topological space is an interior point of a set S if there is some open neighborhood of p which is contained in S. { in'tir·ē·ər 'point }

intermediate value theorem If $f(x)$ is a continuous real-valued function on the closed interval from a to b, then, for any y between the least upper bound and the greatest lower bound of the values of f, there is an x between a and b with $f(x) = y$. { ,in·tər'mēd·ē·ət ¦val·yü 'thir·əm }

intermediate vertex A vertex in an s-t network that is neither the source nor the terminal. { ,in·tər¦mēd·ē·ət 'vər,teks }

internally stable set *See* independent set. { in,tərn·əl·ē ,stā·bəl 'set }

internally tangent circles Two circles, one of which is inside the other, that have a single point in common. { in,tərn·əl·ē ,tan·jənt 'sər·kəlz }

internal operation For a set S, a function whose domain is a set of members of S or a set of ordered sequences of members of S, and whose range is a subset of S. { in,tərn·əl ,äp·ə'rā·shən }

internal tangent For two circles, each exterior to the other, a line that is tangent to both circles and that separates them. { in,tərn·əl 'tan·jənt }

interpolation A process used to estimate an intermediate value of one (dependent) variable which is a function of a second (independent) variable when values of the dependent variable corresponding to several discrete values of the independent variable are known. { in,tər·pə'lā·shən }

interquartile range The distance between the top of the lower quartile and the bottom of the upper quartile of a distribution. { ¦in·tər'kwor,tīl ,rānj }

intersection 1. The point, or set of points, that is common to two or more geometric configurations. **2.** For two sets, the set consisting of all elements common to both of the sets. Also known as meet. **3.** For two fuzzy sets A and B, the fuzzy set whose membership function has a value at any element x that is the minimum of the values of the membership functions of A and B at x. **4.** The intersection of two Boolean matrices A and B, with the same number of rows and columns, is the Boolean matrix whose element c_{ij} in row i and column j is the intersection of corresponding elements a_{ij} in A and b_{ij} in B. { ,in·tər'sek·shən }

interval A set of numbers which consists of those numbers that are greater than one fixed number and less than another, and that may also include one or both of the end numbers. { 'in·tər·vəl }

interval estimate An estimate which specifies a range of values for a population parameter. { 'in·tər·vəl ,es·tə·mət }

interval estimation A technique that expresses uncertainty about an estimate by defining an interval, or range of values, and indicates the certain degree of confidence with which the population parameter will fall within the interval. { 'in·tər·vəl ,es·tə,mā·shən }

interval measurement A method of measuring quantifiable data that assumes an exact knowledge of the quantitative difference between the objects being scaled. Also known as cardinal measurement. { 'in·tər·vəl ,mezh·ər·mənt }

interval of convergence The interval consisting of the real numbers for which a specified power series possesses a limit. { 'in·tər·vəl əv kən'vər·jəns }

interval scale A rule or system for assigning numbers to objects in such a way that the difference between any two objects is reflected in the difference in the numbers assigned to them; used in interval measurement. { 'in·tər·vəl ‚skāl }

intrinsic equations of a curve The equations describing the radius of curvature and torsion of a curve as a function of arc length; these equations determine the curve up to its position in space. Also known as natural equations of a curve. { in'trin· sik i¦kwā·zhənz əv ə 'kərv }

intrinsic geometry of a surface The description of the intrinsic properties of a surface. { in'trin·sik jē'äm·ə·trē əv ə 'sərfəs }

intrinsic property 1. For a curve, a property that can be stated without reference to the coordinate system. **2.** For a surface, a property that can be stated without reference to the surrounding space. { in'trin·sik 'präp·ərd·ē }

invariance *See* invariant property. { in'ver·ē·əns }

invariant 1. An element x of a set E is said to be invariant with respect to a group G of mappings acting on E if $g(x) = x$ for all g in G. **2.** A subset F of a set E is said to be invariant with respect to a group G of mappings acting on E if $g(x)$ is in F for all x in F and all g in G. **3.** For an algebraic equation, an expression involving the coefficients that remains unchanged under a rotation or translation of the coordinate axes in the cartesian space whose coordinates are the unknown quantities. { in'ver·ē·ənt }

invariant function A function f on a set S is said to be invariant under a transformation T of S into itself if $f(Tx) = f(x)$ for all x in S. { in'ver·ē·ənt 'fəŋk·shən }

invariant measure A Borel measure m on a topological space X is invariant for a transformation group (G,X,π) if for all Borel sets A in X and all elements g in G, $m(A_g) = m(A)$, where A_g is the set of elements equal to $\pi(g,x)$ for some x in A. { in'ver·ē·ənt 'mezh·ər }

invariant property A mathematical property of some space unchanged after the application of any member from some given family of transformations. Also known as invariance. { in'ver·ē·ənt 'präp·ərd·ē }

invariant subgroup *See* normal subgroup. { in'ver·ē·ənt 'səb‚grùp }

invariant subspace For a bounded operator on a Banach space, a closed linear subspace of the Banach space such that the operator takes any point in the subspace to another point in the subspace. { in‚ver·ē·ənt 'səb‚spās }

inverse 1. The additive inverse of a real or complex number a is the number which when added to a gives 0; the multiplicative inverse of a is the number which when multiplied with a gives 1. **2.** The inverse of a fractional ideal I of an integral domain R is the set of all elements x in the quotient field K of R such that xy is in I for all y in I. **3.** For a set S with a binary operation $x \cdot y$ that has an identity element e, the inverse of a member, x, of S is another member, $\bar{x}$, of S for which $x \cdot \bar{x} = \bar{x} \cdot x = e$. { 'in‚vərs }

inverse cosecant *See* arc cosecant. { ¦in‚vərs kō'sē‚kant }

inverse cosine *See* arc cosine. { ¦in‚vərs 'kō‚sīn }

inverse cotangent *See* arc cotangent. { ¦in‚vərs kō'tan·jənt }

inverse curves A pair of curves such that every point on one curve is the inverse point of some point on the other curve, with respect to a fixed circle. { 'in‚vərs 'kərvz }

inverse element In a group G the inverse of an element g is the unique element g^{-1} such that $g \cdot g^{-1} = g^{-1} \cdot g = e$, where $\cdot$ denotes the group operation and e is the identity element. { 'in‚vərs 'el·ə·mənt }

inverse function An inverse function for a function f is a function g whose domain is the range of f and whose range is the domain of f with the property that both f composed with g and g composed with f give the identity function. { 'in‚vərs 'fəŋk·shən }

inverse function theorem If f is a continuously differentiable function of euclidean n-space to itself and at a point x_0 the matrix with the entry $(\partial f_i/\partial x_j)x_0$ in the ith row and jth column is nonsingular, then there is a continuously differentiable

inverse hyperbolic cosecant

function $g(y)$ defined in a neighborhood of $f(x_0)$ which is an inverse function for $f(x)$ at all points near x_0. { 'in,vərs 'fəŋk·shən ,thir·əm }

inverse hyperbolic cosecant *See* arc-hyperbolic cosecant. { ¦in,vərs ,hī·pər,bäl·ik kō'sē,kant }

inverse hyperbolic cosine *See* arc-hyperbolic cosine. { ¦in,vərs ,hī·pər,bäl·ik 'kō,sīn }

inverse hyperbolic cotangent *See* arc-hyperbolic cotangent. { ¦in,vərs ,hī·pər,bäl·ik kō'tan·jənt }

inverse hyperbolic function An inverse function of a hyperbolic function; that is, an arc-hyperbolic sine, arc-hyperbolic cosine, arc-hyperbolic tangent, arc-hyperbolic cotangent, arc-hyperbolic secant, or arc-hyperbolic cosecant. Also known as anti-hyperbolic function; arc-hyperbolic function. { ¦in,vərs ,hī·pər,bäl·ik 'fəŋk·shən }

inverse hyperbolic secant *See* arc-hyperbolic secant. { ¦in,vərs ,hī·pər,bäl·ik 'sē,kant }

inverse hyperbolic sine *See* arc-hyperbolic sine. { ¦in,vərs ,hī·pər,bäl·ik 'sīn }

inverse hyperbolic tangent *See* arc-hyperbolic tangent. { ¦in,vərs ,hī·pər,bäl·ik 'tan·jənt }

inverse image *See* pre-image. { ¦in,vərs 'im·ij }

inverse implication The implication that results from replacing both the antecedent and the consequent of a given implication with their negations. { 'in,vərs ,im·plə'kā·shən }

inverse logarithm *See* antilogarithm. { 'in,vərs 'läg·ə,ri_th·əm }

inversely proportional quantities Two variable quantities whose product remains constant. { in¦vərs·lē prə¦pór·shən·əl 'kwän·əd·ēz }

inverse-mapping theorem The theorem that the inverse of a linear, one-to one, continuous mapping between two Banach spaces or two Fréchet spaces is also continuous. { ¦in,vərs 'map·iŋ ,thir·əm }

inverse matrix The inverse of a nonsingular matrix A is the matrix A^{-1} where $A \cdot A^{-1} = A^{-1} \cdot A = I$, the identity matrix. { 'in,vərs 'mā·triks }

inverse operator The inverse of an operator L is the operator which is the inverse function of L. { 'in,vərs 'äp·ə,rād·ər }

inverse points A pair of points lying on a diameter of a circle or sphere such that the product of the distances of the points from the center equals the square of the radius. { 'in,vərs 'póins }

inverse probability principle *See* Bayes' theorem. { ¦in,vərs ,präb·ə'bil·əd·ē ,prin·sə·pəl }

inverse ranks Ranking responses to treatments from largest response to smallest response. { 'in,vərs 'raŋks }

inverse ratio The reciprocal of the ratio of two quantities. Also known as reciprocal ratio. { ¦in,vərs 'rā·shō }

inverse relation For a relation R, the inverse relation R^{-1} is the relation such that the ordered pair (x,y) belongs to R^{-1} if and only if (y,x) belongs to R. { ¦in,vərs ri'lā·shən }

inverse secant *See* arc secant. { ¦in,vərs 'sē,kant }

inverse sine *See* arc sine. { ¦in,vərs 'sīn }

inverse substitution A substitution that precisely nullifies the effect of a given substitution. { ¦in,vərs ,səb·stə'tü·shən }

inverse tangent *See* arc tangent. { ¦in,vərs 'tan·jənt }

inverse trigonometric function An inverse function of a trigonometric function; that is, an arc sine, arc cosine, arc tangent, arc cotangent, arc secant, or arc cosecant. Also known as antitrigonometric function. { ¦in,vərs ,trig·ə·nə,me·trik 'fəŋk·shən }

inverse variation **1.** A relationship between two variables wherein their product is equal to a constant. **2.** An equation or function expressing such a relationship. { 'in,vərs ,ver·e'ā·shən }

inversion **1.** Given a point O lying in a plane or in space, a mapping of the plane or of space, excluding the point O, into itself in which every point is mapped into its inverse point with respect to a circle or sphere centered at O. **2.** The interchange of two adjacent members of a sequence. { in'vər·zhən }

invertible element An element x of a groupoid with a unit element e for which there is an element $\bar{x}$ such that $x \cdot \bar{x} = \bar{x} \cdot x = e$. { in‚vərd·ə·bəl 'el·ə·mənt }

invertible ideal A fractional ideal I of an integral domain R such that R is equal to the set of elements of the form xy, where x is in I and y is in the inverse of I. { in¦vərd· ə·bəl i'dēl }

involute 1. A curve produced by any point of a perfectly flexible inextensible thread that is kept taut as it is wound upon or unwound from another curve. **2.** A curve that lies on the tangent surface of a given space curve and is orthogonal to the tangents to the given curve. **3.** A surface for which a given surface is one of the two surfaces of center. { ¦in·və¦lüt }

involution 1. Any transformation that is its own inverse. **2.** In particular, a correspondence between the points on a line that is its own inverse, given algebraically by $x' = (ax + b)/(cx - a)$, where $a^2 + bc \neq 0$. **3.** A correspondence between the lines passing through a given point on a plane such that corresponding lines pass through corresponding points of an involution of points on a line. { ‚in·və'lü·shən }

irrational algebraic expression An algebraic expression that cannot be written as a quotient of polynomials. { i¦rash·ən·əl ‚al·jə¦brā·ik ik'spresh·ən }

irrational equation An equation having an unknown raised to some fractional power. Also known as radical equation. { i'rash·ən·əl i'kwā·zhən }

irrational number A number which is not the quotient of two integers. { i'rash·ən·əl 'nəm·bər }

irrational radical A radical that is not equivalent to a rational number or expression. { i'räsh·ən·əl 'rad·ə·kəl }

irreducible element An element x of a ring which is not a unit and such that every divisor of x is improper. { ‚ir·ə'düs·ə·bəl 'el·ə·mənt }

irreducible equation An equation that is equivalent to one formed by setting an irreducible polynomial equal to zero. { ‚ir·ə'dü·sə·bəl i'kwā·zhən }

irreducible function *See* irreducible polynomial. { ‚ir·ə'dü·sə·bəl 'fəŋk·shən }

irreducible lambda expression A lambda expression that cannot be converted to a reduced form by a sequence of applications of the renaming and reduction rules. { ‚ir·ə'dü·sə·bəl 'lam·də ik‚spresh·ən }

irreducible module A module whose only submodules are the module itself and the module that consists of the element 0. { ‚ir·i¦dü·sə·bəl 'mäj·əl }

irreducible polynomial A polynomial is irreducible over a field K if it cannot be written as the product of two polynomials of lesser degree whose coefficients come from K. Also known as irreducible function. { ‚ir·ə'dü·sə·bəl ‚päl·ə'nō·mē·əl }

irreducible representation of a group A representation of a group as a family of linear operators of a vector space V where there is no proper closed subspace of V invariant under these operators. { ‚ir·ə'dü·sə·bəl ‚rep·rə·zən'tā·shən əv ə 'grüp }

irreducible tensor A tensor that cannot be written as the inner product of two tensors of lower degree. { ‚ir·ə'dü·sə·bəl 'ten·sər }

irrotational vector field A vector field whose curl is identically zero; every such field is the gradient of a scalar function. Also known as lamellar vector field. { ¦ir· ə'tā·shən·əl ‚fēld }

isarithm *See* isopleth. { 'ī·sə‚rith‧əm }

isochrone *See* semicubical parabola. { 'ī·sə‚krōn }

isochronous curve A curve with the property that the time for a particle to reach a lowest point on the curve if it starts from rest and slides without friction does not depend on the particle's starting point. { ī¦sä·krə·nəs 'kərv }

isogonal conjugates *See* isogonal lines. { ī¦säg·ən·əl 'kän·jə·gəts }

isogonal lines Lines that pass through the vertex of an angle and make equal angles with the bisector of the angle. Also known as isogonal conjugates. { ī¦säg·ən· əl 'līnz }

isogonal transformation A mapping of the plane into itself which leaves the magnitudes of angles between intersecting lines unchanged but may reverse their sense. { ī¦säg·ən·əl ‚tranz·fər'mā·shən }

isogram See isopleth. { 'ī·sə,gram }

isolated point 1. A point p in a topological space is an isolated point of a set if p is in the set and there is a neighborhood of p which contains no other points of the set. **2.** A point that satisfies the equation for a plane curve C but has a neighborhood that includes no other point of C. Also known as acnode; hermit point. { 'ī·sə,lād·əd 'point }

isolated set A set consisting entirely of isolated points. { 'ī·sə,lād·əd 'set }

isolated subgroup An isolated subgroup of a totally ordered Abelian group G is a subgroup of G which is also a segment of G. { |īs·ə,lād·əd 'səb,grüp }

isolated vertex A vertex of a graph that has no edges incident to it. { |ī·sə,lād·əd 'vər,teks }

isometric forms Two bilinear forms f and g on vector spaces E and F for which there exists a linear isomorphism of E onto F such that $f(x,y) = g(\sigma x, \sigma y)$ for all x and y in E. { |ī·sə'me·trik 'fórmz }

isometric spaces Two spaces between which an isometry exists. { |ī·sə'me·trik 'spā·səs }

isometry 1. A mapping f from a metric space X to a metric space Y where the distance between any two points of X equals the distance between their images under f in Y. **2.** A linear isomorphism σ of a vector space E onto itself such that, for a given bilinear form g, $g(\sigma x, \sigma y) = g(x,y)$ for all x and y in E. { ī'säm·ə·trē }

isometry class A set consisting of all bilinear forms (on vector spaces over a given field) which are isometric to a given form. { ī'säm·ə·trē ,klas }

isomorphic systems Two algebraic structures between which an isomorphism exists. { ,ī·sə|mór·fik 'sis·təmz }

isomorphism A one to one function of an algebraic structure (for example, group, ring, module, vector space) onto another of the same type, preserving all algebraic relations; its inverse function behaves likewise. { |ī·sə|mór,fiz·əm }

isomorphism problem For two simple graphs with the same numbers of vertices and edges, the problem of determining whether there exist correspondences between these vertices and edges such that there is an edge between two vertices in one graph if and only if there is an edge between the corresponding vertices in the other. { ,ī·sə'mór,fiz·əm ,präb·ləm }

isoperimetric figures Figures whose perimeters are equal. { |ī·sō,per·ə'me·trik 'fig·yərz }

isoperimetric inequality The statement that the area enclosed by a plane curve is equal to or less than the square of its perimeter divided by 4π. { ,ī·sə,per·ə|me·trik ,in·i'kwäl·əd·ē }

isoperimetric problem In the calculus of variations this problem deals with finding a closed curve in the plane which encloses the greatest area given its length as fixed. { |ī·sō,per·ə'me·trik 'präb·ləm }

isopleth The straight line which cuts the three scales of a nomograph at values satisfying some equation. Also known as index line. { 'ī·sə,pleth }

isoptic The locus of the intersection of tangents to a given curve that meet at a specified constant angle. { ī'säp·tik }

isosceles spherical triangle A spherical triangle that has two equal sides. { ī|säs·ə,lēz |sfer·ə·kəl 'trī,aŋ·gəl }

isosceles triangle A triangle with two sides of equal length. { ī'säs·ə,lēz 'trī,aŋ·gəl }

isotropy group For an operation of a group G on a set S, the isotropy group of an element s of S is the set of elements g in G such that $gs = s$. { ī'sä,trō·pē ,grüp }

isthmus See bridge. { 'is·məs }

iterated integral An integral over an area or volume designated to be performed by successive integrals over line segments. { 'īd·ə,rād·əd 'int·ə·grəl }

iteration See iterative method. { ,īd·ə'rā·shən }

iterative method Any process of successive approximation used in such problems as numerical solution of algebraic equations, differential equations, or the interpolation of the values of a function. Also known as iteration. { 'īd·ə,rād·iv 'meth·əd }

iterative process A process for calculating a desired result by means of a repeated cycle of operations, which comes closer and closer to the desired result; for example, the arithmetical square root of N may be approximated by an iterative process using additions, subtractions, and divisions only. { 'ĭd·ə,rād·iv 'prä·səs }

Itô's formula *See* stochastic chain rule. { 'ē,tōz ,fȯr·myə·lə }

Itô's integral *See* stochastic integral. { 'ē,tōz ,int·ə·grəl }

J

Jacobian The Jacobian of functions $f_i(x_1, x_2, \ldots, x_n)$, $i = 1, 2, \ldots, n$, of real variables x_i is the determinant of the matrix whose ith row lists all the first-order partial derivatives of the function $f_i(x_1, x_2, \ldots, x_n)$. Also known as Jacobian determinant. { jə'kō·bē·ən }

Jacobian determinant *See* Jacobian. { jə'kō·bē·ən di'tər·mə·nənt }

Jacobian elliptic function For m a real number between 0 and 1, and u a real number, let ϕ be that number such that

$$\int_0^\phi d\theta/(1 - m \sin^2 \theta)^{1/2} = u;$$

the 12 Jacobian elliptic functions of u with parameter m are sn $(u|m) = \sin \phi$, cn $(u|m) = \cos \phi$, dn $(u|m) = (1 - m \sin^2 \phi)^{1/2}$, the reciprocals of these three functions, and the quotients of any two of them. { jə'kō·bē·ən ə¦lip·tik 'fəŋk·shən }

Jacobian matrix The matrix used to form the Jacobian. { jə'kō·bē·ən 'mā·triks }

Jacobi canonical matrix A form to which any matrix can be reduced by a collineatory transformation, with zeros below the principal diagonal and characteristic roots as elements of the principal diagonal. { jə¦kōb·ē kə¦nän·ə·kəl 'mā·triks }

Jacobi condition In the calculus of variations, a differential equation used to study the extremals in a variational problem. { jə'kō·bē kən,dish·ən }

Jacobi polynomials Polynomials that are constructed from the hypergeometric function and satisfy the differential equation $(1 - x^2)y'' + [\beta - \alpha - (\alpha + \beta + 2)x]y' + n(\alpha + \beta + n + 1)y = 0$, where n is an integer and α and β are constants greater than -1; in certain cases these generate the Legendre and Chebyshev polynomials. { jə'kō·bē ,päl·ə'nō·mē·əlz }

Jacobi's method 1. A method of determining the eigenvalues of a Hermitian matrix. **2.** A method for finding a complete integral of the general first-order partial differential equation in two independent variables; it involves solving a set of six ordinary differential equations. { jə'kō·bēz ,meth·əd }

Jacobi's theorem The proposition that a periodic, analytic function of a complex variable is simply periodic or doubly periodic. { jə'kō·bēz ,thir·əm }

Jacobi's transformations Transformations of Jacobian elliptic functions to other Jacobian elliptic functions given by change of parameter and variable. { jə'kō·bēz ,tranz·fər'mā·shənz }

Jensen's inequality 1. A general inequality satisfied by a convex function

$$f\left(\sum_{i=1}^n a_i x_i\right) \leq \sum_{i=1}^n a_i f(x_i)$$

where the x_i are any numbers in the region where f is convex and the a_i are nonnegative numbers whose sum is equal to 1. **2.** If $a_1, a_2, \ldots, a_n$ are positive numbers and $s > t > 0$, then $(a_1{}^s + a_2{}^s + \cdots + a_n{}^s)^{1/s}$ is less than or equal to $(a_1{}^t + a_2{}^t + \cdots + a_n{}^t)^{1/t}$. { 'jen·sənz ,in·i'kwäl·ədē }

join 1. The join of two elements of a lattice is their least upper bound. **2.** *See* union. { jȯin }

join-irreducible member A member, A, of a lattice or ring of sets such that, if A is equal to the join of two other members, B and C, then A equals B or A equals C. { ¦jȯin ‚ir·i‚dü·sə·bəl 'mem·bər }

joint distribution For two random variables Z and W, the distribution which gives the probability that $Z = z$ and $W = w$ for all values z and w of Z and W respectively. { 'jȯint ‚dis·trə¦byü·shən }

joint marginal distribution The distribution obtained by summing the joint distribution of three random variables over all possible values of one of these variables. { 'jȯint ¦mär·jən·əl ‚dis·trə'byü·shən }

joint variation The relation of a variable x to two other variables y and z wherein x is proportional to the product of y and z. { ¦jȯint ‚ver·ē'ā·shən }

Jordan algebra A nonassociative algebra over a field in which the products satisfy the Jordan identity $(xy)x^2 = x(yx^2)$. { zhȯr'dän ‚al·jə·brə }

Jordan arc *See* simple arc. { zhȯr'dän ‚ärk }

Jordan condition A condition for the convergence of a Fourier series of a function f at a number x, namely, that there be a neighborhood of x on which f is of bounded variation. { zhȯr'dän kən¦dish·ən }

Jordan content For a set whose exterior Jordan content and interior Jordan content are equal, the common value of these two quantities. Also known as content. { zhȯr'dän ‚kän‚tent }

Jordan curve A simple closed curve in the plane, that is, a curve that is closed, connected, and does not cross itself. { zhȯr'dän ‚kərv }

Jordan curve theorem The theorem that in the plane every simple closed curve separates the plane into two parts. { zhȯr'dän ‚kərv ‚thir·əm }

Jordan form A matrix that has been transformed into a Jordan matrix is said to be in Jordan form. { zhȯr'dän ‚fȯrm }

Jordan-Hölder theorem The theorem that for a group any two composition series have the same number of subgroups listed, and both series produce the same quotient groups. { zhȯr'dän 'hül·dər ‚thir·əm }

Jordan matrix A matrix whose elements are equal and nonzero on the principal diagonal, equal to 1 on the diagonal immediately above, and equal to 0 everywhere else. { zhȯr'dän ‚mā·triks }

J-shaped distribution A frequency distribution that is extremely asymmetrical in that the initial (or final) frequency group contains the highest frequency, with succeeding frequencies becoming smaller (or larger) elsewhere; the shape of the curve roughly approximates the letter "J" lying on its side. { ¦jā ‚shāpt ‚dis·trə'byü·shən }

judgment sample Sample selection in which personal views or opinions of the individual doing the sampling enter into the selection. { 'jəj·mənt ‚sam·pəl }

Julia set For a polynomial, p, with degree greater than 1, the Julia set of p is the boundary of the set of complex numbers, z, such that the sequence $p(z)$, $p^2(z)$, $..., p_n(z), ...$ is bounded, where $p^2(z) = p[p(z)]$, and so forth. { 'jül·yə ‚set }

jump discontinuity A point a where for a real-valued function $f(x)$ the limit on the left of $f(x)$ as x approaches a and the limit on the right both exist but are distinct. { 'jəmp dis‚känt·ən'ü·əd·ē }

jump function A function used to represent a sampled data sequence arising in the numerical study of linear difference equations. { 'jəmp ‚fəŋk·shən }

Jung's theorem The theorem that a set of diameter 1 in an n-dimensional Euclidean space is contained in a closed ball of radius $[n/(2n + 2)]^{1/2}$.

K

Kakeya problem The problem of finding the plane figure of least area within which a unit line segment can be moved continuously so as to return to its original position with its end points reversed; in fact, there is no such minimum area. { kä 'kä·ə ˌpräb·ləm }

kampyle of Eudoxus A plane curve whose equation in Cartesian coordinates x and y is $x^4 = a^2(x^2 + y^2)$, where a is a constant. { kam'pīl əv yü'däk·səs }

kappa curve A plane curve whose equation in Cartesian coordinates x and y is $(x^2 + y^2) y^2 = a^2x^2$, where a is a constant. Also known as Gutschoven's curve. { 'kap·ə ˌkərv }

Kármán swirling flow problem The problem of describing fluid motion above a rotating infinite plane disk when the fluid at infinity does not rotate. { 'kär,män ¦swir·liŋ ¦flō ˌpräb·ləm }

Karmarkar's algorithm A method for solving linear programming problems that has a polynomial time bound and appears to be faster than the simplex method for many complex problems. { ¦kär·mə,kärz 'al·gə,rith·əm }

Karush-Kuhn-Tucker conditions A system of equations and inequalities which the solution of a nonlinear programming problem must satisfy when the objective function and the constraint functions are differentiable. { ¦kär·əsh ¦kyün 'tək·ər kən,dish·ənz }

kei function A function that is expressed in terms of modified Bessel functions of the second kind in a manner similar to that in which the bei function is expressed in terms of Bessel functions. { 'kī ˌfəŋk·shən }

Kekeya needle problem The problem of finding the smallest area of a plane region in which a line segment of unit length can be continuously moved so that it returns to its original position after turning through 360°. { kä,kē·ə 'nēd·əl ˌpräb·ləm }

Kempe chain A subgraph of a graph whose vertices have been colored, consisting of vertices which have been assigned a given color or colors and arcs connecting pairs of such vertices. { 'kem·pə ˌchān }

Kendall's rank correlation coefficient A statistic used as a measure of correlation in nonparametric statistics when the data are in ordinal form. Also known as Kendall's tau. { ¦ken·dəlz ¦raŋk ˌkä·rə'lā·shən ˌkō·ə,fish·ənt }

Kendall's tau *See* Kendall's rank correlation coefficient. { ¦ken·dəlz 'tȯ }

keratoid A plane curve whose equation in Cartesian coordinates x and y is $y^2 = x^2y + x^5$. { 'ker·ə,tȯid }

keratoid cusp A cusp of a curve which has one branch of the curve on each side of the common tangent. Also known as single cusp of the first kind. { 'ker·ə,tȯid ˌkəsp }

ker function A function that is expressed in terms of modified Bessel functions of the second kind in a manner similar to that in which the ber function is expressed in terms of Bessel functions. { 'ker ˌfəŋk·shən }

kernel 1. For any mapping f from a group A to a group B, the kernel of f, denoted ker f, is the set of all elements a of A such that $f(a)$ equals the identity element of B. **2.** For a homomorphism h from a group G to a group H, this consists of all elements of G which h sends to the identity element of H. **3.** For Fredholm and Volterra integral equations, this is the function $K(x,t)$. **4.** For an integral

Killing's equations

transform, the function $K(x,t)$ in the transformation which sends the function $f(x)$ to the function $\int K(x,t)f(t)dt = F(x)$. **5.** *See* null space. { 'kərn·əl }

Killing's equations The equations for an isometry-generating vector field in a geometry. { 'kil·iŋz i‚kwā·zhənz }

Killing vector An element of a vector field in a geometry that generates an isometry. { 'kil·iŋ ‚vek·tər }

Kirkman triple system A resolvable balanced incomplete block design with block size k equal to 3. { ‚kərk·mən ‚trip·əl 'sis·təm }

Klein bottle The nonorientable surface having only one side with no inside or outside; it resembles a bottle pulled into itself. { 'klīn ‚bäd·əl }

Kleinian group A group of conformal mappings of a Riemann surface onto itself which is discontinuous at one or more points and is not discontinuous at more than two points. { 'klī·nē·ən ‚grüp }

Klein's four-group The noncyclic group of order four. { ‚klīnz 'fȯr ‚grüp }

knapsack problem The problem, given a set of integers $\{A_1, A_2, \ldots, A_n\}$ and a target integer B, of determining whether a subset of the A_i can be selected without repetition so that their sum is the target B. { 'nap‚sak ‚präb·ləm }

knot In the general case, a knot consists of an embedding of an n-dimensional sphere in an $(n + 2)$-dimensional sphere; classically, it is an interlaced closed curve, homeomorphic to a circle. { nät }

knot theory The topological and algebraic study of knots emphasizing their classification and how one may be continuously deformed into another. { 'nät ‚thē·ə·rē }

Kobayashi potential A solution of Laplace's equation in three dimensions constructed by superposition of the solutions obtained by separation of variables in cylindrical coordinates. { ‚kō·bī'yä·shē pə‚ten·chəl }

Koch curve A fractal which can be constructed by a recursive procedure; at each step of this procedure every straight segment of the curve is divided into three equal parts and the central piece is then replaced by two similar pieces. { 'kōk ‚kərv }

Koebe function The analytic function $k(z) = z(1 - z)^{-2} = z + 2z^2 + 3z^3 + \cdots$, that maps the unit disk onto the entire complex plane minus the part of the negative real axis to the left of $-1/4$. { 'kā·bē ‚faŋk·shən }

Kolmogorov consistency conditions For each finite subset F of the real numbers or integers, let P_F denote a probability measure defined on the Borel subsets of the cartesian product of $k(F)$ copies of the real line indexed by elements in F, where $k(F)$ denotes the number of elements in F; the family $\{P_F\}$ of measures satisfy the Kolmogorov consistency conditions if given any two finite sets F_1 and F_2 with F_1 contained in F_2, the restriction of P_{F2} to those sets which are independent of the coordinates in F_2 which are not in F_1 coincides with P_{F1}. { ‚kȯl·mə'gȯ·rȯf kən'sis·tən·sē kən‚dish·ənz }

Kolmogorov inequalities For each integer k let X_k be a random variable with finite variance σ_k and suppose $\{X_k\}$ is an independent sequence which is uniformly bounded by some constant c; then for every $\epsilon > 0$, and integer n,

$$1 - (\epsilon + 2c)^2 \Big/ \sum_{k=1}^{n} \sigma_k^2 \leq \text{Prob} \{\max_{k \leq n} |S_k + ES_k| \geq \epsilon\}$$

and

$$\frac{1}{\epsilon^2} \sum_{k=1}^{n} \sigma_k^2 \geq \text{Prob} \{\max_{k \leq n} |S_k + ES_k| \geq \epsilon\};$$

here $S_k = \sum_{i=1}^{k} X_i$ and ES_k denotes the expected value of S_k. { ‚kȯl·mə'gȯ·rȯf ‚in·i'kwäl·əd·ēz }

Kolmogorov-Sinai invariant An isomorphism invariant of measure-preserving transformations; if T is a measure-preserving transformation on a probability space, the Kolmogorov-Sinai invariant is the least upper bound of the set of entropies of T

given each finite partition of the probability space. Also known as entropy of a transformation. { ˌkȯl·mə'gȯ·rȯf 'sī͵nī in͵ver·ē·ənt }

Kolmogorov-Smirnov test A procedure used to measure goodness of fit of sample data to a specified population; critical values exist to test goodness of fit. { ˌkȯl· məˌgȯr·əf 'smir͵nȯf ͵test }

Kolmogorov space *See* T_0 space. { ˌkȯl·mə'gȯr·əf ͵spās }

König-Egerváry theorem The theorem that, for a matrix in which each entry is either 0 or 1, the largest number of 1's that can be chosen so that no two selected 1's lie in the same row or column equals the smallest number of rows and columns that must be deleted to eliminate all the 1's. { ˈkərn·ik 'e·ger͵vär·yi ͵thir·əm }

Königsberg bridge problem The problem of walking across seven bridges connecting four landmasses in a specified manner exactly once and returning to the starting point; this is the original problem which gave rise to graph theory. { ˈkərn·iks͵bərg 'brij ͵präb·ləm }

König's theorem The theorem that the largest possible number of edges in a matching of a bipartite graph equals the smallest possible number of edges in an edge cover of that graph. { 'kər·nigz ͵thir·əm }

Krawtchouk polynomials Families of polynomials which are orthogonal with respect to binomial distributions. { ˈkräv͵chək ͵päl·ə'nō·mē·əlz }

Krein-Milman property The property of some topological vector spaces that any bounded closed convex subset is the closure of the convex span of its extreme points. { 'krīn 'mil·mən ͵präp·ərd·ē }

Krein-Milman theorem The theorem that in a locally convex topological vector space, any compact convex set K is identical with the intersection of all convex sets containing the extreme points of K. { 'krīn 'mil·mən ͵thir·əm }

Kronecker delta The function or symbol δ_{ij} dependent upon the subscripts i and j which are usually integers; its value is 1 if $i = j$ and 0 if $i \neq j$. { 'krō·nek·ər ͵del·tə }

Kronecker product Given two different representations of the same group, their Kronecker product is a representation of the group constructed by taking direct products of matrices from the respective representations. { 'krō·nek·ər ͵präd·əkt }

K theory The study of the mathematical structure resulting from associating an abelian group $K(X)$ with every compact topological space X in a geometrically natural way, with the aid of complex vector bundles over X. Also known as topological K theory. { 'kā ͵thē·ə·rē }

Kuratowski closure-complementation problem The problem of showing that at most 14 distinct sets can be obtained from a subset of a topological space by repeated operations of closure and complementation. { ͵kùr·ə͵tȯf·skē ͵klō·zhər ͵käm·plə·men'tā·shən ͵präb·ləm }

Kuratowski graphs Two graphs which appear in Kuratowski's theorem, the complete graph K_5 with five vertices and the bipartite graph $K_{3,3}$. { ͵kùr·ə'təv͵skē ͵grafs }

Kuratowski's lemma Each linearly ordered subset of a partially ordered set is contained in a maximal linearly ordered subset. { ͵kùr·ə'tȯv·skēz 'lem·ə }

Kuratowski's theorem The proposition that a graph is nonplanar if and only if it has a subgraph which is either a Kuratowski graph or a subdivision of a Kuratowski graph. { ͵kùr·ə'təv͵skēz ͵thir·əm }

Kureppa number A number of the form $!n = 0! + 1! + \cdots + (n - 1)!$, where n is a positive integer. { kù'rep·ə ͵nəm·bər }

kurtosis The extent to which a frequency distribution is concentrated about the mean or peaked; it is sometimes defined as the ratio of the fourth moment of the distribution to the square of the second moment. { kər'tō·səs }

L

labeled graph A graph whose vertices are distinguished by names. { 'lā·bəld ˌgraf }

lacunary space A region in the complex plane that lies entirely outside the domain of a particular monogenic analytic function. { ləˌkü·nə·rē 'spās }

lag correlation The strength of the relationship between two elements in an ordered series, usually a time series, where one element lags a specific number of places behind the other elements. { 'lag ˌkä·rəˌlā·shən }

Lagrange-Helmholtz equation *See* Helmholtz equation. { lə'gränj 'helmˌhōlts iˌkwā·zhən }

Lagrange's formula *See* mean value theorem. { lə'grän·jəz ˌför·myə·lə }

Lagrange's theorem In a group of finite order, the order of any subgroup must divide the order of the entire group. { lə'grän·jəz ˌthir·əm }

Lagrangian multipliers A technique whereby potential extrema of functions of several variables are obtained.   Also known as undetermined multipliers. { lə'grän·jē·ən 'məlˌtəˌplī·ərz }

Laguerre polynomials A sequence of orthogonal polynomials which solve Laguerre's differential equation for positive integral values of the parameter. { lə'ger ˌpälˌə'nō·mē·əlz }

Laguerre's differential equation The equation $xy'' + (1 - x)y' + \alpha y = 0$, where α is a constant. { lə'gerz ˌdif·əˈren·chəl iˈkwā·zhən }

lambda calculus A mathematical formalism to model the mathematical notion of substitution of values for bound variables. { 'lam·də ˌkal·kyə·ləs }

lambda expression An expression used to define a function in the lambda calculus; for example, the function $f(x) = x + 1$ is defined by the expression $\lambda x(x + 1)$. { 'lam·də ikˌspresh·ən }

Lamé functions Functions that arise when Laplace's equation is separated in ellipsoidal coordinates. { lä'mā ˌfəŋk·shənz }

lamellar vector field *See* irrotational vector field. { lə'mel·ər 'vek·tər ˌfēld }

Lamé polynomials Polynomials which result when certain parameters of Lamé functions assume integral values, and which are used to express physical solutions of Laplace's equation in ellipsoidal coordinates. { lä'mā ˌpäl·ə'nō·mē·əlz }

Lamé's equations A general collection of second-order differential equations which have five regular singularities. { lä'māz iˌkwā·zhənz }

Lamé's relations Six independent relations which when satisfied by the covariant metric tensor of a three-dimensional space provide necessary and sufficient conditions for the space to be Euclidean. { lä'māz riˌlā·shənz }

Lamé wave functions Functions which arise when the wave equation is separated in ellipsoidal coordinates.   Also known as ellipsoidal wave functions. { lä'mā 'wāv ˌfəŋk·shənz }

Lanczos's method A transformation method for diagonalizing a matrix in which the matrix used to transform the original matrix to triple-diagonal form is formed from a set of column vectors that are determined by a recursive process. { 'länˌchōz·əs ˌmeth·əd }

language theory A branch of automata theory which attempts to formulate the grammar of a language in mathematical terms; it has been applied to automatic language translation and to the construction of higher-level programming languages and

Laplace operator

systems such as the propositional calculus, nerve networks, sequential machines, and programming schemes. { 'laŋ·gwij ,thē·ə·rē }

Laplace operator The linear operator defined on differentiable functions which gives for each function the sum of all its nonmixed second partial derivatives. Also known as Laplacian. { lə'pläs ,äp·ə,rād·ər }

Laplace's equation The partial differential equation which states that the sum of all the nonmixed second partial derivatives equals 0; the potential functions of many physical systems satisfy this equation. { lə'pläs·əz i,kwā·zhən }

Laplace's expansion An expansion by means of which the determinant of a matrix may be computed in terms of the determinants of all possible smaller square matrices contained in the original. { lə'pläs·əz ik,span·chən }

Laplace's measure of dispersion The expected value of the absolute value of the difference between a random variable and its mean. { lə'pläs·əz ¦mezh·ər əv di'spər·zhən }

Laplace transform For a function $f(x)$ its Laplace transform is the function $F(y)$ defined as the integral over x from 0 to ∞ of the function $e^{-yx}f(x)$. { lə'pläs 'trans,form }

Laplacian *See* Laplace operator. { lə'pläs·ē·ən }

Laspeyre's index A weighted aggregate price index with base-year quantity weights. Also known as base-year method. { lä'perz ,in,deks }

latent root *See* eigenvalue. { 'lāt·ənt 'rüt }

lateral area The area of a surface with the bases (if any) excluded. { 'lad·ə·rəl 'er·ē·ə }

lateral face The lateral face for a prism or pyramid is any edge or face which is not part of a base. { 'lad·ə·rəl 'fās }

Latin rectangle An $r \times n$ matrix, with n equal to or greater than r in which each row is a permutation of the numbers 1, 2, . . ., n, and no number appears in a column more than once. { 'lat·ən 'rek,taŋ·gəl }

Latin square An $n \times n$ square array of n different symbols, each symbol appearing once in each row and once in each column; these symbols prove useful in ordering the observations of an experiment. { 'lat·ən 'skwer }

lattice A partially ordered set in which each pair of elements has both a greatest lower bound and least upper bound. { 'lad·əs }

latus rectum The length of a chord through the focus and perpendicular to the axis of symmetry in a conic section. { ¦lad·əs 'rek·təm }

Laurent expansion An infinite series in which an analytic function $f(z)$ defined on an annulus about the point z_0 may be expanded, with nth term $a_n(z - z_0)^n$, n ranging from $-\infty$ to ∞, and $a_n = 1/(2\pi i)$ times the integral of $f(t)/(t - z_0)^{n+1}$ along a simple closed curve interior to the annulus. Also known as Laurent series. { lȯ'rän ik,span·chən }

Laurent series *See* Laurent expansion. { lȯ'ränz ,sir·ēz }

law of contradiction A principle of logic whereby a proposition cannot be both true and false. { ,lȯ əv ,kän·trə'dik·shən }

law of cosines Given a triangle with angles A, B, and C and sides a, b, c opposite these angles respectively: $a^2 = b^2 + c^2 - 2bc \cos A$. { 'lȯ əv 'kō,sīnz }

law of exponents Any of the laws $a^m a^n = a^{m+n}$, $a^m/a^n = a^{m-n}$, $(a^m)^n = a^n$, $(ab)^n = a^n b^n$, $(a/b)^n = a^n/b^n$; these laws are valid when m and n are any integers, or when a and b are positive and m and n are any real numbers. Also known as exponential law. { 'lȯ əv ik'spō·nəns }

law of large numbers The law that if, in a collection of independent identical experiments, $N(B)$ represents the number of occurrences of an event B in n trials, and p is the probability that B occurs at any given trial, then for large enough n it is unlikely that $N(B)/n$ differs from p by very much. Also known as Bernoulli theorem. { 'lȯ əv ¦lärj 'nəm·bərz }

law of quadrants 1. The law that any angle of a right spherical triangle (except a right angle) and the side opposite it are in the same quadrant. **2.** The law that when two sides of a right spherical triangle are in the same quadrant the third side is in the first quadrant, and when two sides are in different quadrants the third side is in the second quadrant.   { 'lȯ əv 'kwäd·rəns }

law of signs The product or quotient of two numbers is positive if the numbers have the same sign, negative if they have opposite signs.   { 'lȯ əv 'sīnz }

law of sines Given a triangle with angles A, B, and C and sides a, b, c opposite these angles respectively: sin A/a = sin B/b = sin C/c.   { 'lȯ əv 'sīnz }

law of species The law that one-half the sum of two angles in a spherical triangle and one-half the sum of the two opposite sides are of the same species, in that they are both acute or both obtuse angles.   { 'lȯ əv 'spē̦shēz }

law of the excluded middle A principle of logic whereby a proposition is either true or false but cannot be both true and false.   Also known as principle of dichotomy.   { ¦lȯ əv <u>th</u>ē ik̦sklüd·əd 'mid·əl }

law of tangents Given a triangle with angles A, B, and C and sides a, b, c opposite these angles respectively: $(a - b)/(a + b) =$ [tan $\frac{1}{2}(A - B)$]/[tan $\frac{1}{2}(A + B)$].   { 'lȯ əv 'tan·jəns }

law of the mean *See* mean value theorem.   { ¦lȯ əv <u>th</u>ə 'mēn }

lcm *See* least common multiple.

leading zeros Zeros preceding the first nonzero integer of a number.   { 'lēd·iŋ 'zir·ōz }

leaf of Descartes *See* folium of Descartes.   { 'lēf əv dā'kärt }

least common denominator The least common multiple of the denominators of a collection of fractions.   { 'lēst 'käm·ən di'näm·ə̦näd·ər }

least common multiple The least common multiple of a set of quantities (for example, numbers or polynomials) is the smallest quantity divisible by each of them.   Abbreviated lcm.   { 'lēst 'käm·ən 'məl·tə·pəl }

least-squares estimate An estimate obtained by the least-squares method.   { ¦lēst 'skwerz ̦es·tə·mət }

least-squares method A technique of fitting a curve close to some given points which minimizes the sum of the squares of the deviations of the given points from the curve.   { ¦lēst 'skwerz ̦meth·əd }

least upper bound The least upper bound of a subset A of a set S with ordering < is the smallest element of S which is greater than or equal to every element of A. Abbreviated lub.   Also known as supremum (sup).   { ¦lēst ¦əp·ər 'bau̇nd }

least-upper-bound axiom The statement that any set of real numbers that has an upper bound also has a least upper bound.   { ¦lēst ̦əp·ər ¦bau̇nd 'ak·sē·əm }

Lebesgue exterior measure A measure whose value on a set S is the greatest lower bound of the Lebesgue measures of open sets that contain S.   Also known as exterior measure; outer measure.   { lə'beg ik̦stir·ē·ər 'mezh·ər }

Lebesgue integral The generalization of Riemann integration of real valued functions, which allows for integration over more complicated sets, existence of the integral even though the function has many points of discontinuity, and convergence properties which are not valid for Riemann integrals.   { lə'beg ̦int·ə·grəl }

Lebesgue interior measure A measure whose value on a set S is the least upper bound of the Lebesgue measures of the closed sets contained in S.   Also known as inner measure; interior measure.   { lə'beg in̦tir·ē·ər 'mezh·ər }

Lebesgue measure A measure defined on subsets of euclidean space which expresses how one may approximate a set by coverings consisting of intervals.   { lə'beg ̦mezh·ər }

Lebesgue number The Lebesgue number of an open cover of a compact metric space X is a positive real number so that any subset of X whose diameter is less than this number must be completely contained in a member of the cover.   { lə'beg ̦nəm·bər }

Lebesgue-Stieltjes integral A Lebesgue integral of the form

$$\int_b^a f(x) \, d\phi(x)$$

where ϕ is of bounded variation; if $\phi(x) = x$, it reduces to the Lebesgue integral

left-continuous function

of $f(x)$; if $\phi(x)$ is differentiable, it reduces to the Lebesgue integral of $f(x)\phi'(x)$.
{ lə'beg 'stēlt·yəs ˌint·ə·grəl }

left-continuous function A function $f(x)$ of a real variable is left-continuous at a point c if $f(x)$ approaches $f(c)$ as x approaches c from the left, that is, $x < c$ only. { 'left kən'tin·yə·wəs 'fəŋk·shən }

left coset A left coset of a subgroup H of a group G is a subset of G consisting of all elements of the form ah, where a is a fixed element of G and h is any element of H. { ¦left 'käs·ət }

left-hand derivative The limit of the difference quotient $[f(x) - f(c)]/[x - c]$ as x approaches c from the left, that is, $x < c$ only. { 'left ˌhand də'riv·əd·iv }

left-handed coordinate system 1. A three-dimensional rectangular coordinate system such that when the thumb of the left hand extends in the positive direction of the first (or x) axis, the fingers fold in the direction in which the second (or y) axis could be rotated about the first axis to coincide with the third (or z) axis.   **2.** A coordinate system of a Riemannian space which has negative scalar density function.   { 'left ¦hand·əd kō'órd·ən·ət ˌsis·təm }

left-handed curve A space curve whose torsion is positive at a given point.   Also known as sinistrorse curve; sinistrorsum.   { 'left ¦hand·əd 'kərv }

left-hand limit *See* limit on the left.   { 'left ¦hand 'lim·ət }

left identity In a set in which a binary operation · is defined, an element e with the property $e \cdot a = a$ for every element a in the set.   { ¦left i'den·ə·dē }

left inverse For a set S with a binary operation $x \cdot y$ that has an identity element e, the left inverse of a member, x, of S is another member, $\bar{x}$, of S for which $\bar{x} \cdot x = e$.   { ¦left in¦vərs }

left-invertible element An element x of a groupoid with a unit element e for which there is an element $\bar{x}$ such that $\bar{x} \cdot x = e$.   { ¦left in,vərd·ə·bəl 'el·ə·məns }

left module A module over a ring in which the product of a member x of the module and a member a of the ring is written ax.   { 'left ˌmäj·əl }

leg Either side adjacent to the right angle of a right triangle.   { leg }

Legendre equation The second-order linear homogeneous differential equation $(1 - x^2)y'' - 2xy' + \nu(\nu + 1)y = 0$, where ν is real and nonnegative.   { lə'zhän·drə i,kwā·zhən }

Legendre function Any solution of the Legendre equation.   { lə'zhän·drə ˌfəŋk·shən }

Legendre polynomials A collection of orthogonal polynomials which provide solutions to the Legendre equation for nonnegative integral values of the parameter.   { lə'zhän·drə ˌpäl·i'nō·mē·əlz }

Legendre's symbol The symbol $(c|p)$, where p is an odd prime number, and $(c|p)$ is equal to 1 if c is a quadratic residue of p, and is equal to -1 if c is not a quadratic residue of p.   { lə'zhän·drəz ˌsim·bəl }

Legendre transformation A mathematical procedure in which one replaces a function of several variables with a new function which depends on partial derivatives of the original function with respect to some of the original independent variables.   Also known as Legendre contact transformation.   { lə'zhän·drə ˌtranz·fər'mā·shən }

Leibnitz's rule A formula to compute the nth derivative of the product of two functions f and g:

$$d^n(f \cdot g)/dx^n = \sum_{k=0}^{n} \binom{n}{k} d^{n-k} f/dx^{n-k} \cdot d^k g/dx^k$$

where $\binom{n}{k} = n!/(n - k)! \, k!$ { 'līb,nit·səz ˌrül }

Leibnitz's test If the sequence of positive numbers a_n approaches zero monotonically, then the series

$$\sum_{n=1}^{\infty} (-1)^n a_n$$

is convergent. { 'līb,nit·səz ,test }

lemma A mathematical fact germane to the proof of some theorem. { 'lem·ə }

lemma of duBois-Reymond A continuous function $f(x)$ is constant in the interval (a,b) if for certain functions g whose integral over (a,b) is zero, the integral over (a,b) of f times g is zero. { 'lem·ə əv dyüb'wä rä'mōn }

lemniscate of Bernoulli The locus of points (x,y) in the plane satisfying the equation $(x^2 + y^2)^2 = a^2 (x^2 - y^2)$ or, in polar coordinates (r,θ), the equation $r^2 = a^2 \cos 2\theta$. { lem'nis·kət əv ber'nü·ē }

lemniscate of Gerono *See* eight curve. { lem'nis·kət əv je'rän·ō }

length of a curve A curve represented by $x = x(t)$, $y = y(t)$ for $t_1 \leq t \leq t_2$, with $x(t_1) = x_1$, $x(t_2) = x_2$, $y(t_1) = y_1$, $y(t_2) = y_2$, has length from (x_1,y_1) to (x_2,y_2) given by the integral from t_1 to t_2 of the function $\sqrt{(dx/dt)^2 + (dy/dt)^2}$. { 'leŋkth əv ə 'kərv }

leptokurtic distribution A distribution in which the ratio of the fourth moment to the square of the second moment is greater than 3, which is the value for a normal distribution; it appears to be more heavily concentrated about the mean, or more peaked, than a normal distribution. { ¦lep·tə¦kərd·ik ,di·strə'byü·shən }

level In factorial experiments, the quantitative or qualitative intensity at which a particular value of a factor is held fixed during an experiment. { 'lev·əl }

level of significance For a test, the probability of false rejection of the null hypothesis. Also known as significance level. { 'lev·əl əv sig'nif·i·kəns }

Levi-Civita symbol A symbol $\epsilon_{i,j,\ldots,s}$ where $i, j, \ldots, s$ are n indices, each running from 1 to n; the symbol equals zero if any two indices are identical, and 1 or -1 otherwise, depending on whether $i, j, \ldots, s$ form an even or an odd permutation of $1, 2, \ldots, n$. { 'lā·vē chē·vē'tä ,sim·bəl }

lexicographic order Given sets A and B with a common ordering $<$, one defines an ordering between all sequences (finite or infinite) of elements of A and of elements of B by $(a_1,a_2, \ldots) < (b_1,b_2, \ldots)$ if either $a_i = b_i$ for every i, or $a_n < b_n$, where n is the first place in which they differ; this is the way words are ordered in a dictionary. { ¦lek·sə·kō¦graf·ik 'òr·dər }

l'Hôpital's cubic *See* Tschirnhausen's cubic. { lō·pē·tälz 'kyü·bik }

l'Hôpital's rule A rule useful in evaluating indeterminate forms: if both the functions $f(x)$ and $g(x)$ and all their derivatives up to order $(n - 1)$ vanish at $x = a$, but the nth derivatives both do not vanish or both become infinite at $x = a$, then

$$\lim_{x\to a} f(x)/g(x) = f^{(n)}(a)/g^{(n)}(a),$$

$f^{(n)}$ denoting the nth derivative. { lō·pē·tälz ,rül }

l'Huilier's equation An equation used in the solution of a spherical triangle, involving tangents of various functions of its angles and sides. { lə·wē'yäz i,kwā·zhən }

Liapunov function *See* Lyapunov function. { 'lyä·pù·nòf ,fəŋk·shən }

Lie algebra The algebra of vector fields on a manifold with additive operation given by pointwise sum and multiplication by the Lie bracket. { 'lē ,al·jə·brə }

Lie bracket Given vector fields X,Y on a manifold M, their Lie bracket is the vector field whose value is the difference between the values of XY and YX. { 'lē ,brak·ət }

Lie group A topological group which is also a differentiable manifold in such a way that the group operations are themselves analytic functions. { 'lē ,grüp }

lifting 1. Given a fiber bundle $(\bar{X},B,p)$ and a continuous map of a topological space $\bar{Y}$ to B, $g:\bar{Y} \to B$, lifting entails finding a continuous map $\bar{g}:\bar{Y} \to \bar{X}$ such that the function g is the composition $p - \bar{g}$. **2.** *See* translation. { 'lift·iŋ }

likelihood The likelihood of a sample of independent values of $x_1, x_2, \ldots, x_n$, with $f(x)$ the probability function, is the product $f(x_1) - f(x_2) - \cdots - f(x_n)$. { 'līk·lē,hùd }

likelihood ratio The probability of a random drawing of a specified sample from a population, assuming a given hypothesis about the parameters of the population, divided by the probability of a random drawing of the same sample, assuming that the parameters of the population are such that this probability is maximized. { 'līk·lē,hu̇d ,rā·shō }

likelihood ratio test A procedure used in hypothesis testing based on the ratio of the values of two likelihood functions, one derived from the hypothesis being tested and one without the constraints of the hypothesis under test. { ¦līk·lē,hu̇d 'rā·shō ,test }

like terms *See* similar terms. { ¦līk ¦tərmz }

limaçon The locus of points of the plane which in polar coordinates (r,θ) satisfy the equation $r = a \cos \theta + b$. Also known as Pascal's limacon. { 'lim·ə,sän *or* ,lim·ə'sōn }

limit 1. A function $f(x)$ has limit L as x tends to c if given any positive number ϵ (no matter how small) there is a positive number δ such that if x is in the domain of f, x is not c, and $|x - c| < \delta$, then $|f(x) - L| < \epsilon$; written

$$\lim_{x \to c} f(x) = L$$

2. A sequence $\{a_n : n = 1, 2, \ldots\}$ has limit L if given a positive number ϵ (no matter how small), there is a positive integer N such that for all integers n greater than N, $|a_n - L| < \epsilon$. { 'lim·ət }

limit cycle For a differential equation, a closed trajectory C in the plane (corresponding to a periodic solution of the equation) where every point of C has a neighborhood so that every trajectory through it spirals toward C. { 'lim·ət ,sīk·əl }

limit inferior Also known as lower limit. **1.** The limit inferior of a sequence whose nth term is a_n is the limit as N approaches infinity of the greatest lower bound of the terms a_n for which n is greater than N; denoted by

$$\liminf_{n \to \infty} a_n \text{ or } \underline{\lim_{n \to \infty}} a_n$$

2. The limit inferior of a function f at a point c is the limit as ϵ approaches zero of the greatest lower bound of $f(x)$ for $|x - c| < \epsilon$ and $x \neq c$; denoted by

$$\liminf_{x \to c} f(x) \text{ or } \underline{\lim_{n \to c}} f(x)$$

3. For a sequence of sets, the set consisting of all elements that belong to all but a finite number of the sets in the sequence. Also known as restricted limit. { 'lim·ət in¦fir·ē·ər }

limit on the left The limit on the left of the function f at a point c is the limit of f at c which would be obtained if only values of x less than c were taken into account; more precisely, it is the number L which has the property that for any positive number ϵ there is a positive number δ so that if x is the domain of f and $0 < (c - x) < \delta$, then $|f(x) - L| < \epsilon$; denoted by

$$\lim_{x \to c^-} f(x) = L \text{ or } f(c^-) = L$$

Also known as left-hand limit. { 'lim·ət ȯn thə 'left }

limit on the right The limit on the right of the function $f(x)$ at a point c is the limit of f at c which would be obtained if only values of x greater than c were taken into account; more precisely, it is the number L which has the property that for any positive number ϵ there is a positive number δ so that if x is in the domain of f and $0 < (x - c) < \delta$, then $|f(x) - L| < \epsilon$; denoted by

$$\lim_{x \to c^+} f(x) = L \text{ or } f(c^+) = L$$

Also known as right-hand limit. { 'lim·ət ȯn thə 'rīt }

limit point *See* cluster point. { 'lim·ət ,pȯint }

limits of integration The end points of the interval over which a function is being integrated. { 'lim·əts əv ,int·ə'grā·shən }

limit superior Also known as upper limit. **1.** The limit superior of a sequence whose

nth term is a_n is the limit as N approaches infinity of the least upper bound of the terms a_n for which n is greater than N; denoted by

$$\limsup_{n \to \infty} a_n \text{ or } \overline{\lim_{n \to \infty}} a_n$$

2. The limit superior of a function f at a point c is the limit as ϵ approaches zero of the least upper bound of $f(x)$ for $|x - c|$ ϵ and $x \neq c$; denoted by

$$\limsup_{x \to c} f(x) \text{ or } \overline{\lim_{x \to c}} f(x)$$

3. For a sequence of sets, the set consisting of all elements that belong to infinitely many of the sets in the sequence. Also known as complete limit. { 'lim·ət sə'pir·ē·ər }

Lindelöf space A topological space where if a family of open sets covers the space, then a countable number of these sets also covers the space. { 'lin·də,löf ,späs }

Lindelöf theorem The proposition that there is a countable subcover of each open cover of a subset of a space whose topology has a countable base. { 'lin·də,lef ,thir·əm }

line The set of points $(x_1, \ldots, x_n)$ in Euclidean space, each of whose coordinates is a linear function of a single parameter t; $x_i = f_i(t)$. Also known as straight line. { līn }

linear algebra The study of vector spaces and linear transformations. { 'lin·ē·ər 'al·jə·brə }

linear algebraic equation An equation in some algebraic system where the unknowns occur linearly, that is, to the first power. { 'lin·ē·ər ,al·jə,brā·ik i'kwā·zhən }

linear combination A linear combination of vectors $v_1, \ldots, v_n$ in a vector space is any expression of the form $a_1 v_1 + a_2 v_2 + \cdots + a_n v_n$, where the a_i are scalars. { 'lin·ē·ər ,käm·bə'nā·shən }

linear congruence The relation between two quantities that have the same remainder on division by a given integer, where the quantities are polynomials of, at most, the first degree in the variables involved. { 'lin·ē·ər kəŋ'grü·əns }

linear dependence The property of a set of vectors $v, \ldots, v_n$ in a vector space for which there exists a linear combination such that $a_1 v_1 + \cdots + a_n v_n = 0$, and at least one of the scalars a_i is not zero. { 'lin·ē·ər di'pen·dəns }

linear differential equation A differential equation in which all derivatives occur linearly, and all coefficients are functions of the independent variable. { 'lin·ē·ər ,dif·ə,ren·chəl i'kwā·zhən }

linear discriminant function A function, used in conjunction with a set of threshold values in a classification procedure, whose values are linear combinations of the values of selected variables. { ,lin·ē·ər di,skrim·ə·nənt ,fəŋk·shən }

linear element On a surface determined by equations $x = f(u,v)$, $y = g(u,v)$, and $z = h(u,v)$, the element of length ds given by $ds^2 = E\,du^2 + 2F\,du\,dv + G\,dv^2$, where E, F, and G are functions of u and v. { 'lin·ē·ər 'el·ə·mənt }

linear equation A linear equation in the variables $x_1, \ldots, x_n$, and y is any equation of the form $a_1 x_1 + a_2 x_2 + \cdots + a_n x_n = y$. { 'lin·ē·ər i'kwā·zhən }

linear form A homogeneous polynomial of the first degree. { 'lin·ē·ər 'fȯrm }

linear fractional transformations See Möbius transformations. { 'lin·ē·ər ,frak·shən·əl ,tranz·fər'mā·shənz }

linear function See linear transformation. { 'lin·ē·ər 'fəŋk·shən }

linear functional A linear transformation from a vector space to its scalar field. { 'lin·ē·ər 'fəŋk·shən·əl }

linear hypothesis See linear model. { 'lin·ē·ər hī'päth·ə·səs }

linear independence The property of a set of vectors $v_1, \ldots, v_n$ in a vector space where if $a_1 v_1 + a_2 v_2 + \cdots + a_n v_n = 0$, then all the scalars $a_i = 0$. { 'lin·ē·ər ,in·də'pen·dəns }

linear inequalities A collection of relations among variables x_i, where at least one relation has the form $\Sigma_i a_i x_i \geq 0$. { 'lin·ē·ər ,in·i'kwäl·əd·ēz }

linear interpolation A process to find a value of a function between two known values

linearity

under the assumption that the three plotted points lie on a straight line. { 'lin·ē· ər in₁tər·pə'lā·shən }

linearity The property whereby a mathematical system is well behaved (in the context of the given system) with regard to addition and scalar multiplication. { ₁lin·ē'ar· əd·ē }

linearly dependent quantities Quantities that satisfy a homogeneous linear equation in which at least one of the coefficients is not zero. { 'lin·ē·ər·lē di¦pen·dənt 'kwän·tə₁tēz }

linearly disjoint extensions Two extension fields E and F of a field k contained in a common field L, such that any finite set of elements in E that is linearly independent when E is regarded as a vector space over k remains linearly independent when E is regarded as a vector space over F. { ¦lin·ē·ər·lē ¦dis¦jóint ik'sten·chənz }

linearly independent quantities Quantities which do not jointly satisfy a homogeneous linear equation unless all coefficients are zero. { 'lin·ē·ər·lē ₁in·də¦pen·dənt 'kwän· əd·ēz }

linearly ordered set A set with an ordering ≤ such that for any two elements a and b either $a ≤ b$ or $b ≤ a$. Also known as chain; completely ordered set; serially ordered set; simply ordered set; totally ordered set. { 'lin·ē·ər·lē ¦òr·dərd 'set }

linear manifold A subset of a vector space which is itself a vector space with the induced operations of addition and scalar multiplication. { 'lin·ē·ər 'man·ə₁fōld }

linear model A mathematical model in which linear equations connect the random variables and the parameters. Also known as linear hypothesis. { 'lin·ē·ər 'mäd·əl }

linear operator *See* linear transformation. { 'lin·ē·ər 'äp·ə₁rād·ər }

linear order Any order < on a set S with the property that for any two elements a and b in S exactly one of the statements $a < b$, $a = b$, or $b < a$ is true. Also known as complete order; serial order; simple order; total order. { 'lin·ē·ər 'òr· dər }

linear programming The study of maximizing or minimizing a linear function $f(x_1, ..., x_n)$ subject to given constraints which are linear inequalities involving the variables x_i. { 'lin·ē·ər 'prō₁gram·iŋ }

linear regression The straight line running among the points of a scatter diagram about which the amount of scatter is smallest, as defined, for example, by the least squares method. { 'lin·ē·ər ri'gresh·ən }

linear scale *See* uniform scale. { ¦lin·ē·ər 'skāl }

linear space *See* vector space. { 'lin·ē·ər 'spās }

linear span { ₁lin·ē·ər 'span }

linear system A system where all the interrelationships among the quantities involved are expressed by linear equations which may be algebraic, differential, or integral. { 'lin·ē·ər 'sis·təm }

linear topological space *See* topological vector space. { ¦lin·ē·ər ₁täp·ə¦läj·ə·kəl 'spās }

linear transformation A function T defined in a vector space E and having its values in another vector space over the same field, such that if f and g are vectors in E, and c is a scalar, then $T(f + g) = Tf + Tg$ and $T(cf) = c(Tf)$. Also known as homogeneous transformation; linear function; linear operator. { 'lin·ē·ər ₁tranz· fər'mā·shən }

linear trend A first step in analyzing a time series, to determine whether a linear relationship provides a good approximation to the long-term movement of the series; computed by the method of semiaverages or by the method of least squares. { ¦lin·ē·ər 'trend }

line at infinity *See* ideal line. { 'līn at in'fin·əd·ē }

line graph A graph in which successive points representing the value of a variable at selected values of the independent variable are connected by straight lines. { 'līn ₁graf }

line integral 1. For a curve in a vector space defined by $\mathbf{x} = \mathbf{x}(t)$, and a vector function $\mathbf{V}$ defined on this curve, the line integral of $\mathbf{V}$ along the curve is the integral over t of the scalar product of $\mathbf{V}[\mathbf{x}(t)]$ and $d\mathbf{x}/dt$; this is written $\int \mathbf{V} \cdot d\mathbf{x}$. **2.** For a curve which is defined by $x = x(t)$, $y = y(t)$, and a scalar function f depending on x and y, the line integral of f along the curve is the integral over t of $f[x(t),y(t)] \cdot \sqrt{(dx/dt)^2 + (dy/dt)^2}$; this is written $\int f \, ds$, where $ds = \sqrt{(dx)^2 + (dy)^2}$ is an infinitesimal element of length along the curve. **3.** For a curve in the complex plane defined by $z = z(t)$, and a function f depending on z, the line integral of f along the curve is the integral over t of $f[z(t)]$ (dz/dt); this is written $\int f \, dz$. { 'līn ¦int· ə·grəl }

line of curvature A curve on a surface whose tangent lies along a principal direction at each point. { 'līn əv 'kər·və·chər }

line of striction The locus of the central points of the rulings of a given ruled surface. { 'līn əv 'strik·shən }

line of support Relative to a convex region in a plane, a line that contains at least one point of the region but is such that a half-plane on one side of the line contains no points of the region. { 'līn əv sə'pȯrt }

line segment A connected piece of a line. { 'līn ¦seg·mənt }

link relatives method A method for computing indexes by dividing the value of a magnitude in one period by the value in the previous period. { ¦liŋk 'rel·ə·tivz ¦meth·əd }

Liouville function A function $\lambda(n)$ on the positive integers such that $\lambda(1) = 1$, and for $n \geq 2$, $\lambda(n)$ is -1 raised to the number of prime factors of n, with repeated factors counted the number of times they appear. { 'lyü‚vēl ‚fəŋk·shən }

Liouville-Neumann series An infinite series of functions constructed from the given functions in the Fredholm equation which under certain conditions provides a solution. Also known as Neumann series. { 'lyü‚vēl 'nȯi‚män ‚sir·ēz }

Liouville number An irrational number x such that for any integer n there exist integers p and q, with q greater than 1, for which the absolute value of $x - (p/q)$ is less than $1/q^n$. { 'lyü‚vēl ‚nəm·bər }

Liouville's theorem Every function of a complex variable which is bounded and analytic in the entire complex plane must be constant. { 'lyü‚vēlz ‚thir·əm }

Lipschitz condition A function f satisfies such a condition at a point b if $|f(x) - f(b)| \leq K|x - b|$, with K a constant, for all x in some neighborhood of b. { 'lip‚shits kən‚dish·ən }

Lipschitz mapping A function f from a metric space to itself for which there is a positive constant K such that, for any two elements in the space, a and b, the distance between $f(a)$ and $f(b)$ is less than or equal to K times the distance between a and b. { 'lip‚shits ‚map·iŋ }

literal constant A letter denoting a constant. { 'lid·ə·rəl 'kän·stənt }

literal expression An expression or equation in which the constants are represented by letters. { 'lid·ə·rəl ik'spresh·ən }

literal notation The use of letters to denote numbers, known or unknown. { 'lid·ə· rəl nō'tā·shən }

Littlewood conjecture The statement that there exists a number C such that, whenever $n_1, n_2, \ldots, n_N$ are N distinct integers, the integral over x from $-\pi$ to π of the absolute value of the sum from $k = 1$ to $k = N$ of the exponential functions of $in_k x$ is greater than $2\pi \, C \log N$. { 'lid·əl‚wu̇d kən‚jek·chər }

lituus The trumpet-shaped plane curve whose points in polar coordinates (r, θ) satisfy the equation $r^2 = a/\theta$. { 'lich·ə·wəs }

Lobachevski geometry A system of planar geometry in which the euclidean parallel postulate fails; any point p not on a line L has at least two lines through it parallel to L. Also known as Bolyai geometry; hyperbolic geometry. { lō·bə'chef·skē jē'äm·ə·trē }

local algebra An algebra A over a field F which is the sum of the radical of A and the subalgebra consisting of products of elements of F with the multiplicative identity of A. { ¦lō·kəl 'al·jə·brə }

local base For a point x in a topological space, a family of neighborhoods of x such that every neighborhood of x contains a member of the family. Also known as base for the neighborhood system. { ¦lō·kəl 'bās }

local coefficient

local coefficient By using fiber bundles where the fiber is a group, one may generalize cohomology theory for spaces; one uses such bundles as the algebraic base for such a theory and calls the bundle a system of local coefficients. { 'lō·kəl ˌkō·i'fish·ənt }

local coordinate system The coordinate system about a point which is induced when the global space is locally Euclidean. { 'lō·kəl kō'órd·ən·ət ˌsis·təm }

local distortion The absolute value of the derivative of an analytic function at a given point. { 'lō·kəl di'stór·shən }

localization principle The principle that the convergence of the Fourier series of a function at a point depends only on the behavior of the function in some arbitrarily small neighborhood of that point. { ˌlō·kə·lə'zā·shən ˌprin·sə·pəl }

locally arcwise connected topological space A topological space in which every point has an arcwise connected neighborhood, that is, an open set any two points of which can be joined by an arc. { 'lō·kə·lē 'ärk,wīz kə,nek·təd ¦täp·ə¦läj·ə·kəl ¦spās }

locally compact topological space A topological space in which every point lies in a compact neighborhood. { 'lō·kə·lē kəm'pakt ¦täp·ə¦läj·ə·kəl ¦spās }

locally connected topological space A topological space in which every point has a connected neighborhood. { 'lō·kə·lē kə'nek·təd ¦täp·ə¦läj·ə·kəl ¦spās }

locally convex space A Hausdorff topological vector space E such that every neighborhood of any point x belonging to E contains a convex neighborhood of x. { 'lō·kə·lē 'kän,veks ˌspās }

locally Euclidean topological space A topological space in which every point has a neighborhood which is homeomorphic to a Euclidean space. { 'lō·kə·lē yü'klid·ē·ən ¦täp·ə¦läj·ə·kəl ¦spās }

locally finite family of sets A family of subsets of a topological space such that each point of the topological space has a neighborhood that intersects only a finite number of these subsets. { ¦lō·kə·lē ¦fī,nīt 'fam·lē əv 'sets }

locally integrable function A function is said to be locally integrable on an open set S in n-dimensional Euclidean space if it is defined almost everywhere in S and has a finite integral on compact subsets S. { ¦lō·kə·lē ¦int·ə·grə·bəl 'fəŋk·shən }

locally one to one A function is locally one to one if it is one to one in some neighborhood of each point. { 'lō·kə·lē 'wən tə 'wən }

locally trivial bundle A bundle for which each point in the base has a neighborhood U whose inverse image under the projection map is isomorphic to a Cartesian product of U with a space isomorphic to the fibers of the bundle. { ¦lō·kə·lē ¦triv·ē·əl 'bən·dəl }

local maximum A local maximum of a function f is a value $f(c)$ of f where $f(x) \leq f(c)$ for all x in some neighborhood of c; if $f(c)$ is a local maximum, f is said to have a local maximum at c. { 'lō·kəl 'mak·sə·məm }

local minimum A local minimum of a function f is a value $f(c)$ of f where $f(x) \geq f(c)$ for all x in some neighborhood of c; if $f(c)$ is a local minimum, f is said to have a local minimum at c. { 'lō·kəl 'min·ə·məm }

local property A property of an object (such as a space, function, curve, or surface) whose specification is based on the behavior of the object in the neighborhoods of certain points. { 'lō·kəl ˌpräp·ərd·ē }

local quasi-F martingale A stochastic process $\{X_t\}$ such that the process obtained from $\{X_t\}$ by stopping it when it reaches n or $-n$ is a quasi-F martingale for each integer n. { 'lō·kəl 'kwä·zē¦ef 'mart·ən,gäl }

local ring A ring with only one maximal ideal. { 'lō·kəl 'riŋ }

local solution A function which solves a system of equations only in a neighborhood of some point. { 'lō·kəl sə'lü·shən }

located vector An ordered pair of points in n-dimensional Euclidean space. { ¦lō,kād·əd 'vek·tər }

location principle A principle useful in locating the roots of an equation stating that if a continuous function has opposite signs for two values of the independent variable, then it is zero for some value of the variable between these two values. { lō'kā·shən ˌprin·sə·pəl }

locus A collection of points in a Euclidean space whose coordinates satisfy one or more algebraic conditions. { 'lō·kəs }

logarithm 1. The real-valued function log u defined by log $u = v$ if $e^v = u$, e^v denoting the exponential function. Also known as hyperbolic logarithm; Naperian logarithm; natural logarithm. **2.** An analog in complex variables relative to the function e^z. { 'läg·ə,ri<u>th</u>·əm }

logarithmically convex function A function whose logarithm is a convex function. { ,läg·ə¦ri<u>th</u>·mik·lē ¦kän¦veks ,fəŋk·shən }

logarithmic coordinate paper Paper ruled with two sets of mutually perpendicular, parallel lines spaced according to the logarithms of consecutive numbers, rather than the numbers themselves. { 'läg·ə,ri<u>th</u>·mik kō'ȯrd·ən·ət ,pā·pər }

logarithmic coordinates In the plane, logarithmic coordinates are defined by two coordinate axes, each marked with a scale where the distance between two points is the difference of the logarithms of the two numbers. { 'läg·ə,ri<u>th</u>·mik kō'ȯrd·ən·əts }

logarithmic curve A curve whose equation in Cartesian coordinates is $y = \log ax$, where a is greater than 1. { 'läg·ə,ri<u>th</u>·mik 'kərv }

logarithmic derivative The logarithmic derivative of a function $f(z)$ of a real (complex) variable is the ratio $f'(z)/f(z)$, that is, the derivative of $\log f(z)$. { 'läg·ə,ri<u>th</u>·mik də'riv·əd·iv }

logarithmic differentiation A technique often helpful in computing the derivatives of a differentiable function $f(x)$; set $g(x) = \log f(x)$ where $f(x) \neq 0$, then $g'(x) = f'(x)/f(x)$, and if there is some other way to find $g'(x)$, then one also finds $f'(x)$. { 'läg·ə,ri<u>th</u>·mik ,dif·ə,ren·chē'ā·shən }

logarithmic distribution A frequency distribution whose value at any integer $n = 1, 2, \ldots$ is $\lambda^n/(-n) \log (1 - \lambda)$, where λ is fixed. { 'läg·ə,ri<u>th</u>·mik ,dis·trə'byü·shən }

logarithmic equation An equation which involves a logarithmic function of some variable. { 'läg·ə,ri<u>th</u>·mik i'kwā·zhən }

logarithmic scale A scale in which the distances that numbers are at from a reference point are proportional to their logarithms. { 'läg·ə,ri<u>th</u>·mik 'skāl }

logarithmic series The expansion of the natural logarithm of $1 + x$ in a Maclaurin series; namely, $x - x^2/2 + x^3/3 - \cdots$. { ¦läg·ə,ə,ri<u>th</u>·mik 'sir·ēz }

logarithmic spiral The spiral plane curve whose points in polar coordinates (r, θ) satisfy the equation log $r = a\theta$. Also known as equiangular spiral. { 'läg·ə,ri<u>th</u>·mik 'spī·rəl }

logarithmic transformation The replacement of a variate y with a new variate $z = \log y$ or $z = \log (y + c)$, where c is a constant; this operation is often performed when the resulting distribution is normal, or if the resulting relationship with another variable is linear. { 'läg·ə,ri<u>th</u>·mik ,tranz·fər'mā·shən }

logarithmic trigonometric function The logarithm of the corresponding trigonometric function. { 'läg·ə,ri<u>th</u>·mik 'trig·ə·nə,me·trik ,fəŋk·shən }

logic The subject that investigates, formulates, and establishes principles of valid reasoning. { 'läj·ik }

logical addition The additive binary operation of a Boolean algebra. { 'läj·ə·kəl ə'dish·ən }

logical connectives Symbols which link mathematical statements; these symbols represent the terms "and," "or," "implication," and "negation." { 'läj·ə·kəl kə'nek·tivz }

logical function See propositional function.. { 'läj·ə·kəl 'fəŋk·shən }

logically equivalent statements Two statements that are equivalent because of their logical form rather than their mathematical content. { ,läj·i·klē i,kwiv·ə·lənt 'stāt·məns }

logical multiplication The multiplicative binary operation of a Boolean algebra. { 'läj·ə·kəl məl·tə·plə'kā·shən }

logistic curve 1. A type of growth curve, representing the size of a population y as a function of time t: $y = k/(1 + e^{-kbt})$, where k and b are positive constants. Also known as Pearl-Reed curve. **2.** More generally, a curve representing a function of the form $y = k/(1 + e^{-kbt})$, where c is a constant and $f(t)$ is some function of time. { lə'jis·tik 'kərv }

lognormal distribution A probability distribution in which the logarithm of the parameter has a normal distribution. { 'läg,nȯr·məl ,di·strə'byü·shən }

long division 1. Division of numbers in which the divisor contains more than one digit. **2.** Division of algebraic quantities in which the divisor contains more than one term. { ,lȯŋ di'vizh·ən }

long radius The distance from the center of a regular polygon to a vertex. { 'lȯŋ ,rād·ē·əs }

long run frequency The ratio of the number of occurrences of an event in a large number of trials to the number of trials. { 'lȯŋ ,rən 'frē·kwən·sē }

long-time trend *See* secular trend. { ¦lȯŋ ,tīm 'trend }

lookahead tree *See* game tree. { 'lùk·ə,hed ,trē }

loop A line which begins and ends at the same point of the graph. { lüp }

Lorentz group The group of all Lorentz transformations of euclidean four-space with composition as the operation. { 'lȯr,ens ,grüp }

Lorentz transformation Any linear transformation of Euclidean four space which preserves the quadratic form $q(x,y,z,t) = t^2 - x^2 - y^2 - z^2$. { 'lȯr,ens ,tranz·fər,mā·shən }

Lorenz curve A graph for showing the concentration of ownership of economic quantities such as wealth and income; it is formed by plotting the cumulative distribution of the amount of the variable concerned against the cumulative frequency distribution of the individuals possessing the amount. { 'lȯr,ens ,kərv }

loss function In decision theory, the function, dependent upon the decision and the true underlying distributions, which expresses the loss produced in taking the decision. { 'lȯs ,fəŋk·shən }

lower bound 1. A lower bound of a subset A of a set S is a point of S which is smaller than every element of A. **2.** A lower bound on a function f with values in a partially ordered set S is an element of S which is smaller than every element in the range of f. { 'lō·ər ¦baùnd }

lower limit *See* limit inferior. { ¦lō·ər 'lim·ət }

lower semicontinuous function A real-valued function $f(x)$ is lower semicontinuous at a point x_0 if, for any small positive number ϵ, $f(x)$ is always greater that $f(x_0) - \epsilon$ for all x in some neighborhood of x_0. { ¦lō·ər ,sem·ē·kən'tin·yə·wəs ,fəŋk·shən }

loxodromic spiral A curve on a surface of revolution which cuts the meridians at a constant angle other than 90°. { ¦lak·sə¦dräm·ik 'spī·rəl }

lub *See* least upper bound.

Lucas numbers The terms of the Fibonacci sequence whose first two terms are 1 and 3. { 'lü·kəs ,nəm·bərz }

lune A section of a plane bounded by two circular arcs, or of a sphere bounded by two great circles. { lün }

lune of Hippocrates 1. A section of the plane, bounded by two circular arcs, whose area equals that of a polygon used in constructing the circles. **2.** One of a small finite number of sections of the plane, each bounded by two circular arcs, such that the sum of their areas equals that of a polygon used in constructing the circles. { ¦lün əv hip'äk·rə,tēz }

Luzin's theorem Given a measurable function f which is finite almost everywhere in a euclidean space, then for every number $\epsilon > 0$ there is a continuous function g which agrees with f, except on a set of measure less than ϵ. Also spelled Lusin's theorem. { ,lü·zēnz ,thir·əm }

Lyapunov function A function of a vector and of time which is positive-definite and has a negative-definite derivative with respect to time for nonzero vectors, is identically zero for the zero vector, and approaches infinity as the norm of the vector approaches infinity; used in determining the stability of control systems. Also spelled Liapunov function. { lē'ap·ə,nȯf ,fəŋk·shən }

M

m *See* milli-.

MacDonald functions *See* modified Hankel functions. { mək'dän·əld ˌfəŋk·shənz }

Machin's formula The formula $\pi/4 = 4$ arc tan $(1/5)$ − arc tan $(1/239)$, which has been used to compute the value of π. { 'mā·chənz ˌför·myə·lə }

Maclaurin-Cauchy test *See* Cauchy's test for convergence. { mə'klör·ən kō'shē ˌtest }

Maclaurin expansion The power series representation of a function arising from Maclaurin's theorem. { mə'klör·ən ik,span·chən }

Maclaurin series The power series in the Maclaurin expansion. { mə'klör·ən ˌsir·ēz }

Maclaurin's theorem The theorem giving conditions when a function, which is infinitely differentiable, may be represented in a neighborhood of the origin as an infinite series with nth term $(1/n!) \cdot f^{(n)}(0) \cdot x^n$, where $f^{(n)}$ denotes the nth derivative. { mə'klör·ənz ˌthir·əm }

magic square **1.** A square array of integers where the sum of the entries of each row, each column, and each diagonal is the same. **2.** A square array of integers where the sum of the entries in each row and each column (but not necessarily each diagonal) is the same. Also known as semimagic square. { 'maj·ik 'skwer }

magnitude *See* absolute value. { 'mag·nəˌtüd }

main diagonal *See* principal diagonal. { ¦mān dī'ag·ən·əl }

main effect The effect of the change in level of one factor in a factorial experiment measured independently of other variables. { ¦mān i'fekt }

major arc The longer of the two arcs produced by a secant of a circle. { 'mā·jər 'ärk }

major axis The longer of the two axes with respect to which an ellipse is symmetric. { 'mā·jər 'ak·səs }

majority A logic operator having the property that if P, Q, R are statements, then the function (P, Q, R, . . .) is true if more than half the statements are true, or false if half or less are true. { mə'jär·əd·ē }

Mandelbrot dimensionality *See* fractal dimensionality. { ¦män·dəlˌbröt di,men·shə'nal·əd·ē }

Mandelbrot set The set of complex numbers, c, for which the sequence $s_0, s_1, \ldots$ is bounded, where $s_0 = 0$, and $s_{n+1} = s_n{}^2 + c$. { ¦män·dəlˌbröt ˌset }

manifold A topological space which is locally Euclidean; there are four types: topological, piecewise linear, differentiable, and complex, depending on whether the local coordinate systems are obtained from continuous, piecewise linear, differentiable, or complex analytic functions of those in Euclidean space; intuitively, a surface. { 'man·əˌföld }

Mann-Whitney test A procedure used in nonparametric statistics to determine whether the means of two populations are equal. { ¦man 'wit·nē ˌtest }

mantissa The positive decimal part of a common logarithm. { man'tis·ə }

map *See* mapping. { map }

mapping **1.** Any function or multiple-valued relation. Also known as map. **2.** In topology, a continuous function. { 'map·iŋ }

marginal probability Probability expressed by the two conditional probability distributions which arise from the joint distribution of two random variables. { 'mär·jən·əl ˌpräb·ə'bil·əd·ē }

mark The name or value given to a class interval; frequently, the value of the midpoint or the integer nearest the midpoint. { märk }

Markov chain A Markov process whose state space is finite or countably infinite. { 'mar₁kȯf ₁chān }

Markov inequality If x is a random variable with probability P and expectation E, then, for any positive number a and positive integer n, $P(|x| \geq a) \leq E(|x|^n/a^n)$. { 'mar₁kȯf ₁in·i'kwäl·əd·ē }

Markov process A stochastic process which assumes that in a series of random events the probability of an occurrence of each event depends only on the immediately preceding outcome. { 'mär₁kȯf prä·səs }

marriage theorem The proposition that a family of n subsets of a set S with n elements is a system of distinct representatives for S if any k of the subsets, $k = 1, 2, \ldots$, n, together contain at least k distinct elements. Also known as Hall's theorem. { 'mar·ij ₁thir·əm }

martingale A sequence of random variables $x_1, x_2, \ldots$, where the conditional expected value of x_{n+1} given $x_1, x_2, \ldots, x_n$ equals x_n. { 'märt·ən₁gāl }

Mascheroni's constant See Euler's constant. { ₁mäsk·ə'rō₁nēz 'kän·stənt }

match See biconditional operation. { mach }

matched groups Groups of individuals or objects chosen so that the mean values (or some other characteristic) of some variable are the same for all the groups, in order to minimize the variation due to this variable. { ¦macht 'grüps }

matched pairs The design of an experiment for paired comparison in which the assignment of subjects to treatment or control is not completely at random, but the randomization is restricted to occur separately within each pair. { 'macht 'perz }

matching A set of edges in a graph, no two of which have a vertex in common. Also known as independent edge set. { 'mach·iŋ }

matching distribution The distribution of number of matches obtained if N tickets labeled 1 to N are drawn at random one at a time and laid in a row, and a match is counted when a ticket's label matches its position. { 'mach·iŋ ₁di·strə'byü·shən }

material implication See implication. { mə¦tir·ē·əl ₁im·plə'kā·shən }

mathematical analysis See analysis. { ¦math·ə¦mad·ə·kəl ə'nal·ə·səs }

mathematical induction A general method of proving statements concerning a positive integral variable: if a statement is proven true for $x = 1$, and if it is proven that, if the statement is true for $x = 1, \ldots, n$, then it is true for $x = n + 1$, it follows that the statement is true for any integer. Also known as complete induction; method of infinite descent; proof by descent. { ¦math·ə¦mad·ə·kəl in'dək·shən }

mathematical logic The study of mathematical theories from the viewpoint of model theory, recursive function theory, proof theory, and set theory. { ¦math·ə¦mad·ə·kəl 'läj·ik }

mathematical model 1. A mathematical representation of a process, device, or concept by means of a number of variables which are defined to represent the inputs, outputs, and internal states of the device or process, and a set of equations and inequalities describing the interaction of these variables. **2.** A mathematical theory or system together with its axioms. { ¦math·ə¦mad·ə·kəl 'mäd·əl }

mathematical probability The ratio of the number of mutually exclusive, equally likely outcomes of interest to the total number of such outcomes when the total is exhaustive. Also known as a priori probability. { ¦math·ə¦mad·ə·kəl ₁präb·ə'bil·əd·ē }

mathematical programming See optimization theory. { ¦math·ə¦mad·ə·kəl 'prō₁gram·iŋ }

mathematical system A structure formed from one or more sets of undefined objects, various concepts which may or may not be defined, and a set of axioms relating these objects and concepts. { ¦math·ə¦mad·ə·kəl 'sis·təm }

mathematical table A listing of the values of a function of one or several variables at a series of values of the arguments, usually equally spaced. { ¦math·ə¦mad·ə·kəl 'tā·bəl }

Mathieu equation A differential equation of the form $y'' + (a + b \cos 2x)y = 0$, whose solution depends on periodic functions. { ma'tyü i‚kwā·zhən }

Mathieu functions Any solution of the Mathieu equation which is periodic and an even or odd function. { ma'tyü ‚fəŋk·shənz }

matrix A rectangular array of numbers or scalars from a vector space. { 'mā·triks }

matrix algebra An algebra whose elements are matrices and whose operations are addition and multiplication of matrices. { 'mā·triks 'al·jə·brə }

matrix calculus The treatment of matrices whose entries are functions as functions in their own right with a corresponding theory of differentiation; this has application to the study of multidimensional derivatives of functions of several variables. { 'mā·triks 'kal·kyə·ləs }

matrix element One of the set of numbers which form a matrix. { 'mā·triks ‚el·ə·mənt }

matrix game A game involving two persons, which gives rise to a matrix representing the amount received by the two players. Also known as rectangular game. { 'mā·triks ‚gām }

matrix of a linear transformation A unique matrix A, such that for a specified linear transformation L from one vector space to another, and for specified finite bases in each space, L applied to a vector is equal to A times that vector. { 'mā·triks əv ə 'lin·ē·ər ‚tranz·fər'mā·shən }

matrix theory The algebraic study of matrices and their use in evaluating linear processes. { 'mā·triks ‚thē·ə·rē }

max See maximum. { maks }

max-flow min-cut theorem See Ford-Fulkerson theorem. { ‚maks¦flō ‚min'kət ‚thir·əm }

maximal chain A sequence of $n + 1$ subsets of a set of n elements, such that the first member of the sequence is the empty set and each member of the sequence is a proper subset of the next one. { ¦mak·sə·məl 'chān }

maximal element See maximal member. { 'mak·sə·məl 'el·ə·mənt }

maximal ideal An ideal I in a ring R which is not equal to R, and such that there is no ideal containing I and not equal to I or R. { ¦mak·sə·məl ī'dēl }

maximal independent set An independent set of vertices of a graph which is not a proper subset of another independent set. { ¦mak·sə·məl ‚in·də‚pen·dənt 'set }

maximal member In a partially ordered set a maximal member is one for which no other element follows it in the ordering. Also known as maximal element. { 'mak·sə·məl 'mem·bər }

maximal planar graph A planar graph to which no new arcs can be added without forcing crossings and hence violating planarity. { 'mak·sə·məl ¦plān·ər ‚graf }

maximax criterion In decision theory, one of several possible prescriptions for making a decision under conditions of uncertainty; it prescribes the strategy which will maximize the maximum possible profit. { 'mak·sə‚maks krī‚tir·ē·ən }

maxim criterion One of several prescriptions for making a decision under conditions of uncertainty; it prescribes the strategy which will maximize the minimum profit. Also known as maximin criterion. { 'mak·səm krī‚tir·ē·ən }

maximin 1. The maximum of a set of minima. **2.** In the theory of games, the largest of a set of minimum possible gains, each representing the least advantageous outcome of a particular strategy. { 'mak·sə‚min }

maximin criterion See maxim criterion. { 'mak·sə‚min krī‚tir·ē·ən }

maximizing a function Finding the largest value assumed by a function. { 'mak·sə‚mīz·iŋ ə 'fəŋk·shən }

maximum The maximum of a real-valued function is the greatest value it assumes. Abbreviated max. { 'mak·sə·məm }

maximum cardinality matching See maximum matching. { ‚mak·sə·məm ‚kärd·ən'al·əd·ē ‚mach·iŋ }

maximum flow problem The problem of finding a feasible flow in an s-t network with the largest possible flow value for a given weight function. { ¦mak·sə·məm 'flō ‚präb·ləm }

maximum independent set An incident set of vertices of a graph such that there is no other independent set with more vertices. { ¦mak·sə·məm ‚in·də‚pen·dənt 'set }

maximum likelihood method A technique in statistics where the likelihood distribution is so maximized as to produce an estimate to the random variables involved. { 'mak·sə·məm 'līk·lē‚hud ‚meth·əd }

maximum matching A matching of edges in a graph such that no other matching has a greater number of edges. Also known as maximum cardinality matching. { ¦mak·sə·məm 'mach·iŋ }

maximum-minimum principle See min-max theorem. { 'mak·sə·məm 'min·ə·məm ‚prin·sə·pəl }

maximum-modulus theorem For a complex analytic function in a closed bounded simply connected region its modulus assumes its maximum value on the boundary of the region. { 'mak·sə·məm 'mäj·ə·ləs ‚thir·əm }

maximum-value theorem The theorem that there is a point in the domain of a real-valued function at which the function has its greatest value if this domain is compact. { ¦mak·sə·məm 'val·yü ‚thir·əm }

meager set A set that is a countable union of nowhere-dense sets. Also known as set of first category. { ¦mē·gər 'set }

mean A single number that typifies a set of numbers, such as the arithmetic mean, the geometric mean, or the expected value. Also known as mean value. { mēn }

mean curvature Half the sum of the principal curvatures at a point on a surface. { ¦mēn 'kər·və·chər }

mean deviation See average deviation. { 'mēn ‚dē·vē'ā·shən }

mean difference The average of the absolute values of the $n(n - 1)/2$ differences between pairs of elements in a statistical distribution that has n elements. { 'mēn 'dif·rəns }

mean evolute The envelope of the planes that are orthogonal to the normals of a given surface and cut the normals halfway between the centers of principal curvature of the surface. { ¦mēn 'ev·ə‚lüt }

mean proportional For two numbers a and b, a number x, such that $x/a = b/x$. { 'mēn prə'pór·shən·əl }

mean rank method A method of handling data which has the same observed frequency occurring at two or more consecutive ranks; it consists of assigning the average of the ranks as the rank for the common frequency. { 'mēn 'raŋk ‚meth·əd }

mean square The arithmetic mean of the squares of the differences of a set of values from some given value. { ¦mēn 'skwer }

mean-square deviation A measure of the extent to which a collection $v_1, v_2, \ldots, v_n$ of numbers is unequal; it is given by the expression $(1/n)[(v_1 - \bar{v})^2 + \cdots + (v_n - \bar{v})^2]$, where $\bar{v}$ is the mean of the numbers. { 'mēn 'skwer dē·vē'ā·shən }

mean-square error The residual or error sum of squares divided by the number of degrees of freedom of the sum; gives an estimate of the error or residual variance. { ¦mēn ¦skwer 'er·ər }

mean terms The second and third terms of a proportion. { 'mēn 'tərmz }

mean value 1. For a function $f(x)$ defined on an interval (a,b), the integral from a to b of $f(x)\, dx$ divided by $b - a$. 2. See mean. { 'mēn 'val·yü }

mean value theorem The proposition that, if a function $f(x)$ is continuous on the closed interval $[a,b]$ and differentiable on the open interval (a,b), then there exists x_0, $a < x_0 < b$, such that $f(b) - f(a) = (b - a)f'(x_0)$. Also known as first law of the mean; Lagrange's formula; law of the mean. { 'mēn 'val·yü ‚thir·əm }

measurable function 1. A real valued function f defined on a measurable space X, where for every real number a all those points x in X for which $f(x) \geq a$ form a measurable set. 2. A function on a measurable space to a measurable space such that the inverse image of a measurable set is a measurable set. { 'mezh·rə·bəl 'fəŋk·shən }

measurable set A member of the sigma-algebra of subsets of a measurable space. { 'mezh·rə·bəl 'set }

measurable space A set together with a sigma-algebra of subsets of this set. { 'mezh·rə·bəl 'spās }

measure A nonnegative real valued function m defined on a sigma-algebra of subsets of a set S whose value is zero on the empty set, and whose value on a countable union of disjoint sets is the sum of its values on each set. { 'mezh·ər }

measure of location A statistic, such as the mean, median, quartile, or mode; it has the property for the mean that if a constant is added to each value the same constant must also be added to the location measure. { ¦mezh·ər əv lō'kā·shən }

measure-preserving transformation A transformation T of a measure space S into itself such that if E is a measurable subset of S then so is $T^{-1}E$ (the set of points mapped into E by T) and the measure of $T^{-1}E$ is then equal to that of E. { ¦mezh·ər pri¦zərv·iŋ ¸tranz·fər'mā·shən }

measure space A set together with a sigma-algebra of subsets of the set and a measure defined on this sigma-algebra. { 'mezh·ər ¸spās }

measure theory The study of measures and their applications, particularly the integration of mathematical functions. { 'mezh·ər ¸thē·ə·rē }

measure zero 1. A set has measure zero if it is measurable and the measure of it is zero. **2.** A subset of Euclidean n-dimensional space which has the property that for any positive number ϵ there is a covering of the set by n-dimensional rectangles such that the sum of the volumes of the rectangles is less than ϵ. { 'mezh·ər ¸zir·ō }

mechanic's rule A rule for estimating the square root of a number x whereby an estimate e is made of $\sqrt{x}$, a new estimate is made by taking the quantity $e' = (1/2)[e + (x/e)]$, and this procedure is repeated as many times as required to achieve the desired accuracy. { mi'kan·iks ¦rül }

median 1. Any line in a triangle which joins a vertex to the midpoint of the opposite side. **2.** The line that joins the midpoints of the nonparallel sides of a trapezoid. Also known as midline. An average of a series of quantities or values; specifically, the quantity or value of that item which is so positioned in the series, when arranged in order of numerical quantity or value, that there are an equal number of items of greater magnitude and lesser magnitude. { 'mē·dē·ən }

median point The point at which all three medians of a triangle intersect. { 'med·ē·ən ¸pòint }

meet The meet of two elements of a lattice is their greatest lower bound. { mēt }

meet-irreducible member A member, A, of a lattice or ring of sets such that, if A is equal to the meet of two other members, B and C, then A equals B or A equals C. { ¦mēt ¸ir·i'dü·sə·bəl ¸mem·bər }

Meijer transform The Meijer transform of a function $f(x)$ is the function $F(y)$ defined as the integral from 0 to ∞ of $\sqrt{xy}K_n(xy)f(x)dx$ where K_n is a modified Bessel function. { 'mā·ər ¸tranz¸fòrm }

Mellin transform The transform $F(s)$ of a function $f(t)$ defined as the integral over t from 0 to ∞ of $f(t)t^{s-1}$. { me'lēn ¸tranz¸fòrm }

member 1. An individual object that belongs to a set. Also known as element. **2.** For an equation, the expression on either side of the equality sign. { 'mem·bər }

membership function The characteristic function of a fuzzy set, which assigns to each element in a universal set a value between 0 and 1. { 'mem·bər¸ship ¸fəŋk·shən }

ménage number One of the numbers M_n that count the number of ways, once n wives are seated in alternate seats about a circular table, that their husbands can be seated in the seats between them so that no husband sits next to his wife. { mā'näzh ¸nəm·bər }

ménage problem See problème des ménages. { mā'näzh ¸präb·ləm }

Menelaus' theorem If ABC is a triangle and PQR is a straight line that cuts AB, CA, and the extension of BC at P, Q, and R respectively, then $(AP/PB)(CQ/QA)(BR/CR) = 1$. { ¦men·ə¦lā·əs ¸thir·əm }

Menger's theorem A theorem in graph theory which states that if G is a connected graph and A and B are disjoint sets of points of G, then the minimum number of points whose deletion separates A and B is equal to the maximum number of disjoint paths between A and B { 'meŋ·ərz ¸thir·əm }

mensuration The measurement of geometric quantities; for example, length, area, and volume. { ˌmen·səˈrā·shən }

meridian section The intersection of a surface of revolution with a plane that contains the axis of revolution. { məˈrid·ē·ən ˌsek·shən }

meromorphic function A function of complex variables which is analytic in its domain of definition save at a finite number of points which are poles. { ¦mer·ə¦mȯr·fik 'fəŋk·shən }

Mersenne number A number of the form $2^p - 1$, where p is a prime number. { mərˈsen ¦nəm·bər }

Mersenne prime A Mersenne number that is also a prime number. { mərˈsen ¦prīm }

mesh *See* fineness. { mesh }

mesokurtic distribution A distribution in which the ratio of the fourth moment to the square of the second moment equals 3, which is the value for a normal distribution. { ˌmes·ə¦kərd·ik ˌdi·strə'byü·shən }

metacompact space A topological space with the property that every open covering F is associated with a point-finite open covering G, such that every element of G is a subset of an element of F. { ˌmed·ə¦käm¸pakt 'spās }

metamathematics The study of the principles of deductive logic as they are used in mathematical logic. { ¦med·ə¸math·ə'mad·iks }

method of exhaustion A method of finding areas and volumes by finding an increasing or decreasing sequence of sets whose areas or volumes are known and less than or greater than the desired area or volume, and then showing that the area or volume between the boundaries of the approximating sets and the boundary of the set to be measured approaches zero (is exhausted). { ˌmeth·əd əv ig'zȯs·chən }

method of infinite descent *See* mathematical induction. { ¦meth·əd əv ¦in·fə·nət di'sent }

method of moments A method of estimating the parameters of a frequency distribution by first computing as many moments of the distribution as there are parameters to be estimated and then using a function that relates the parameters to moments. { ¦meth·əd əv 'mō·məns }

method of moving averages A series of averages where each average is the mean value of the time series over a fixed interval of time, and where all possible averages of the length are included in the analysis; used to smooth data in a time series. { ¦meth·əd əv ¦müv·iŋ 'av·rij·əz }

method of semiaverages A method for providing a quick estimate of a linear regression line, in which data are divided into two equal sets and the means of the two sets or two other points representative of each set are determined and a straight line drawn through them. { ¦meth·əd əv 'sem·ē¸av·rij·əz }

metric A real valued "distance" function on a topological space X satisfying four rules: for x, y, and z in X, the distance from x to itself is zero; the distance from x to y is positive if x and y are different; the distance from x to y is the same as the distance from y to x; and the distance from x to y is less than or equal to the distance from x to z plus the distance from z to y (triangle inequality). { 'me·trik }

metric space Any topological space which has a metric defined on it. { 'me·trik 'spās }

metric tensor A second rank tensor of a Riemannian space whose components are functions which help define magnitude and direction of vectors about a point. Also known as fundamental tensor. { 'me·trik 'ten·sər }

metrizable space A topological space on which can be defined a metric whose topological structure is equivalent to the original one. { mə'trīz·ə·bəl 'spās }

Meusnier's theorem A theorem stating that the curvature of a surface curve equals the curvature of the normal section through the tangent to the curve divided by the cosine of the angle between the plane of this normal section and the osculating plane of the curve. { mən'yāz ˌthir·əm }

micro- A prefix representing 10^{-6}, or one-millionth. { 'mī·krō }

micromicro- *See* pico-. { ¦mī·krō¦mī·krō }

middleware Software that allows different computer programs used in a corporate network to work together. { 'mid·əl¸wer }

midline *See* median. { 'mid,līn }

midpoint The midpoint of a line segment is the point which separates the segment into two equal parts. { 'mid,pȯint }

mil A unit of angular measure which, due to nonuniformity of usage, may have any one of three values: 0.001 radian or approximately $0.0572958°$; 1/6400 of a full revolution or $0.05625°$; 1/1000 of a right angle or $0.09°$. { mil }

milli- A prefix representing 10^{-3}, or one-thousandth. Abbreviated m. { 'mil·ē }

million The number 10^6, or 1,000,000. { 'mil·yən }

Milne method A technique which provides numerical solutions to ordinary differential equations. { 'miln ,meth·əd }

minimal element *See* minimal member. { 'min·ə·məl ¦el·ə·mənt }

minimal equation 1. An algebraic equation whose zeros define a minimal surface. **2.** *See* reduced characteristic equation. { 'min·ə·məl i'kwā·zhən }

minimal member In a partially ordered set, a minimal member is one for which no other element precedes it in the ordering. Also known as minimal element. { 'min·ə·məl ¦mem·bər }

minimal polynomial The polynomial of least degree which both divides the characteristic polynomial of a matrix and has the same roots. { 'min·ə·məl ,päl·ə'nō·mē·əl }

minimal surface A surface that has assumed a geometric configuration of least area among those into which it can readily deform. { 'min·ə·məl 'sər·fəs }

minimal transformation group A transformation group such that every orbit is dense in the phase space. { ¦min·ə·məl ,tranz·fər'mā·shən ,grüp }

minimax 1. The minimum of a set of maxima. **2.** In the theory of games, the smallest of a set of maximum possible losses, each representing the most unfavorable outcome of a particular strategy. { 'min·ə,maks }

minimax criterion A concept in game theory and decision theory which requires that losses or expected losses associated with a variable that can be controlled be minimized, and thus maximizes the losses or expected losses associated with the variable that cannot be controlled. { 'min·ə,maks krī,tir·ē·ən }

minimax estimator A random variable obtained by applying the minimax criterion to a risk function associated with a loss function. { 'min·ə,maks ,es·tə,mād·ər }

minimax technique *See* min-max technique. { 'min·ē,maks tek,nēk }

minimax theorem A theorem of games that the lowest maximum expected loss in a two-person zero-sum game equals the highest minimum expected gain. { 'min·ə,maks ,thir·əm }

minimization The determination of the simplest expression of a Boolean function equivalent to a given one. { ,min·ə·mə'zā·shən }

minimum The least value that a real valued function assumes. { 'min·ə·məm }

minimum cut For an *s-t* network, an *s-t* cut whose weight has the minimum possible value. { 'min·ə·məm ¦kət }

minimum dominating vertex set A dominating vertex set such that there is no other dominating vertex set with fewer vertices. { ¦min·ə·məm¦däm·ə,nād·iŋ 'vər,teks ,set }

minimum edge cover An edge cover of a graph such that there is no other edge cover with fewer vertices. { ,min·ə·məm 'ej ,kəv·ər }

minimum-modulus theorem The theorem that a nonvanishing, complex analytic function in a closed, bounded, simply connected region assumes its minimum absolute value on the boundary of the region. { ¦min·ə·məm 'mäj·ə·ləs ,thir·əm }

minimum-variance estimator An estimator that possesses the least variance among the members of a defined class of estimators. { ¦min·i·məm 'ver·ē·əns ,es·tə,mād·ər }

minimum vertex cover A vertex cover in a graph such that there is no other vertex cover with fewer vertices. { ,min·ə·məm 'vər,teks ,kəv·ər }

Minkowski distance function Relative to a convex body with the origin O in its interior, the function whose value at a point P is the distance ratio OP/OQ, where Q is the point of the convex body on the ray OP that is furthest from O. { miŋ'kȯf·skē *or* min'kaů·skē 'dis·təns ,fəŋk·shən }

Minkowski's inequality

Minkowski's inequality 1. An inequality involving powers of sums of sequences of real or complex numbers, a_k and b_k:

$$\left[\sum_{k=1} |a_k + b_k|^s \right]^{1/s} \leq \left[\sum_{k=1} |a_k|^s \right]^{1/s} + \left[\sum_{k=1}^{\infty} |b_k|^s \right]^{1/s}$$

provided $s \geq 1$. **2.** An inequality involving powers of integrals of real or complex functions, f and g, over an interval or region R:

$$\left[\int_R |f(x) + g(x)|^s \, dx \right]^{1/s} \leq \left[\int_R |f(x)|^s \, dx \right]^{1/s} + \left[\int_R |g(x)|^s \, dx \right]^{1/s}$$

provided $s \geq 1$ and the integrals involved exist. { miŋ'kȯf·skēz ‚in·i'kwäl·əd·ē }

min-max technique A method of approximation of a function f by a function g from some class where the maximum of the modulus of $f - g$ is minimized over this class. Also known as Chebyshev approximation; minimax technique. { 'min 'maks tek‚nēk }

min-max theorem The theorem that provides information concerning the nth eigenvalue of a symmetric operator on an inner product space without necessitating knowledge of the other eigenvalues. Also known as maximum-minimum principle. { 'min 'maks ‚thir·əm }

minor The minor of an entry of a matrix is the determinant of the matrix obtained by removing the row and column containing the entry. Also known as cofactor; complementary minor. { 'mīn·ər }

minor arc The smaller of the two arcs on a circle produced by a secant. { 'mīn·ər 'ärk }

minor axis The smaller of the two axes of an ellipse. { 'mīn·ər 'ak·səs }

minuend The quantity from which another quantity is to be subtracted. { 'min·yə‚wend }

minus A minus B means that the quantity B is to be subtracted from the quantity A. { 'mī·nəs }

minus sign *See* subtraction sign. { 'mī·nəs ‚sīn }

minute A unit of measurement of angle that is equal to 1/60 of a degree. Symbolized '. Also known as arcmin. { 'min·ət }

mirror plane of symmetry *See* plane of mirror symmetry. { 'mir·ər 'plān əv 'sim·ə·trē }

Mirsky's theorem The theorem that, in a finite partially ordered set, the maximum cardinality of a chain is equal to the minimum number of disjoint antichains into which the partially ordered set can be partitioned. { 'mir·skēz ‚thir·əm }

Mittag-Leffler's theorem A theorem that enables one to explicitly write down a formula for a meromorphic complex function with given poles; for a function $f(z)$ with poles at $z = z_i$, having order m_i and principal parts $\sum_{j=1}^{m_i} a_{ij} (z - z_i)^{-j}$, the formula is $f(z) = \sum_i \left[\sum_{j=1}^{m_i} a_{ij} (z - z_i)^{-j} + p_i(z) \right] + g(z)$ where the $p_i(z)$ are polynomials, $g(z)$ is an entire function, and the series converges uniformly in every bounded region where $f(z)$ is analytic. { 'mi‚täk 'lef·lərz ‚thir·əm }

mixed-base notation A computer number system in which a single base, such as 10 in the decimal system, is replaced by two number bases used alternately, such as 2 and 5. { 'mikst ¦bās nō'tā·shən }

mixed-base number A number in mixed-base notation. Also known as mixed-radix number. { 'mikst ¦bās 'nəm·bər }

mixed decimal Any decimal plus an integer. { 'mikst 'des·məl }

mixed graph A graph in which directions are associated with some arcs but not with others. { ¦mikst 'graf }

mixed model 1. A model having both determinate and stochastic elements in its equations. **2.** A model having both difference and differential equations. **3.** A model containing both endogenous and exogenous elements. **4.** In analysis of variance for a two-way layout, the combined rows and columns. { 'mikst ‚mäd·əl }

mixed number The sum of an integer and a fraction. { 'mikst 'nəm·bər }

mixed partial derivative A partial derivative whose differentiations are with respect to two or more different variables. { ¦mikst 'pär·shəl də¦riv·əd·iv }

mixed radix Pertaining to a numeration system using more than one radix, such as the biquinary system. { 'mikst 'rā·diks }

mixed-radix number *See* mixed-base number. { 'mikst ¦rā·diks 'nəm·bər }

mixed sampling The use of two or more methods of sampling; for example, in multistage sampling, if samples are drawn at random at one stage and drawn by a systematic method at another. { ¦mikst 'sam·pliŋ }

mixed strategy A method of playing a matrix game in which the player attaches a probability weight to each of the possible options, the probability weights being nonnegative numbers whose sum is unity, and then operates a chance device that chooses among the options with probabilities equal to the corresponding weights. A concept in game theory which allows a player more than one choice of action which is determined by a chance mechanism. { ¦mikst 'strad·ə·jē }

mixed surd A surd containing a rational factor or term, as well as irrational numbers. { ¦mikst 'sərd }

mixed tensor A tensor with both contravariant and covariant indices. { ¦mikst 'ten·sər }

mixing transformation A function of a measure space which moves the measurable sets in such a manner that, asymptotically as regards measure, any measurable set is distributed uniformly throughout the space. { 'mik·siŋ ˌtranz·fər'mā·shən }

Möbius band The nonorientable surface obtained from a rectangular strip by twisting it once and then gluing the two ends. Also known as Möbius strip. { 'mər·bē·əs ˌband }

Möbius function The function μ of the positive integers where $\mu(1) = 1$, $\mu(n) = (-1)^r$ if n factors into r distinct primes, and $\mu(n) = 0$ otherwise; also, $\mu(n)$ is the sum of the primitive nth roots of unity. { 'mər·bē·əs ˌfəŋk·shən }

Möbius strip *See* Möbius band. { 'mər·bē·əs ˌstrip }

Möbius transformations These are the most commonly used conformal mappings of the complex plane; their form is $f(z) = (az + b)/(cz + d)$ where the real numbers a, b, c, and d satisfy $ad - bc \neq 0$. Also known as bilinear transformations; homographic transformations; linear fractional transformations. { 'mər·bē·əs ˌtranz·fər'mā·shənz }

modal class The class that contains more individuals than any other class in a statistical distribution. { 'mōd·əl ˌklas }

mode The most frequently occurring member of a set of numbers. { mōd }

model theory The general qualitative study of the structure of a mathematical theory. { 'mäd·əl ˌthē·ə·rē }

modern algebra The study of algebraic systems such as groups, rings, modules, and fields. { 'mäd·ərn 'al·jə·brə }

modified Bessel equation The differential equation $z^2 f''(z) + zf'(z) - (z^2 + n^2)f(z) = 0$, where z is a variable that can have real or complex values and n is a real or complex number. { ¦mōd·ə¦fīd 'bes·əl iˌkwā·zhən }

modified Bessel function of the first kind *See* modified Bessel function. { ¦mäd·ə¦fīd ¦bes·əl ˌfəŋk·shən əv thə 'fərst ˌkīnd }

modified Bessel function of the second kind *See* modified Hankel function. { ¦mäd·ə¦fīd ¦bes·əl ˌfəŋk·shən əv thə 'sek·ənd ˌkīnd }

modified Bessel functions The functions defined by $I_\nu(x) = \exp(-i\nu\pi/2)\, J_\nu(ix)$, where J_ν is the Bessel function of order ν, and x is real and positive. Also known as modified Bessel function of the first kind. { 'mäd·ə¦fīd 'bes·əl ˌfəŋk·shənz }

modified exponential curve The equation resulting when a constant is added to the exponential curve equation; used to estimate trend in a nonlinear time series. { ¦mäd·ə¦fīd ˌek·spə¦nen·chəl 'kərv }

modified Hankel functions The functions defined by $K_\nu(x) = (i\pi/2)\exp(i\nu\pi/2) H_\nu^{(1)}(ix)$, where $H_\nu^{(1)}$ is the first Hankel function of order ν, and x is real and

modified mean

positive. Also known as modified Bessel function of the second kind. { 'mäd·ə,fīd 'häŋk·əl ,fəŋk·shənz }

modified mean A mean computed after elimination of observations judged to be atypical. { ¦mäd·ə,fīd 'mēn }

modular lattice A lattice with the property that, if x is equal to or greater than z, then for any element y, the greatest lower bound of x and v equals the least upper bound of w and z, where v is the least upper bound of y and z and w is the greatest lower bound of x and y. { ,mäj·ə·lər 'lad·əs }

module A vector space in which the scalars are a ring rather than a field. { 'mäj·ül }

modulo **1.** A group G modulo a subgroup H is the quotient group G/H of cosets of H in G. **2.** A technique of identifying elements in an algebraic structure in such a manner that the resulting collection of identified objects is the same type of structure. { 'mäj·ə,lō }

modulo N Two integers are said to be congruent modulo N (where N is some integer) if they have the same remainder when divided by N. { 'mäj·ə,lō 'en }

modulo N arithmetic Calculations in which all integers are replaced by their remainders after division by N (where N is some fixed integer.) { 'mäj·ə,lō ¦en ə'rith·mə·tik }

modulus **1.** The modulus of a logarithm with a given base is the factor by which a logarithm with a second base must be multiplied to give the first logarithm. **2.** *See* absolute value. { 'mäj·ə·ləs }

modulus of a congruence A number a, such that two specified numbers b and c give the same remainder when divided by a; b and c are then said to be congruent, modulus a (or congruent, modulo a). { 'mäj·ə·ləs əv ə kən'grü·əns }

modulus of continuity For a real valued continuous function f, this is the function whose value at a real number r is the maximum of the modulus of $f(x) - f(y)$ where the modulus of $x - y$ is less than r; this function is useful in approximation theory. { 'mäj·ə·ləs əv ,känt·ən'ü·əd·ē }

molding surface A surface generated by a plane curve as its plane rolls without slipping over a cylinder. { 'mōld·iŋ ,sər·fəs }

moment The nth moment of a distribution $f(x)$ about a point x_0 is the expected value of $(x - x_0)^n$, that is, the integral of $(x - x_0)^n df(x)$, where $df(x)$ is the probability of some quantity's occurrence; the first moment is the mean of the distribution, while the variance may be found in terms of the first and second moments. { 'mō·mənt }

moment generating function For a frequency function $f(x)$, a function $\phi(t)$ that is defined as the integral from $-\infty$ to ∞ of $\exp(tx) f(x)dx$, and whose derivatives evaluated at $t = 0$ give the moments of f. { ¦mō·mənt ¦jen·ə,rād·iŋ 'fəŋk·shən }

moment problem The problem of finding a distribution whose moments have specified values, or of determining whether such a distribution exists. { 'mō·mənt ,präb·ləm }

Monge form An equation of a surface of the form $z = f(x,y)$, where x, y, and z are cartesian coordinates. { 'mônzh ,fôrm }

Monge's theorem For three coplanar circles, and for radii of these circles which are parallel to each other, the three outer centers of similitude of the circles taken in pairs lie on a single straight line, and any two inner centers of similitude lie on a straight line with one of the outer centers. { 'mōnzh·əz ,thir·əm }

monic equation A polynomial equation with integer coefficients, where the coefficient of the term of highest degree is +1. { ¦mō·nik i'kwā·zhən }

monic polynomial A polynomial in which the coefficient of the term of highest degree is +1 and the coefficients of the other terms are integers. { ¦mō·nik ,päl·ə'nō·mē·əl }

monodromy theorem If a complex function is analytic at a point of a bounded simply connected domain and can be continued analytically along every curve from the point, then it represents a single-valued analytic function in the domain. { 'män·ə,drō·mē ,thir·əm }

monogenic analytic function An analytic function whose domain of definition has been extended directly or indirectly by analytic continuation as far as theoretically possible. { ¦män·ə‚jen·ik ‚an·ə¦lid·ik 'fəŋk·shən }

monoid A semigroup which has an identity element. { 'mä‚nȯid }

monomial A polynomial of degree one. { mə'nō·mē·əl }

monomial factor A single factor that can be divided out of every term in a given expression. { mə'nō·mē·əl ‚fak·tər }

monomino *See* square. { mə'näm·ə·nō }

monotone convergence theorem The integral of the limit of a monotone increasing sequence of nonnegative measurable functions is equal to the limit of the integrals of the functions in the sequence. { 'män·ə‚tōn kən'vər·jəns ‚thir·əm }

monotone decreasing function *See* monotone nonincreasing function. { ¦män·ə‚tōn di¦krēs·iŋ 'fəŋk·shən }

monotone decreasing sequence A sequence of real numbers in which each term is equal to or less than the preceding term. { ¦män·ə‚tōn di¦krēs·iŋ 'sē·kwəns }

monotone function A function which is either monotone nondecreasing or monotone nonincreasing. Also known as monotonic function. { ‚män·ə‚tōn ‚fəŋk·shən }

monotone increasing function *See* monotone nondecreasing function. { ¦män·ə‚tōn in¦krēs·iŋ 'fəŋk·shən }

monotone increasing sequence A sequence of real numbers in which each term is equal to or greater than the preceding term. { ‚män·ə‚tōn in'krēs·iŋ ‚sē·kwəns }

monotone nondecreasing function A function which never decreases, that is, if $x \leq y$ then $f(x) \leq f(y)$. Also known as monotone increasing function; monotonically nondecreasing function. { ‚män·ə‚tōn ¦nän·di'krēs·iŋ ‚fəŋk·shən }

monotone nondecreasing sequence 1. A sequence, $\{S_n\}$, of real numbers that never decreases; that is, $S_{n+1} \geq S_n$ for all n. **2.** A sequence of real-valued functions, $\{f_n\}$, defined on the same domain, D, that never decreases; that is, $f_{n+1}(x) \geq f_n(x)$ for all n and for all x in D.

monotone nonincreasing function A function which never increases, that is, if $x \leq y$ then $f(x) \geq f(y)$. Also known as monotone decreasing function; monotonically nonincreasing function. { ‚män·ə‚tōn ¦nän·in'krēs·iŋ ‚fəŋk·shən }

monotone nonincreasing sequence 1. A sequence, $\{S_n\}$, of real numbers that never increases; that is, $S_{n+1} \leq S_n$ for all n. **2.** A sequence of real-valued functions, $\{f_n\}$, defined on the same domain, D, that never increases; that is, $f_{n+1}(x) \leq f_n(x)$ for all n and for all x in D.

monotone sequence 1. A sequence of real numbers that is monotone-nondecreasing or monotone-nonincreasing. **2.** A sequence of real-valued functions, defined on the same domain, that is either monotone-nondecreasing or monotone-nonincreasing. { 'män·ə‚tōn ¦sē·kwəns }

monotonically nondecreasing function *See* monotone nondecreasing function. { ¦män·ə¦tän·ik·lē ¦nän·di'krēs·iŋ ‚fəŋk·shən }

monotonically nonincreasing function *See* monotone nonincreasing function. { ¦män·ə¦tän·ik·lē ¦nän·in'krēs·iŋ ‚fəŋk·shən }

monotonic decreasing function *See* monotone nonincreasing function. { ‚män·ə¦tän·ik di¦krēs·iŋ 'fəŋk·shən }

monotonic function *See* monotone function. { ¦män·ə¦tän·ik 'fəŋk·shən }

monotonic system of sets *See* nested sets. { ¦män·ə¦tän·ik ¦sis·təm əv 'sets }

Monte Carlo method A technique which obtains a probabilistic approximation to the solution of a problem by using statistical sampling techniques. { 'män·tē 'kär·lō ‚meth·əd }

Moore-Smith convergence Convergence of a net to a point x in a topological space, in the sense that for each neighborhood of x there is an element a of the directed system that indexes the net such that, if b is also an element of this directed system and $b \geq a$, then x_b (the element indexed by b) is in this neighborhood. { 'mur 'smith kən'vər·jəns }

Moore-Smith set *See* directed set. { ¦mur 'smith ‚set }

Moore space A topological space that has a sequence of coverings by open sets, such that each member of the sequence is a subcollection of the previous one, and such that, for any two points, x and y, of an open set S in the space, there is an open covering in the sequence such that the closure of any member of this covering that includes x is a subset of S and does not include y. { 'mür ,spās }

Morera's theorem If a function of a complex variable is continuous in a simply connected domain D, and if the integral of the function about every simply connected curve in D vanishes, then the function is analytic in D. { mȯ'rer·əz ,thir·əm }

morphism The class of elements which together with objects form a category; in most cases, morphisms are functions which preserve some structure on a set. { 'mȯr,fiz·əm }

Morse theory The study of differentiable mappings of differentiable manifolds, which by examining critical points shows how manifolds can be constructed from one another. { 'mȯrs ,thē·ə·rē }

Morse-Thue sequence A sequence of binary digits defined by the number of 1's modulo 2 in successive integers when written in binary notation: 01101001 . . . . { ¦mȯrs 'thü ,sē·kwəns }

most powerful test If two tests have the same level of significance, then the test with a smaller-size type II error is the most powerful test of the two at that significance level. { 'mōst 'pau̇·ər·fu̇l 'test }

moving totals The sum of the year's figures and those of some years before and after it. { 'müv·iŋ 'tōd·əlz }

moving trihedral For a space curve, a configuration consisting of the tangent, principal normal, and binormal of the curve at a variable point on the curve. { 'müv·iŋ trī'hē·drəl }

Muller method A method for finding zeros of a function $f(x)$, in which one repeatedly evaluates $f(x)$ at three points, x_1, x_2, and x_3, fits a quadratic polynomial to $f(x_1)$, $f(x_2)$, and $f(x_3)$, and uses x_2, x_3, and the root of this quadratic polynomial nearest to x_3 as three new points to repeat the process. { 'məl·ər ,meth·əd }

multicollinearity A concept in regression analysis describing the situation where, because of the high degree of correlation between two or more independent variables, it is not possible to separate accurately the effect of each individual independent variable upon the dependent variable. { ,məl·tē·kō,lin·ē'ar·əd·ē }

multidimensional derivative The generalized derivative of a function of several variables which is usually represented as a matrix involving the various partial derivatives of the function. { ¦məl·tə·di'men·shən·əl də'riv·əd·iv }

multifoil A plane figure consisting of congruent arcs of a circle arranged around a regular polygon, with the end points of each arc located at the midpoints of adjacent sides of the polygon, and the tangents to the arcs at these points perpendicular to the sides. { 'məl·tē,fȯil }

multigraph 1. A graph with no loops. 2. A graph that may have more than one edge joining a particular pair of vertices. { 'məl·tə,graf }

multilinear algebra The study of functions of several variables which are linear relative to each variable. { 'məl·tə,lin·ē·ər 'al·jə·brə }

multilinear form A multilinear form of degree n is a polynomial expression which is linear in each of n variables. { 'məl·tə,lin·ē·ər 'fȯrm }

multilinear function A function of several variables that is a linear function of each variable when the other variables are given fixed values. { ¦məl·tə,lin·ē·ər 'fəŋk·shən }

multimodal distribution A frequency distribution that has several relative maxima. { ,məl·tē¦mōd·əl ,di·strə'byü·shən }

multinomial An algebraic expression which involves the sum of at least two terms. { ¦məl·tə¦nō·mē·əl }

multinomial distribution The joint distribution of the set of random variables which are the number of occurrences of the possible outcomes in a sequence of multinomial trials. { ¦məl·tə¦nō·mē·əl ,di·strə'byü·shən }

multinomial theorem The rule for expanding $(x_1 + x_2 + \cdots + x_m)^n$, where m and n are positive integers; a generalization of the binomial theorem. { ,məl·tə¦nō·mē·əl 'thir·əm }

multinomial trials Unrelated trials with more than two possible outcomes the probabilities of which do not change from trial to trial. { ˌməl·tə¦nō·mē·əl 'trīlz }

multiphase sampling A sampling method in which certain items of information are drawn from the whole units of a sample and certain other items of information are taken from the subsample. { ¦məl·tə‚fāz 'sam·pliŋ }

multiple The product of a number or quantity by an integer. { 'məl·tə·pəl }

multiple coefficient of determination A statistic that measures the proportion of total variation which is explained by the regression line; computed by taking the square root of the coefficient of multiple correlation. { ˌməl·tə·pəl ‚kō·ə¦fish·ənt əv di‚tər·mə'nā·shən }

multiple edges *See* parallel edges. { ‚məl·tə·pəl 'ej·əs }

multiple integral An integral over a subset of n-dimensional space. { 'məl·tə·pəl 'int·ə·grəl }

multiple linear correlation An index for estimating the strength of the linear relationship between one dependent variable and two or more independent variables. { 'məl·tə·pəl ¦lin·ē·ər ‚kär·ə'lā·shən }

multiple linear regression A technique for determining the linear relationship between one dependent variable and two or more independent variables. { ¦məl·tə·pəl ¦lin·ē·ər ri'gresh·ən }

multiple point A point of a curve through which passes more than one arc of the curve. { 'məl·tə·pəl 'pȯint }

multiple root A polynomial $f(x)$ has c as a multiple root if $(x - c)^n$ is a factor for some $n > 1$. Also known as repeated root. { 'məl·tə·pəl 'rüt }

multiple stratification Division of a population into two or more parts with respect to two or more variables. { 'məl·tə·pəl ‚strad·ə·fə'kā·shən }

multiple-valued A relation between sets is multiple-valued if it associates to an element of one more than one element from the other; sometimes functions are allowed to be multiple-valued. { 'məl·tə·pəl 'val‚yüd }

multiple-valued logic A form of logic in which statements can have values other than the two values "true" and "false." { 'məl·tə·pəl ¦val‚yüd 'läj·ik }

multiplicand If a number x is to be multiplied by a number y, then x is called the multiplicand. { ‚məl·tə·pli'kand }

multiplication Any algebraic operation analogous to multiplication of real numbers. { ‚məl·tə·pli'kā·shən }

multiplication formula An equation that expresses a function of a multiple of a quantity in terms of functions of the quantity itself and possibly functions of other multiples of the quantity. { ‚məl·tə·plə'kā·shən ‚fȯr·myə·lə }

multiplication on the left *See* premultiplication. { ‚məl·tə·pli'kā·shən ȯn thə 'left }

multiplication on the right *See* postmultiplication. { ‚məl·tə·pli'kā·shən ȯn thə 'rīt }

multiplication sign The symbol $\times$ or $\cdot$, used to indicate multiplication. Also known as times sign. { ‚məl·tə·pli'kā·shən ‚sīn }

multiplicative identity In a mathematical system with an operation of multiplication, denoted $\times$, an element 1 such that $1 \times e = e \times 1 = e$ for any element e in the system. { ‚məl·tə'plik·əd·iv ī'den·əd·ē }

multiplicative inverse In a mathematical system with an operation of multiplication, denoted $\times$, the multiplicative inverse of an element e is an element $\bar{e}$ such that $e \times \bar{e} = \bar{e} \times e = 1$, where 1 is the multiplicative identity. { ‚məl·tə'plik·əd·iv 'in‚vərs }

multiplicative number-theoretic function A number theoretic function, f, which has the properties that mn is in its range whenever m and n are, and that $f(mn) = f(m)f(n)$ whenever m and n are relatively prime. { ‚məl·tə¦plik·əd·iv ‚nəm·bər ‚thē·ə¦red·ik 'fəŋk·shən }

multiplicative subset A subset S of a commutative ring such that if x and y are in S then so is xy. { ‚məl·tə'plik·əd·iv 'səb‚set }

multiplicity 1. A root of a polynomial $f(x)$ has multiplicity n if $(x - a)^n$ is a factor of $f(x)$ and n is the largest possible integer for which this is true. **2.** The geometric multiplicity of an eigenvalue λ of a linear transformation T is the dimension of the null space of the transformation $T - \lambda I$, where I denotes the identity transformation. **3.** The algebraic multiplicity of an eigenvalue λ of a linear transformation T on a finite-dimensional vector space is the multiplicity of λ as a root of the characteristic polynomial of T. { ˌməl·tə'plis·əd·ē }

multiplier If a number x is to be multiplied by a number y, then y is called the multiplier. { 'məl·tə,plī·ər }

multiply connected region An open set in the plane which has holes in it. { 'məl·tə·plē kə¦nek·təd 'rē·jən }

multiply perfect number An integer such that the sum of all its factors is a multiple of the integer itself. { ˌməl·tə·plē ¦pər·fikt 'nəm·bər }

multistage sampling A sampling method in which the population is divided into a number of groups or primary stages from which samples are drawn; these are then divided into groups or secondary stages from which samples are drawn, and so on. { ¦məl·tə,stāj 'sam·pliŋ }

multivariate analysis The study of random variables which are multidimensional. { ¦məl·tē'ver·ē·ət ə'nal·ə·səs }

multivariate distribution For two or more random variables, X_1, X_2, . . ., and X_n, the distribution which gives the probability that $X_1 = x_1$, $X_2 = x_2$, . . ., and $X_n = x_n$ for all values, x_1, x_2, . . ., and x_n, of X_1, X_2, . . ., and X_n respectively. { ¦məl·tē'ver·ē·ət 'dis·trə'byü·shən }

mutually exclusive events Two or more events such that the occurrence of any one makes impossible the occurrence of any of the others. { ¦myü·chə·lē ik¦sklü·siv i'vens }

N

nabla *See* del operator. { 'nab·lə }

Nakayama's lemma The proposition that, if R is a commutative ring, I is an ideal contained in all maximal ideals of R, and M is a finitely generated module over R, and if $IM = M$, where IM denotes the set of all elements of the form am with a in I and m in M, then $M = 0$. { ,nä·kä,yä·mäz 'lem·ə }

NAND A logic operator having the characteristic that if P, Q, R, . . . are statements, then the NAND of P, Q, R, . . . is true if at least one statement is false, false if all statements are true. Derived from NOT-AND. Also known as sheffer stroke. { nand }

nano- A prefix representing 10^{-9}, which is 0.000000001 or one-billionth of the unit adjoined. { 'nan·ō }

Naperian logarithm *See* logarithm. { nā'pir·ē·ən 'läg·ə,rith·əm }

Napierian logarithm *See* logarithm. { nā'pir·ē·ən 'läg·ə,rith·əm }

Napier's analogies Formulas which enable one to study the relationships between the sides and the angles of a spherical triangle. { 'nā·pē·ərz ə'nal·ə·jēz }

Napier's rules Two rules which give the formulas necessary in the solution of right spherical triangles. { 'nā·pē·ərz ,rülz }

nappe One of the two parts of a conical surface defined by the vertex. { nap }

n-ary composition A function that associates an element of a set with every sequence of n elements of the set. { 'en·ə·rē ,käm·pə'zish·ən }

n-ary tree A rooted tree in which each vertex has at most n successors. { 'en·ə·rē ,trē }

natural boundary Those points of the boundary of a region where an analytic function is defined through which the function cannot be continued analytically. { 'nach·rəl 'baun·drē }

natural equations of a curve *See* intrinsic equations of a curve. { ,nach·rəl i,kwā·zhənz əv ə 'kərv }

natural function A trigonometric function, as opposed to its logarithm. { 'nach·rəl 'fəŋk·shən }

natural logarithm *See* logarithm. { 'nach·rəl 'läg·ə,rith·əm }

natural number One of the integers 1, 2, 3, ... { 'nach·rəl 'nəm·bər }

navel point *See* umbilical point. { 'nā·vəl ,point }

n-cell A set that is homeomorphic either with the set of points in n-dimensional Euclidean space $(n = 1, 2, . . .)$ whose distance from the origin is less than unity, or with the set of points whose distance from the origin is less than or equal to unity. { 'en ,sel }

n-colorable graph A graph whose nodes can be colored using one of n colors on each node in such a way that no edge connects a pair of nodes with the same color. { ¦en ,kəl·ə·rə·bəl 'graf }

n-connected graph A connected graph for which the removal of n points is required to disconnect the graph. { 'en kə,nek·təd 'graf }

n-dimensional space A vector space whose basis has n vectors. { 'en di'men·shən·əl 'spās }

nearly isometric spaces Two Banach spaces, A and B, such that for any numbers $c < 1$ and $d > 1$ there is a bijective mapping, f, from A to B such that the norm of $f(x)$ divided by the norm of x lies in the interval $[c,d]$. { ¦nir·lē ‚ī·sə‚me·trik 'spās·əz }

near ring An algebraic system with two binary operations called multiplication and addition; the system is a group (not necessarily commutative) relative to addition, and multiplication is associative, and is left-distributive with respect to addition, that is, $x(y + z) = xy + xz$ for any x, y, and z in the near ring. { ¦nir ¦riŋ }

necessary condition A mathematical statement that must be true if a given statement is true. { ‚nes·ə‚ser·ē kən'dish·ən }

negation The negation of a proposition P is a proposition which is true if and only if P is false; this is often written ~ P. Also known as denial. { nə'gā·shən }

negative angle The angle subtended by moving a ray in the clockwise direction. { 'neg·əd·iv 'aŋ·gəl }

negative binomial distribution The distribution of a negative binomial random variable. Also known as Pascal distribution. { ¦neg·əd·iv bī¦nō·mē·əl ‚di·strə'byü·shən }

negative correlation A relation between two quantities such that when one increases the other decreases. { 'neg·əd·iv ‚kär·ə'lā·shən }

negative integer The additive inverse of a positive integer relative to the additive group structure of the integers. { 'neg·əd·iv 'int·ə·jər }

negative number A real number that is less than 0. { ¦neg·əd·iv 'nəm·bər }

negative part For a real-valued function f, this is the function, denoted f^-, for which $f^-(x) = f(x)$ if $f(x) \le 0$ and $f^-(x) = 0$ if $f(x) > 0$. { 'neg·əd·iv 'pärt }

negative pedal 1. The negative pedal of a curve with respect to a point O is the envelope of the line drawn through a point P of the curve perpendicular to OP. Also known as first negative pedal. **2.** Any curve that can be derived from a given curve by repeated application of the procedure specified in the first definition. { 'neg·əd·iv 'ped·əl }

negative series A series whose terms are all negative real numbers. { 'neg·əd·iv 'sir‚ēz }

negative sign The symbol −, used to indicate a negative number. { 'neg·əd·iv ¦sīn }

negative skewness Skewness in which the mean is smaller than the mode. { 'neg·əd·iv 'skü·nəs }

negative with respect to a measure A set A is negative with respect to a signed measure m if, for every measurable set B, the intersection of A and B, $A \cap B$, is measurable and $m(A \cap B) \le 0$. { ¦neg·əd·iv with ri¦spekt tü ə 'mezh·ər }

neighborhood of a point A set in a topological space which contains an open set which contains the point; in Euclidean space, an example of a neighborhood of a point is an open (without boundary) ball centered at that point. { 'nā·bər‚hu̇d əv ə 'pȯint }

Neil's parabola The graph of the equation $y = ax^{3/2}$, where a is a constant. { 'nēlz pə'rab·ə·lə }

nephroid An epicycloid for which the diameter of the fixed circle is two times the diameter of the rolling circle. { 'ne‚frȯid }

nephroid of Freeth See Freeth's nephroid. { 'ne‚frȯid əv 'frēth }

nested intervals A sequence of intervals, each of which is contained in the preceding interval. { 'nes·təd 'in·tər·vəlz }

nested sets A family of sets where, given any two of its sets, one is contained in the other. Also known as monotonic system of sets. { 'nes·təd 'sets }

net 1. A set whose members are indexed by elements from a directed set; this is a generalization of a sequence. Also known as Moore-Smith sequence. **2.** A nondegenerate partial plane satisfying the parallel axiom. { net }

net flow The net flow at a vertex in an s-t network is the outflow at that vertex minus the inflow there. { ¦net 'flō }

network The name given to a graph in applications in management and the engineering sciences; to each segment linking points in the graph, there is usually associated a direction and a capacity on the flow of some quantity. { 'net‚wərk }

Neumann boundary condition The boundary condition imposed on the Neumann problem in potential theory. { 'nȯi‚män 'bau̇n·drē kən‚dish·ən }

Neumann function 1. One of a class of Bessel functions arising in the study of the solutions to Bessel's differential equation. **2.** A harmonic potential function in potential theory occurring in the study of Neumann's problem. { 'nói,män ,fəŋk·shən }

Neumann line The generalization of the concept of a line occurring in Neumann's study of continuous geometry. { 'nói,män ,līn }

Neumann problem The determination of a harmonic function within a finite region of three-dimensional space enclosed by a closed surface when the normal derivatives of the function on the surface are specified. { 'nói,män ,präb·ləm }

Neumann series *See* Liouville-Neumann series. { 'nói,män ,sir·ēz }

Newton-Cotes formulas Approximation formulas for the integral of a function along a small interval in terms of the values of the function and its derivatives. { 'nüt·ən 'kōts ,fór·myə·ləz }

Newton-Raphson formula If c is an approximate value of a root of the equation $f(x) = 0$, then a better approximation is the number $c - [f(c)/f'(c)]$. { 'nüt·ən 'raf·sən ,fór·myə·lə }

Newton's identity The identity $C(n,r)C(r,k) = C(n,k) C(n - k, r - k)$, where, in general, $C(n,r)$ is the number of distinct subsets of r elements in a set of n elements (the binomial coefficient). { nüt·ənz ī'dendə,dē }

Newton's inequality For any set of n numbers ($n = 0, 1, 2, \ldots$), the inequality $p_{r-1}p_{r+1} \leqq p_r^2$, for $1 \leqq r < n$, where p_r is the average value of the terms constituting the rth elementary symmetric function of the numbers. { ¦nüt·ənz ,in·i'kwäl·əd·ē }

Newton's method A technique to approximate the roots of an equation by the methods of the calculus. { 'nüt·ənz ,meth·əd }

Newton's square-root method A technique for the estimation of the roots of an equation exhibiting faster convergence than Newton's method; this involves calculus methods and the square-root function. { 'nüt·ənz 'skwer ,rüt ,meth·əd }

Neyman-Pearson theory A theory that determines what is the best test to use to examine a statistical hypothesis. { 'nā·mən 'pir·sən ,thē·ə·rē }

nilmanifold The factor space of a connected nilpotent Lie group by a closed subgroup. { ¦nil'man·ə,fōld }

nilpotent An element of some algebraic system which vanishes when raised to a certain power. { ¦nil'pōt·ənt }

nilradical For an ideal, I, in a ring, R, the set of all elements, a, in R for which a_n is a member of I for some positive integer n. Also known as radical. { ,nil'rad·ə·kəl }

n-net A finite net in which n lines pass through each point. { 'en ,net }

node *See* crunode. { nōd }

Noetherian module A module in which every ascending sequence of submodules has only a finite number of distinct members. { ,nō·ə¦thir·ē·ən 'mäj·əl }

Noetherian ring A ring is Noetherian on left ideals (or right ideals) if every ascending sequence of left ideals (or right ideals) has only a finite number of distinct members. { ,nō·ə'thir·ē·ən'riŋ }

nominal scale measurement A method for sorting objects into categories according to some distinguishing characteristic and attaching a name or label to each category; considered the weakest type of measurement. { ¦näm·ə·nəl ¦skāl 'mezh·ər·mənt }

nomogram *See* nomograph. { 'näm·ə,gram }

nomograph A chart which represents an equation containing three variables by means of three scales so that a straight line cuts the three scales in values of the three variables satisfying the equation. Also known as abac; alignment chart; nomogram. { 'näm·ə,graf }

nonagon A nine-sided polygon. Also known as enneagon. { 'nän·ə,gän }

nonahedron A polyhedron with nine faces. { ,nō·nə'hē·drən }

nonassociative algebra A generalization of the concept of an algebra; it is a nonassociative ring R which is a vector space over a field F satisfying $a(xy) = (ax)y = x(ay)$ for all a in F and x and y in R. { ,nän·ə¦sō·shəd·əv 'al·jə·brə }

nonassociative ring A generalization of the concept of a ring; it is an algebraic system with two binary operations called addition and multiplication such that the system is a commutative group relative to addition, and multiplication is distributive with respect to addition, but multiplication is not assumed to be associative. { ˌnän·əˈsō·shəd·əv ˈriŋ }

nonatomic Boolean algebra A Boolean algebra in which there is no element x with the property that if $y \cdot x = y$ for some y, then $y = 0$. { ˌnän·əˈtäm·ik ˈbü·lē·ən ˈal·jə·brə }

nonatomic measure space A measure space in which no point has positive measure. { ˌnän·əˈtäm·ik ˈmezh·ər ˌspās }

noncentral chi-square distribution The distribution of the sum of squares of independent normal random variables, each with unit variance and nonzero mean; used to determine the power function of the chi-square test. { ˌnänˈsen·trəl ˈkī ˌskwer ˌdis·trəˈbyü·shən }

noncentral distribution A distribution of random variables which is not normal. { ˈnän ˌsen·trəl ˌdi·strəˈbyü·shən }

noncentral F distribution The distribution of the ratio of two independent random variables, one with a noncentral chi-square distribution and one with a central chi-square distribution; used to determine the power of the F test in the analysis of variance. { ˌnänˈsen·trəl ˈef ˌdis·trəˈbyü·shən }

noncentral quadric A quadric surface that does not have a point about which the surface is symmetrical; namely, an elliptic or hyperbolic paraboloid, or a quadric cylinder. { ˈnänˌsen·trəl ˈkwäˌdrik }

noncentral t distribution A particular case of a noncentral F distribution; used to test the power of the t test. { ˌnänˈsen·trəl ˈtē ˌdis·trəˈbyü·shən }

noncritical region In testing hypotheses, the set of values leading to acceptance of the null hypothesis. { ˌnänˈkrit·ə·kəl ˈrē·jən }

nondegenerate plane In projective geometry, a plane in which to every line L there are at least two distinct points that do not lie on L, and to every point p there are at least two distinct lines which do not pass through p. { ˌnän·diˈjen·ə·rət ˈplān }

nondenumerable set A set that cannot be put into one-to-one correspondence with the positive integers or any subset of the positive integers. { ˌnän·diˈnüm·rə·bəl ˈset }

nondifferentiable programming The branch of nonlinear programming which does not require the objective and constraint functions to be differentiable. { ˌnänˌdif·əˈren·chə·bəl ˈprōˌgram·iŋ }

nondimensional parameter *See* dimensionless number. { ˌnän·diˈmen·chən·əl pəˈram·əd·ər }

non-Euclidean geometry A geometry in which one or more of the axioms of Euclidean geometry are modified or discarded. { ˌnän·yüˈklid·ē·ən jēˈäm·ə·trē }

nonexpansive mapping A function f from a metric space to itself such that, for any two elements in the space, a and b, the distance between $f(a)$ and $f(b)$ is not greater than the distance between a and b. { ˌnän·ikˌspan·siv ˈmap·iŋ }

nonholonomic constraint One of a nonintegrable set of differential equations which describe the restrictions on the motion of a system. { ˌnänˌhäl·əˈnäm·ik kən ˈstränt }

nonlinear equation An equation in variables $x_1, \ldots, x_n, y$ which cannot be put into the form $a_1x_1 + \cdots + a_nx_n = y$. { ˈnänˌlin·ē·ər iˈkwā·zhən }

nonlinear programming A branch of applied mathematics concerned with finding the maximum or minimum of a function of several variables, when the variables are constrained to yield values of other functions lying in a certain range, and either the function to be maximized or minimized, or at least one of the functions whose value is constrained, is nonlinear. { ˈnänˌlin·ē·ər ˈprōˌgram·iŋ }

nonlinear regression *See* curvilinear regression. { ˈnänˌlin·ē·ər riˈgresh·ən }

nonlinear system A system in which the interrelationships among the quantities involved are expressed by equations, some of which are not linear. { ˈnänˌlin·ē·ər ˈsis·təm }

nonnegative semidefinite *See* positive semidefinite. { ˌnänˈneg·əd·iv ˌsem·iˈdef·ə·nət }

nonomino One of the 1285 plane figures that can be formed by joining nine unit squares along their sides. { nō′näm·ə·nō }

nonorientable surface *See* one-sided surface. { ˌnän¸ȯr·ē¦ent·ə·bəl ′sər·fəs }

nonparametric statistics A class of statistical methods applicable to a large set of probability distributions used to test for correlation, location, independence, and so on. { ¦nän ¦par·ə′me·trik stə′tis·tiks }

nonperiodic decimal *See* nonrepeating decimal. { ˌnän¸pir·ē¸äd·ik ′des·məl }

nonprobabilistic sampling A process in which some criterion other than the laws of probability determines the elements of the population to be included in the sample. { ˌnän¸präb·ə¦lis·tik ′sam·pliŋ }

nonrecurring decimal *See* nonrepeating decimal. { ˌnän·ri·kər·iŋ ′des·məl }

nonremovable discontinuity A point at which a function is not continuous or is undefined, and cannot be made continuous by being given a new value at the point. { ¦nän·ri′müv·ə·bəl dis¸känt·ən′ü·əd·ē }

nonrepeating decimal An infinite decimal that fails to have any finite block of digits that eventually repeats indefinitely. Also known as nonperiodic decimal; nonrecurring decimal. { ˌnän·ri¸pēd·iŋ ′des·məl }

nonresidue A nonresidue of m of order n, where m and n are integers, is an integer a such that $x^n = a + bm$, where x and b are integers, has no solution. { ˌnän′rez· ə¸dü }

nonsense correlation A correlation between two variables that is not due to any causal relationship, but to the fact that each variable is correlated with a third variable, or to random sampling fluctuations. Also known as illusory correlation. { ′nän¸sens ¸kär·ə¸lā·shən }

nonsingular matrix A matrix which has an inverse; equivalently, its determinant is not zero. { ′nän¸siŋ·gyə·lər ′mā·triks }

nonsingular transformation A linear transformation which has an inverse; equivalently, it has null space kernel consisting only of the zero vector. { ′nän¸siŋ·gyə·lər ¸tranz· fər′mā·shən }

nonsquare Banach space A Banach space in which there are no nonzero elements, x and y, that satisfy the equation $\|x + y\| = \|x - y\| = 2\|x\| = 2\|y\|$. { ˌnän¸skwer ′bä¸näk ¸spās }

nonstandard numbers A generalization of the real numbers to include infinitesimal and infinite quantities by considering equivalence classes of infinite sequences of numbers. Also known as hyperreal numbers. { ˌnän¸stan·dərd ′nəm·bərz }

nonterminal vertex A vertex in a rooted tree that has at least one successor. { ¦nän′tər· mən·əl ′vər¸teks }

nonterminating continued fraction A continued fraction that has an infinite number of terms. { ˌnän¦tər·mə¸nād·iŋ kən¦tin·yüd ′frak·shən }

nonterminating decimal A decimal for which there is no digit to the right of the decimal point such that all digits farther to the right are zero. { ˌnän¦tər·mə¸nād· iŋ ′des·məl }

nontrivial solution A solution of a set of homogeneous linear equations in which at least one of the variables has a value different from zero. { ¦nän′triv·ē·əl sə′lü·shən }

NOR A logic operator having the property that if P, Q, R, . . . are statements, then the NOR of P, Q, R, . . . is true if all statements are false, false if at least one statement is true. Derived from NOT-OR. Also known as Peirce stroke relationship. { nȯr }

norm 1. A scalar valued function on a vector space with properties analogous to those of the modulus of a complex number; namely: the norm of the zero vector is zero, all other vectors have positive norm, the norm of a scalar times a vector equals the absolute value of the scalar times the norm of the vector, and the norm of a sum is less than or equal to the sum of the norms. **2.** For a matrix, the square root of the sum of the squares of the moduli of the matrix entries. **3.** For a quaternion, the product of the quaternion and its conjugate. **4.** *See* absolute value. { nȯrm }

normal bundle If A is a manifold and B is a submanifold of A, then the normal bundle

normal curvature

of B in A is the set of pairs (x,y), where x is in B, y is a tangent vector to A, and y is orthogonal to B. { ¦nȯr·məl ¦bən·dəl }

normal curvature The normal curvature at a point on a surface is the curvature of the normal section to the point. { 'nȯr·məl 'kər·və·chər }

normal curve *See* Gaussian curve. { 'nȯr·məl 'kərv }

normal density function A normally distributed frequency distribution of a random variable x with mean e and variance σ is given by $(1/\sqrt{2\sigma})$ exp $[-(x-e)^2/\sigma^2]$. { 'nȯr·məl 'den·səd·ē ˌfəŋk·shən }

normal derivative The directional derivative of a function at a point on a given curve or surface in the direction of the normal to the curve or surface. { 'nȯr·məl di'riv·əd·iv }

normal distribution A commonly occurring probability distribution that has the form

$$(1/\sigma\sqrt{2\pi}) \int_{-\infty}^{u} \exp(-u^2/2)du$$
$$u = (x-e)/\sigma$$

where e is the mean and σ is the variance. Also known as Gauss' error curve; Gaussian distribution. { 'nȯr·məl ˌdi·strə'byü·shən }

normal divisor *See* normal subgroup. { 'nȯr·məl di'vīz·ər }

normal equations The set of equations arising in the least squares method whose solutions give the constants that determine the shape of the estimated function. { ¦nȯr·məl i'kwā·zhənz }

normal extension An algebraic extension of K of a field k, contained in the algebraic closure $\bar{k}$ of k, such that every injective homomorphism of K into $\bar{k}$, inducing the identity on k, is an automorphism of K. { 'nȯr·məl ik'sten·chən }

normal family A family of complex functions analytic in a common domain where every sequence of these functions has a subsequence converging uniformly on compact subsets of the domain to an analytic function on the domain or to $+\infty$. { 'nȯr·məl 'fam·lē }

normal function *See* normalized function. { 'nȯr·məl 'fəŋk·shən }

normalize To multiply a quantity by a suitable constant or scalar so that it then has norm one; that is, its norm is then equal to one. To carry out a normal transformation on a variate. { 'nȯr·məˌlīz }

normalized function A function with norm one; the norm is usually given by an integral $(\int |f|^p d\mu)^{1/p}$, $1 \le p < \infty$. Also known as normal function. { 'nȯr·məˌlīzd 'fəŋk·shən }

normalized standard scores A procedure in which each set of original scores is converted to some standard scale under the assumption that the distribution of scores approximates that of a normal. { ¦nȯr·məˌlīzd ¦stan·dərd 'skȯrz }

normalized support function The function that results from restricting the domain of the independent variable of the support function to the unit sphere. { ¦nȯr·məˌlīzd sə'pȯrt ˌfəŋk·shən }

normalized variate A variate to which a normal transformation has been applied and which therefore has a normal distribution. { ¦nȯr·məˌlīzd 'ver·ē·ət }

normalizer The normalizer of a subset S of a group G is the subgroup of G consisting of all elements x such that xsx^{-1} is in S whenever s is in S. { 'nȯr·məˌlīz·ər }

normally distributed observations Any set of observations whose histogram looks like the normal curve. { 'nȯr·mə·lē di'strib·yəd·əd ˌäb·zər'vā·shənz }

normal map A planar map in which no more than three regions meet any one point and no region completely encloses another. Also known as regular map. { 'nȯr·məl 'map }

normal matrix A matrix is normal if multiplying it on the right by its adjoint is the same as multiplying it on the left. { 'nȯr·məl 'mā,triks }

normal number A number whose expansion with respect to a given base (not necessarily 10) is such that all the digits occur with equal frequency, and all blocks of digits of the same length occur equally often. { 'nȯr·məl 'nəm·bər }

normal operator A linear operator where composing it with its adjoint operator in either order gives the same result. Also known as normal transformation.   { 'nȯr·məl 'äp·ə‚rād·ər }

normal pedal curve The normal pedal curve of a given curve C with respect to a fixed point P is the locus of the foot of the perpendicular from P to the normal to C.   { 'nȯr·məl 'ped·əl ‚kərv }

normal plane For a point P on a curve in space, the plane passing through P which is perpendicular to the tangent to the curve at P.   { 'nȯr·məl 'plān }

normal probability paper Graph paper with the abscissa ruled in uniform increments and the ordinate ruled in such a way that the plot of a cumulative normal distribution is a straight line.   { ¦nȯr·məl ‚präb·ə'bil·əd·ē ‚pā·pər }

normal section Relative to a surface, this is a planar section produced by a plane containing the normal to a point.   { 'nȯr·məl 'sek·shən }

normal series A normal series of a group G is a normal tower of subgroups of G, G_0, $G_1, \ldots, G_n$, in which $G_0 = G$ and G_n is the trivial group containing only the identity element.   { 'nȯr·məl 'sir‚ēz }

normal space A topological space in which any two disjoint closed sets may be covered respectively by two disjoint open sets.   { 'nȯr·məl 'spās }

normal subgroup A subgroup N of a group G where every expression $g^{-1}ng$ is in N for every g in G and every n in N. Also known as invariant subgroup; normal divisor.   { 'nȯr·məl 'səb‚grüp }

normal to a curve The normal to a curve at a point is the line perpendicular to the tangent line at the point.   { 'nȯr·məl tü ə 'kərv }

normal to a surface The normal to a surface at a point is the line perpendicular to the tangent plane at that point.   { 'nȯr·məl tü ə 'sər·fəs }

normal tower A tower of subgroups, $G_0, G_1, \ldots, G_n$, such that each G_{i+1} is normal in G_i, $i = 1, 2, \ldots, n - 1$.   { 'nȯr·məl 'taȯ·ər }

normal transformation *See* normal operator. A transformation on a variate that converts it into a variate which has a normal distribution.   { 'nȯr·məl ‚tranz·fər'mā·shən }

normed linear space A vector space which has a norm defined on it. Also known as normed vector space.   { 'nȯrmd 'lin·ē·ər 'spās }

normed vector space *See* normed linear space.   { 'nȯrmd 'vek·tər 'spās }

NOT-AND *See* NAND.   { 'nät 'and }

notation 1. The use of symbols to denote quantities or operations. **2.** *See* positional notation.   { nō'tā·shən }

NOT function A logical operator having the property that if P is a statement, then the NOT of P is true if P is false, and false if P is true.   { 'nät ‚fəŋk·shən }

NOT-OR *See* NOR.   { 'nät 'ȯr }

nowhere dense set A set in a topological space whose closure has empty interior. Also known as rare set.   { 'nō‚wer 'dens 'set }

n space A vector space over the real numbers whose basis has n vectors.   { 'en ‚spās }

n-sphere The set of all points in $(n + 1)$-dimensional Euclidean space whose distance from the origin is unity, where n is a positive integer.   { 'en ‚sfir }

nuisance parameter A parameter to be estimated by a statistic which arises in the distribution of the statistic under some hypothesis to be tested about the parameter.   { 'nü·səns pə‚ram·əd·ər }

null Indicating that an object is nonexistent or a quantity is zero.   { nəl }

nullary composition The selection of a particular element of a set.   { 'nəl·ə·rē ‚käm·pə'zish·ən }

null geodesic In a Riemannian space, a minimal geodesic curve.   { 'nəl ‚jē·ə'des·ik }

null hypothesis The hypothesis that there is no validity to the specific claim that two variations (treatments) of the same thing can be distinguished by a specific procedure.   { 'nəl hī'päth·ə·səs }

nullity The dimension of the null space of a linear transformation.   { 'nəl·əd·ē }

null matrix The matrix all of whose entries are zero.   { 'nəl 'mā·triks }

null sequence A sequence of numbers or functions which converges to the number zero or the zero function.   { 'nəl 'sē·kwəns }

null set The empty set; the set which contains no elements.   { 'nəl 'set }

null space For a linear transformation, the vector subspace of all vectors which the transformation sends to the zero vector. Also known as kernel. { 'nəl 'spās }

null vector A vector whose invariant length, that is, the sum over the coordinates of the vector space of the product of its covariant component and contravariant component, is equal to zero. { 'nəl 'vek·tər }

number 1. Any real or complex number. **2.** The number of elements in a set is the cardinality of the set. { 'nəm·bər }

number class modulo N The class of all numbers which differ from a given number by a multiple of N. { 'nəm·bər ¦klas ¦mäj·ə·lō 'en }

number field Any set of real or complex numbers that includes the sum, difference, product, and quotient (except division by zero) of any two members of the set. { 'nəm·bər ‚fēld }

number line *See* real line. { 'nəm·bər ‚līn }

number scale Representation of points on a line with numbers arranged in some order. { 'nəm·bər ‚skāl }

number system 1. A mathematical system, such as the real or complex numbers, the quaternions, or the Cayley numbers, that satisfies many of the axioms of the real number system; in general, it is a finite-dimensional vector space over the real numbers with multiplicative operation under which it is an associative or nonassociative division algebra. **2.** *See* numeration system. { 'nəm·bər ‚sis·təm }

number-theoretic function A function whose domain is the set of positive integers. { ¦nəm·bər ‚thē·ə¦red·ik 'fəŋk·shən }

number theory The study of integers and relations between them. { 'nəm·bər 'thē·ə·rē }

numeral A symbol used to denote a number. { 'nüm·rəl }

numeral system *See* numeration system. { 'nüm·rəl ‚sis·təm }

numeration The listing of numbers in their natural order. { ‚nü·mə'rā·shən }

numeration system An orderly method of representing numbers by numerals in which each numeral is associated with a unique number. Also known as number system; numeral system. { ‚nü·mə'rā·shən ‚sis·təm }

numerator In a fraction a/b, the numerator is the quantity a. { 'nü·mə‚rād·ər }

numerical Pertaining to numbers. { nü'mer·i·kəl }

numerical analysis The study of approximation techniques using arithmetic for solutions of mathematical problems. { nü'mer·i·kəl ə'nal·ə·səs }

numerical equation An equation all of whose constants and coefficients are numbers. { nü'mer·i·kəl i'kwā·zhən }

numerical integration The process of using a set of approximate values of a function to calculate its integral to comparable accuracy. { nü'mer·i·kəl ‚int·ə'grā·shən }

numerical range For a linear operator T of a Hilbert space into itself, the set of values assumed by the inner product of Tx with x as x ranges over the set of vectors with norm equal to 1. { nü'mer·i·kəl 'rānj }

numerical tensor A tensor whose components are the same in all coordinate systems. { nü'mer·i·kəl 'ten·sər }

numerical value *See* absolute value. { nü'mer·i·kəl 'val·yü }

n-way analysis A statistical analysis in which n major factors are used to jointly classify the observed values, where n is a positive integer. { 'en ‚wā ə‚nal·ə·səs }

O

obelisk A frustrum of a regular, rectangular pyramid. { 'äb·ə‚lisk }

objective function In nonlinear programming, the function, expressing given conditions for a system, which one seeks to minimize subject to given constraints. { äb'jek·tiv 'fəŋk·shən }

objective probabilities Probabilities determined by the long-run relative frequency of an event. Also known as frequency probabilities. { əb'jek·tiv ‚präb·ə'bil·əd·ēz }

oblate ellipsoid *See* oblate spheroid. { 'ä‚blāt i'lip‚sȯid }

oblate spheroid The surface or ellipsoid generated by rotating an ellipse about one of its axes so that the diameter of its equatorial circle exceeds the length of the axis of revolution. Also known as oblate ellipsoid. { 'ä‚blāt 'sfir‚ȯid }

oblate spheroidal coordinate system A three-dimensional coordinate system whose coordinate surfaces are the surfaces generated by rotating a plane containing a system of confocal ellipses and hyperbolas about the minor axis of the ellipses, together with the planes passing through the axis of rotation. { ¦ō‚blāt sfir¦ȯid·əl kō'ȯrd·ən·ət ‚sis·təm }

oblique angle An angle that is neither a right angle nor a multiple of a right angle. { ə'blēk 'aŋ·gəl }

oblique circular cone A circular cone whose axis is not perpendicular to its base. { ə¦blēk ¦sər·kyə·lər 'kōn }

oblique coordinates Magnitudes defining a point relative to two intersecting nonperpendicular lines, called axes; the magnitudes indicate the distance from each axis, measured along a parallel to the other axis; oblique coordinates are a form of cartesian coordinates. { ə'blēk kō'ȯrd·ən·əts }

oblique lines Lines that are neither perpendicular nor parallel. { ə'blēk 'līnz }

oblique parallelepiped A parallelepiped whose lateral edges are not perpendicular to its bases. { ə¦blēk ‚par·ə‚lel·ə'pī‚ped }

oblique spherical triangle A spherical triangle that has no right angle. { ə¦blēk ¦sfer·ə·kəl 'trī‚aŋ·gəl }

oblique strophoid A plane curve derived from a straight line L and two points called the pole and the fixed point, where the fixed point lies on L but is not the foot of the perpendicular from the pole to the line; it consists of the locus of points on a rotating line L' passing through the pole whose distance from the intersection of L and L' is equal to the distance of this intersection from the fixed point. { ə'blēk 'strō‚fȯid }

oblique triangle A triangle that does not contain a right angle. { ə'blēk 'trī‚aŋ·gəl }

obtuse angle An angle of more than 90° and less than 180°. { äb'tüs 'aŋ·gəl }

obtuse triangle A triangle having one obtuse angle. { äb'tüs 'trī‚aŋ·gəl }

OC curve *See* operating characteristic curve. { ¦ō'sē ‚kərv }

octagon A polygon with eight sides. { 'äk·tə‚gän }

octahedral group The group of motions of three-dimensional space that transform a regular octahedron into itself. { ‚äk·tə'hē·drəl ‚grüp }

octahedron A polyhedron having eight faces, each of which is an equilateral triangle. { ‚ak·tə'hē·drən }

octal Pertaining to the octal number system. { 'äkt·əl }

octal digit The symbol 0, 1, 2, 3, 4, 5, 6, or 7 used as a digit in the octal number system. { 'äkt·əl 'dij·ət }

octal number system A number system in which a number r is written as $n_k n_{k-1} \cdots n_1$ where $r = n_1 8^0 + n_2 8^1 + \cdots + n_k 8^{k-1}$. Also known as octonary number system. { 'äkt·əl 'nəm·bər ,sis·təm }

octant 1. One of the eight regions into which three-dimensional Euclidean space is divided by the coordinate planes of a Cartesian coordinate system. **2.** A unit of plane angle equal to 45° or $\pi/8$ radians. { 'äk·tənt }

octillion 1. The number 10^{27}. **2.** In British and German usage, the number 10^{48}. { äk'til·yən }

octomino One of the 369 plane figures that can be formed by joining eight unit squares along their sides. { äk'täm·ə·nō }

octonary number system See octal number system. { äk¦tän·ə·rē 'nəm·bər ,sis·təm }

octonions See Cayley numbers. { äk'tän·yənz }

odd function A function $f(x)$ is odd if, for every x, $f(-x) = -f(x)$. { 'äd ,fəŋk·shən }

odd number A natural number not divisible by 2. { 'äd 'nəm·bər }

odd permutation A permutation that may be represented as the result of an odd number of transpositions. { 'äd ,pər·myə'tā·shən }

odds ratio The ratio of the probability of occurrence of an event to the probability of the event not occurring. { 'ädz ,rā·shō }

odd vertex A vertex whose degree is an odd number. { 'äd ¦vər,teks }

one-dimensional strain A transformation that elongates or compresses a configuration in a given direction, given by $x' = kx$, $y' = y$, $z' = z$, where k is a constant, when the direction is that of the x axis. { 'wən di,men·chən·əl 'strān }

one-parameter semigroup A semigroup with which there is associated a bijective mapping from the positive real numbers onto the semigroup. { ¦wən pə¦ram·əd·ər 'sem·i,grüp }

one-point compactification The one-point compactification $\bar{X}$ of a topological space X is the union of X with a set consisting of a single element, with the topology of $\bar{X}$ consisting of the open subsets of X and all subsets of $\bar{X}$ whose complements in $\bar{X}$ are closed compact subsets in X. Also known as Alexandroff compactification. { 'wən ,pȯint kəm,pak·tə·fə'kā·shən }

one-sample problem The problem of testing the hypothesis that the average of a sequence of observations or measurement of the same kind has a specified value. { 'wən ,sam·pəl 'präb·ləm }

one-sided limit Either a limit on the left or a limit on the right. { 'wən ,sīd·əd 'lim·ət }

one-sided surface A surface such that an object resting on one side can be moved continuously over the surface to reach the other side without going around an edge; the Möbius band and the Klein bottle are examples. Also known as nonorientable surface. { 'wən ,sīd·əd 'sər·fəs }

one-sided test A test statistic T which rejects a hypothesis only for $T \geq d$ or $T \leq c$ but not for both (here d and c are critical values). { 'wən ,sīd·əd 'test }

one-tail test See one-tailed test. { ¦wən ¦tāl 'test }

one-tailed test A statistical test in which the critical region consists of all values of a test statistic that are less than a given value or greater than a given value, but not both. Also known as one-tail test. { ¦wən ¦tāld 'test }

one-to-one correspondence A pairing between two classes of elements whereby each element of either class is made to correspond to one and only one element of the other class. { ¦wən tə ¦wən ,kär·ə'spän·dəns }

one-valued function See single-valued function. { ¦wən ,val·yüd 'fəŋk·shən }

one-way classification The basis for the simplest case of the analysis of variance; a set of observations are categorized according to values of one variable or one characteristic. { 'wən ,wā ,klas·ə·fə'kā·shən }

open ball In a metric space, an open set about a point x which consists of all points that are less than a fixed distance from x. { 'ō·pən 'bȯl }

open circular region The interior of a circle. { ¦ō·pən 'sər·kyə·lər ,rē·jən }

open covering For a set S in a topological space, a collection of open sets whose union contains S. { 'ō·pən 'kəv·ər·iŋ }

open-ended class The first or last class interval in a frequency distribution having no upper or lower limit. { ¦ō·pən ¦en·dəd 'klas }

open half plane A half plane that does not include any of the line that bounds it. { ¦ō· pən ¦haf 'plān }

open half space A half space that does not include any of the plane that bounds it. { ¦ō·pən ¦haf 'spās }

open interval An open interval of real numbers, denoted by (a,b), consists of all numbers strictly greater than a and strictly less than b. { 'ō·pən 'in·tər·vəl }

open map A function between two topological spaces which sends each open set of one to an open set of the other. { 'ō·pən 'map }

open mapping theorem A continuous linear function between Banach spaces which has closed range must be an open map. { 'ō·pən 'map·iŋ ‚thir·əm }

open n-cell A set that is homeomorphic with the set of points in n-dimensional Euclidean space ($n = 1, 2, \ldots$) whose distance from the origin is less than unity. { ¦ō· pən 'en ‚sel }

open polygonal region The interior of a polygon. { ¦ō·pən pə'lig·ən·əl ‚rē·jən }

open rectangular region The interior of a rectangle. { ¦ō·pən rek'taŋ·gyə·lər ‚rē·jən }

open region *See* domain. { 'ō·pən ‚rē·jən }

open sentence *See* propositional function. { ¦ō·pən 'sent·əns }

open set A set included in a topology; equivalently, a set which is a neighborhood of each of its points; a topology on a space is determined by a collection of subsets which are called open. { 'ō·pən ‚set }

open simplex A modification of a simplex with vertices $p_0, p_1, \ldots, p_n$, in which the points of the simplex, $a_0 p_0 + a_1 p_1 + \cdots + a_n p_n$, are restricted by the condition that each of the coefficients a_i must be greater than 0. { 'ō·pən 'sim‚pleks }

open statement *See* propositional function. { ¦ō·pən 'stāt·mənt }

open triangular region The interior of a triangle. { ¦ō·pən trī'aŋ·gyə·lər ‚rē·jən }

operating characteristic curve In hypothesis testing, a plot of the probability of accepting the hypothesis against the true state of nature. Abbreviated OC curve. { 'äp·ə‚rād·iŋ ‚kar·ik·tə'ris·tik 'kərv }

operation An operation of a group G on a set S is a mapping which associates to each ordered pair (g,s), where g is in G and s is in S, another element in S, denoted gs, such that, for any g,h in G and s in S, $(gh)s = g(hs)$, and $es = s$, where e is the identity element of G. { ‚äp·ə'rā·shən }

operational analysis *See* operational calculus. { ‚äp·ə'rā·shən·əl ə'nal·ə·səs }

operational calculus A technique by which problems in analysis, in particular differential equations, are transformed into algebraic problems, usually the problem of solving a polynomial equation. Also known as operational analysis. { ‚äp·ə'rā· shən·əl 'kal·kyə·ləs }

operations research The mathematical study of systems with input and output from the viewpoint of optimization subject to given constraints. { ‚äp·ə'rā·shənz ri‚sərch }

operator A function between vector spaces. { 'äp·ə‚rād·ər }

operator algebra An algebra whose elements are functions and in which the multiplication of two elements f and g is defined by composition; that is, $(fg)(x) = (f \circ g)(x) = f[g(x)]$. { 'äp·ə‚rād·ər ‚al·jə·brə }

operator theory The general qualitative study of operators in terms of such concepts as eigenvalues, range, domain, and continuity. { 'äp·ə‚rād·ər ‚thē·ə·rē }

oppositely congruent figures Two solid figures, one of which can be made to coincide with the other by a rigid motion in space combined with reflection through a plane. { ¦äp·ə·zət·lē ¦kän‚grü·ənt 'fig·yərz }

opposite rays Two rays that lie on the same or parallel lines but point in opposite directions. { ¦äp·ə·zət 'rāz }

opposite side 1. One of two sides of a polygon with an even number of sides that have the same number of sides between them along either path around the polygon from one of the sides to the other. **2.** For a given vertex of a polygon with an odd number of sides, a side of the polygon that has the same number of sides between it and the vertex along either path around the polygon.   { 'äp·ə·zət 'sīd }

opposite vertices Two vertices of a polygon with an even number of sides that have the same number of sides between them along either path around the polygon from one vertex to the other.   { 'äp·ə·zət 'vərd·ə,sēz }

optimal policy In optimization problems of systems, a sequence of decisions changing the states of a system in such a manner that a given criterion function is minimized.   { 'äp·tə·məl 'päl·ə·sē }

optimal strategy One of the pair of mixed strategies carried out by the two players of a matrix game when each player adjusts strategy so as to minimize the maximum loss that an opponent can inflict.   { 'äp·tə·məl 'strad·ə·jē }

optimal system A system where the variables representing the various states are so determined that a given criterion function is minimized subject to given constraints.   { 'äp·tə·məl 'sis·təm }

optimization The maximizing or minimizing of a given function possibly subject to some type of constraints.   { ,äp·tə·mə'zā·shən }

optimization theory The specific methodology, techniques, and procedures used to decide on the one specific solution in a defined set of possible alternatives that will best satisfy a selected criterion; includes linear and nonlinear programming, stochastic programming, and control theory.   Also known as mathematical programming.   { ,äp·tə·mə'zā·shən ,thē·ə·rē }

optimum allocation A procedure used in stratified sampling to allocate numbers of sample units to different strata to either maximize precision at a fixed cost or minimize cost for a selected level of precision.   { 'äp·tə·məm ,al·ə'kā·shən }

or A logical operation whose result is false (or zero) only if every one of its operands is false, and true (or one) otherwise.   Also known as inclusive or.   { ȯr }

orbit Let G be a group which operates on a set S; the orbit of an element s of S under G is the subset of S consisting of all elements gs where g is in G.   { 'ȯr·bət }

orbit space The orbit space of a G space X is the topological space whose points are equivalence classes obtained by identifying points in X which have the same G orbit and whose topology is the largest topology that makes the function which sends x to its orbit continuous.   { 'ȯr·bət ,spās }

order 1. A differential equation has order n if the derivatives of a function appear up to the nth derivative.   **2.** The number of elements contained within a given group.   **3.** A square matrix with n rows and n columns has order n.   **4.** The number of poles a given elliptic function has in a parallelogram region where it repeats its values.   **5.** A characteristic of infinitesimals used in their comparison.   **6.** For a polynomial, the largest exponent appearing in the polynomial.   **7.** The number of vertices of a graph.   **8.** For a pole of an analytic function, the largest negative power in the function's Laurent expansion about the pole.   **9.** For a zero point z_0 of an analytic function, the integer n such that the function near the pole has the form $g(z)(z - z_0)^n$, where $g(z)$ is analytic at z_0 and does not vanish there.   **10.** For an algebraic curve or surface, the degree of its equation.   **11.** For an algebra, the dimension of the underlying vector space.   **12.** For a branch point of a Riemann surface, the number of sheets of the surface that join at the branch point, minus one.   **13.** *See* ordering.   { 'ȯrd·ər }

ordered field A field with an ordering as a set analogous to the properties of less than or equal for real numbers relative to addition and multiplication.   { 'ȯrd·ərd 'fēld }

ordered *n*-tuple A set of n elements, $x_1, x_2, \ldots, x_n$, written $(x_1, x_2, \ldots, x_n)$, where x_1 is distinguished as first, x_2 as second, and so on.   { ¦ȯrd·ərd 'en, tep·əl }

ordered pair A pair of elements x and y from a set, written (x,y), where x is distinguished as first and y as second.   { 'ȯrd·ərd 'per }

ordered partition For a set A, an ordered sequence whose members are the members of a partition of A.   { ¦ȯrd·ərd pär'tish·ən }

ordered quadruple A set of four elements, distinguished as first, second, third, and fourth.   { ¦ȯrd·ərd kwä'drüp·əl }

ordered rings Rings which have an ordering on them as sets in a manner analogous to the behavior of the usual ordering of the real numbers relative to addition and multiplication. { 'ȯrd·ərd 'riŋz }

ordered triple A set of three elements, written (x,y,z), where x is distinguished as first, y as second, and z as third. { ¦ȯrd·ərd 'trip·əl }

ordering A binary relation, denoted ≤, among the elements of a set such that $a \leq b$ and $b \leq c$ implies $a \leq c$, and $a \leq b$, $b \leq a$ implies $a = b$; it need not be the case that either $a \leq b$ or $b \leq a$. Also known as order; order relation; partial ordering. { 'ȯrd·ə·riŋ }

order of degeneracy See degree of degeneracy. { 'ȯrd·ər əv di¦jen·ə·rə·sē }

order relation See ordering. { 'ȯrd·ər ri,lā·shən }

order statistics Variate values arranged in ascending order of magnitude; for example, first-order statistic is the smallest value of sample observations. { ¦ȯr·dər stə¦tis·tiks }

ordinal number A generalized number which expresses the size of a set, in the sense of "how many" elements. { 'ȯrd·nəl 'nəm·bər }

ordinal scale measurement A method of measuring quantifiable data in nonparametric statistics that is considered to be stronger than nominal scale; it expresses the relationship of order by characterizing objects by relative rank. { ¦ȯrd·nəl ,skāl 'mezh·ər·mənt }

ordinary differential equation An equation involving functions of one variable and their derivatives. { 'ȯrd·ən,er·ē ,dif·ə'ren·chəl i'kwā·zhən }

ordinary generating function See generating function. { ,ȯrd·ən,er·ē 'jen·ə,rād·iŋ ,fəŋk·shən }

ordinary point A point of a curve where a curve does not cross itself and where there is a smoothly turning tangent. Also known as regular point; simple point { 'ȯrd·ən,er·ē 'pȯint }

ordinary singular point A singular point at which the tangents to all branches at the point are distinct. { ¦ȯrd·ən,er·ē ¦siŋ·gyə·lər 'pȯint }

ordinate The perpendicular distance of a point (x,y) of the plane from the x axis. { 'ȯrd·ən·ət }

orientable surface A surface for which an object resting on one side of it cannot be moved continuously over it to get to the other side without going around an edge. { ,ȯr·ē,en·tə·bəl 'sər·fəs }

orientation 1. A choice of sense or direction in a topological space. 2. An ordering $p_0, p_1, \ldots, p_n$ of the vertices of a simplex, two such orderings being regarded as equivalent if they differ by an even permutation. 3. For a simple graph, a directed graph that results from assigning a direction to each of the edges. { ,ȯr·ē·ən'tā·shən }

oriented graph A directed graph in which there is no pair of points a and b such that there is both an arc directed from a to b and an arc directed from b to a. { 'ȯr·ē·ent·əd 'graf }

oriented simplex A simplex for which an order has been assigned to the vertices. { ,ȯr·ē,ent·əd 'sim,pleks }

oriented simplicial complex A simplicial complex each of whose simplexes is an oriented simplex. { ,ȯr·ē,ent·əd sim¦plish·əl 'käm,pleks }

origin The point of a coordinate system at which all coordinate axes meet. { 'är·ə·jən }

Ornstein-Uhlenbeck process A stochastic process used as a theoretical model for Brownian motion. { ¦ȯrn,stīn 'ü·lən,bek ,prä,ses }

orthocenter The point at which the altitudes of a triangle intersect. { ¦ȯr·thō'sen·tər }

orthogonal Perpendicular, or some concept analogous to it. { ȯr'thäg·ən·əl }

orthogonal basis A basis for an inner product space consisting of mutually orthogonal vectors. { ȯr'thäg·ən·əl 'bā·səs }

orthogonal complement In an inner product space, the orthogonal complement of a vector **v** consists of all vectors orthogonal to **v**; the orthogonal complement of a subset S consists of all vectors orthogonal to each vector in S. { ȯr'thäg·ən·əl 'käm·plə·mənt }

orthogonal family See orthogonal system. { ȯr'thäg·ən·əl'fam·lē }

orthogonal functions

orthogonal functions Two real-valued functions are orthogonal if their inner product vanishes. { ȯr'thäg·ən·əl 'fəŋk·shənz }

orthogonal group The group of matrices arising from the orthogonal transformations of a euclidean space. { ȯr'thäg·ən·əl 'grüp }

orthogonality Two geometric objects have this property if they are perpendicular. { ȯr,thäg·ə'nal·əd·ē }

orthogonalization A procedure in which, given a set of linearly independent vectors in an inner product space, a set of orthogonal vectors is recursively obtained so that each set spans the same subspace. { ȯr,thäg·ə·nə·lə'zā·shən }

orthogonal Latin squares Two Latin squares which, when superposed, have the property that the cells contain each of the possible pairs of symbols exactly once. { ȯr¦thäg·ən·əl ¦lat·ən 'skwerz }

orthogonal lines Lines which are perpendicular. { ȯr'thäg·ən·əl 'līnz }

orthogonal matrix A matrix whose inverse and transpose are identical. { ȯr'thäg·ən·əl 'mā·triks }

orthogonal polynomial Orthogonal polynomials are various families of polynomials, which arise as solutions to differential equations related to the hypergeometric equation, and which are mutually orthogonal as functions. { ȯr'thäg·ən·əl päl·ə'nō·mē·əl }

orthogonal projection Also known as orthographic projection. **1.** A continuous linear map P of a Hilbert space H onto a subspace M such that if $\mathbf{h}$ is any vector in H, $\mathbf{h} = P\mathbf{h} + \mathbf{w}$, where $\mathbf{w}$ is in the orthogonal complement of M. **2.** A mapping of a configuration into a line or plane that associates to any point of the configuration the intersection with the line or plane of the line passing through the point and perpendicular to the line or plane. { ȯr'thäg·ən·əl prə'jek·shən }

orthogonal series An infinite series each term of which is the product of a member of an orthogonal family of functions and a coefficient; the coefficients are usually chosen so that the series converges to a desired function. { ȯr¦thäg·ən·əl 'sir,ēz }

orthogonal spaces Two subspaces F and F' of a vector space E with a scalar product g such that $g(x,x') = 0$ for any x in F and x' in F'. { ȯr¦thäg·ən·əl 'spās·əz }

orthogonal sum 1. A vector space E with a scalar product is said to be the orthogonal sum of subspaces F and F' if E is the direct sum of F and F' and if F and F' are orthogonal spaces. **2.** A scalar product g on a vector space E is said to be the orthogonal sum of scalar products f and f' on subspaces F and F' if E is the orthogonal sum of F and F' (in the sense of the first definition) and if $g(x + x', y + y') = f(x,y) + f'(x',y')$ for all x,y in F and x',y' in F'. { ȯr¦thäg·ən·əl 'səm }

orthogonal system 1. A system made up of n families of curves on an n-dimensional manifold in an $(n + 1)$-dimensional Euclidean space, such that exactly one curve from each family passes through every point in the manifold, and, at each point, the tangents to the n curves that pass through that point are mutually perpendicular. **2.** A set of real-valued functions, the inner products of any two of which vanish. Also known as orthogonal family. { ȯr'thäg·ən·əl 'sis·təm }

orthogonal trajectory A curve that intersects all the curves of a given family at right angles. { ȯr'thäg·ən·əl trə'jek·tə·rē }

orthogonal transformation A linear transformation between real inner product spaces which preserves the length of vectors. { ȯr'thäg·ən·əl ,tranz,fȯr'mā·shən }

orthogonal vectors In an inner product space, two vectors are orthogonal if their inner product vanishes. { ȯr'thäg·ən·əl 'vek·tərz }

orthographic projection See orthogonal projection. { ¦ȯr·thə¦graf·ik prə'jek·shən }

orthonormal coordinates In an inner product space, the coordinates for a vector expressed relative to an orthonormal basis. { ¦ȯr·thə¦nȯr·məl kō'ȯrd·ən·əts }

orthonormal functions Orthogonal functions $f_1, f_2, \ldots$ with the additional property that the inner product of $f_n(x)$ with itself is 1. { ¦ȯr·thə¦nȯr·məl 'fəŋk·shənz }

orthonormal vectors A collection of mutually orthogonal vectors, each having length 1. { ¦ȯr·thə¦nȯr·məl 'vek·tərz }

orthoptic The locus of the intersection of tangents to a given curve that meet at a right angle. { ȯr'thäp·tik }

orthotomic The orthotomic of a curve with respect to a point is the envelope of the circles which pass through the point and whose centers lie on the curve. { ˌȯr·thəˈtäm·ik }

oscillating series A series that is divergent but not properly divergent; that is, the partial sums do not approach a limit, or become arbitrarily large or arbitrarily small. { 'äs·ə,lād·iŋ ,sir,ēz }

oscillation 1. The oscillation of a real-valued function on an interval is the difference between its least upper bound and greatest lower bound there. **2.** The oscillation of a real-valued function at a point x is the limit of the oscillation of the function on the interval $[x - e, x + e]$ as e approaches 0. Also known as saltus. { ˌäs·əˈlā·shən }

osculating circle For a plane curve C at a point p, the limiting circle obtained by taking the circle that is tangent to C at p and passes through a variable point q on C, and then letting q approach p. { ¦äs·kyə,lād·iŋ 'sər·kəl }

osculating plane For a curve C at some point p, this is the limiting plane obtained from taking planes through the tangent to C at p and containing some variable point p' and then letting p' approach p along C. { 'äs·kyə,lād·iŋ 'plān }

osculating sphere For a curve C at a point p, the limiting sphere obtained by taking the sphere that passes through p and three other points on C and then letting these three points approach p independently along C. { ¦äs·kyə,lād·iŋ 'sfir }

Ostrogradski's theorem *See* Gauss' theorem. { ˌȯ·strəˈgräd·skēz ˌthir·əm }

outdegree For a vertex, v, in a directed graph, the number of arcs directed from v to other vertices. { 'aùt·di,grē }

outer automorphism Any element of the quotient group formed from the group of automorphisms of a group and the subgroup of inner automorphisms. { 'aùd·ər ¦ȯd·ō'mȯr,fiz·əm }

outer measure 1. A function with the same properties as a measure except that it is only countably subadditive rather than countably additive; usually defined on the collection of all subsets of a given set. **2.** *See* Lebesgue exterior measure. { 'aùd·ər 'mezh·ər }

outer product For any two tensors R and S, a tensor T each of whose indices corresponds to an index of R or an index of S, and each of whose components is the product of the component of R and the component of S with identical values of the corresponding indices. { ¦aùd·ər 'präd·əkt }

outflow The outflow from a vertex in an *s-t* network is the sum of the flows of all the arcs that originate at that vertex. { 'aùt,flō }

outlier In a set of data, a value so far removed from other values in the distribution that its presence cannot be attributed to the random combination of chance causes. { 'aùt,lī·ər }

oval A curve shaped like a section of an egg. { 'ō·vəl }

oval of Cassini An ovallike curve similar to a lemniscate obtained as the locus corresponding to a general type of quadratic equation in two variables x and y; it is expressed as $[(x + a)^2 + y^2] [(x - a)^2 + y^2] = k^4$, where a and k are constants. Also known as Cassinian oval. { 'ō·vəl əv kə'sē·nē }

over a map A map f from a set A to a set L is said to be over a map g from a set B to L if B is a subset of A and the restriction of f to B equals g. { ¦ō·vər ə 'map }

over a set A map f from a set A to a set L is said to be over a set B if B is a subset of both A and L and if the restriction of F to B is the identity map on B. { ¦ō·vər ə 'set }

P

p- *See* pico-.

Paasche's index A weighted aggregate price index with given-year quantity weights. Also known as given-year method. { ¦päs·kəz 'in¦deks }

Pade table A table associated to a power series having in its pth row and qth column the ratio of a polynomial of degree q by one of degree p so that this fraction expanded into a power series agrees with the original up to the $p + q$ term. { 'päd· ə ˌtā·bəl }

p-adic field For a fixed prime number, p, the set of all p-adic numbers, with addition and multiplication defined in a natural way. { ˌpē 'ad·ik ¦fēld }

p-adic integer For a fixed prime number p, a sequence of integers, $x_0, x_1, \ldots$, such that $x_n - x_{n-1}$ divisible by p^n for all $n \geq 0$; two such sequences, x_n and y_n, are considered equal if $x_n - y_n$ is divisible by p^{n+1} for all $n \geq 0$, and the sum and product of two such sequences is defined by term-by-term addition and multiplication. { pē¦adik 'int·i·jər }

p-adic number For a fixed prime number p, a fraction of the form a/p^k, where a is a p-adic integer and k is a nonnegative integer; two such fractions, a/p^k and b/p^m, are considered equal if ap^m and bp^k are the same p-adic integer.

paired comparison A method used where order relations are more easily determined than measurements, such as studying taste preferences; in the comparison of a group of objects, each pair of objects is tested with either one or the other or neither preferred. { ¦perd kəm'par·ə·sən }

pairwise disjoint The property of a collection of sets such that no two members of the collection have any elements in common. { ¦per,wīz dis'jóint }

Pappian plane Any projective plane in which points and lines satisfy Pappus' theorem (third definition). { ¦pap·ē·ən 'plān }

Pappus' theorem 1. The proposition that the area of a surface of revolution generated by rotating a plane curve about an axis in its own plane which does not intersect it is equal to the length of the curve multiplied by the length of the path of its centroid. **2.** The proposition that the volume of a solid of revolution generated by rotating a plane area about an axis in its own plane which does not intersect it is equal to the area multiplied by the length of the path of its centroid. **3.** A theorem of projective geometry which states that if A, B, and C are collinear points and A', B' and C' are also collinear points, then the intersection of AB' with $A'B$, the intersection of AC' with $A'C$, and the intersection of BC' with $B'C$ are collinear. **4.** A theorem of projective geometry which states that if A, B, C, and D are fixed points on a conic and P is a variable point on the same conic, then the product of the perpendiculars from P to AB and CD divided by the product of the perpendiculars from P to AD and BC is constant. { 'pap·əs ˌthir·əm }

parabola The plane curve given by an equation of the form $y = ax^2 + bx + c$. { pə'rab· ə·lə }

parabolic coordinate system 1. A two-dimensional coordinate system determined by a system of confocal parabolas. **2.** A three-dimensional coordinate system whose coordinate surfaces are the surfaces generated by rotating a plane containing a system of confocal parabolas about the axis of symmetry of the parabolas, together with the planes passing through the axis of rotation. { ¦par·ə¦bäl·ik kō'órd·ən·ət ˌsis·təm }

parabolic cylinder A cylinder whose directrix is a parabola. { ¦par·ə¦bäl·ik 'sil·ən·dər }

parabolic cylinder functions Solutions to the Weber differential equation, which results from separation of variables of the Laplace equation in parabolic cylindrical coordinates. { ¦par·ə¦bäl·ik 'sil·ən·dər ˌfəŋk·shənz }

parabolic cylindrical coordinate system A three-dimensional coordinate system in which two of the coordinates depend on the x and y coordinates in the same manner as parabolic coordinates and are independent of the z coordinate, while the third coordinate is directly proportional to the z coordinate. { ¦par·ə¦bäl·ik si¦lin·drə·kəl kō'órd·ən·ət ˌsis·təm }

parabolic differential equation A general type of second-order partial differential equation which includes the heat equation and has the form

$$\sum_{i,j=1}^{n} A_{ij}(\partial^2 u/\partial x_i \partial x_j) + \sum_{i=1}^{n} B_i(\partial u/\partial x_i) + Cu + F = 0$$

where the A_{ij}, B_i, C, and F are suitably differentiable real functions of $x_1, x_2, \ldots, x_n$, and there exists at each point $(x_1, \ldots, x_n)$ a real linear transformation on the x_i which reduces the quadratic form

$$\sum_{i,j=1}^{n} A_{ij} x_i x_j$$

to a sum of fewer than n squares, not necessarily all of the same sign, while the same transformation does not reduce the B_i to 0. Also known as parabolic partial differential equation. { ¦par·ə¦bäl·ik ˌdif·ə'ren·chəl i'kwā·zhən }

parabolic partial differential equation See parabolic differential equation. { ¦par·ə¦bäl·ik ¦pär·shəl ˌdif·ə'ren·chəl i,kwā·zhən }

parabolic point A point on a surface where the total curvature vanishes. { ¦par·ə¦bäl·ik 'póint }

parabolic Riemann surface See parabolic type. { ˌpar·ə,bäl·ik 'rē,män ˌsər·fəs }

parabolic rule See Simpson's rule. { ¦par·ə¦bäl·ik 'rül }

parabolic segment The line segment given by a chord perpendicular to the axis of a parabola. { ¦par·ə¦bäl·ik 'seg·mənt }

parabolic spiral The curve whose equation in polar coordinates is $r^2 = a\theta$. { ¦par·ə¦bäl·ik 'spī·rəl }

parabolic type A type of simply connected Riemann surface that can be mapped conformally on the complex plane, excluding the origin and the point at infinity. Also known as Riemann surface. { ¦par·ə¦bäl·ik ˌtīp }

paraboloid A surface where sections through one of its axes are ellipses or hyperbolas, and sections through the other are parabolas. { pə'rab·ə,lóid }

paraboloidal coordinate system A three-dimensional coordinate system in which the coordinate surfaces form families of confocal elliptic and hyperbolic paraboloids. { pə¦rab·ə¦lóid·əl kō'órd·ən·ət ˌsis·təm }

paraboloid of revolution The surface obtained by rotating a parabola about its axis. { pə'rab·ə,lóid əv ˌrev·ə'lü·shən }

paracompact space A topological space with the property that every open covering F is associated with a locally finite open covering G, such that every element of G is a subset of an element F. { ¦par·ə¦käm,pakt ˌspās }

parallel 1. Lines are parallel in a Euclidean space if they lie in a common plane and do not intersect. **2.** Planes are parallel in a Euclidean three-dimensional space if they do not intersect. **3.** A circle parallel to the primary great circle of a sphere or spheroid. **4.** A curve is parallel to a given curve C if it consists of points that are a fixed distance from C along lines perpendicular to C. { 'par·ə,lel }

parallel axiom The axiom of an affine plane which states that if p and L are a point and line in the plane such that p is not on L, then there exists exactly one line that passes through p and does not intersect L. { ¦par·ə,lel 'ak·sē·əm }

parallel curves Two curves such that one curve is the locus of points on the normals to the other curve at a fixed distance along the normals. { 'par·ə‚lel ‚kərvz }

parallel displacement A vector A at a point P of an affine space is said to be obtained from a vector B at a point Q of the space by a parallel displacement with respect to a curve connecting A and B if a vector $V(X)$ can be associated with each point X on the curve in such a manner that $A = V(P)$, $B = V(Q)$, and the values of V at neighboring points of the curve are parallel as specified by the affine connection. { 'par·ə‚lel di'splās·mənt }

parallel edges Two or more edges that join the same pair of vertices in a graph. Also known as multiple edges. { ¦par·ə‚lel 'ej·əz }

parallelepiped A polyhedron all of whose faces are parallelograms. { ‚par·ə‚lel·ə'pī·pəd }

parallelogram A four-sided polygon with each pair of opposite sides parallel. { ‚par·ə'lel·ə‚gram }

parallelogram law The rule that the sum of two vectors is the diagonal of a parallelogram whose sides are the vectors to be added.

parallelogram of vectors A parallelogram whose sides form two vectors to be added and whose diagonal is the sum of the two vectors. { ‚par·ə'lel·ə‚gram əv 'vek·tərz }

parallelotope A parallelepiped with sides in proportion of 1, 1/2, and 1/4. { ‚par·ə'lel·ə‚tōp }

parallel projection A central projection in which the center of projection is the point at infinity, so that the projectors are parallel; equivalent to an orthogonal projection. { ¦par·ə‚lel prə'jek·shən }

parallel rays 1. Two rays lying on the same line or on parallel lines. **2.** Two rays that lie on the same line or on parallel lines, and point in the same direction. { ¦par·ə‚lel 'rāz }

parallel surfaces Two surfaces such that one surface is the locus of points on the normals to the other curve at a fixed distance along the normals. { ¦par·ə‚lel 'sər·fəs·əz }

parallel vectors 1. Two nonzero vectors such that one vector equals the product of the other vector and a nonzero scalar. **2.** Two nonzero vectors in a vector space over the real numbers such that one vector equals the product of the other vector and a positive number. { ¦par·ə‚lel 'vek·tərz }

parameter An arbitrary constant or variable so appearing in a mathematical expression that changing it gives various cases of the phenomenon represented. { pə'ram·əd·ər }

parameter of distribution For a fixed line on a ruled surface, a quantity whose magnitude is the limit, as a variable line on the surface approaches the fixed line, of the ratio of the minimum distance between the two lines to the angle between them; and whose sign is positive or negative according to whether the motion of the tangent plane to the surface is left- or right-handed as the point of tangency moves along the fixed line in a positive direction. { pə¦ram·əd·ər əv ‚dis·trə'byü·shən }

parametric curves On a surface determined by equations $x = f(u,v)$, $y = g(u,v)$, and $z = h(u,v)$, these are families of curves obtained by setting the parameters u and v equal to various constants. { ¦par·ə¦me·trik 'kərvz }

parametric equation An equation where coordinates of points appear dependent on parameters such as the parametric equation of a curve or a surface. { ¦par·ə¦me·trik i'kwā·zhən }

parity Two integers have the same parity if they are both even or both odd. { 'par·əd·ē }

Parseval's equation The equation which states that the square of the length of a vector in an inner product space is equal to the sum of the squares of the inner products of the vector with each member of a complete orthonormal base for the space. Also known as Parseval's identity; Parseval's relation. { 'pär·sə·vəlz i‚kwä·zhən }

Parseval's identity *See* Parseval's equation. { 'pär·sə·vəlz ī'den·əd·ē }

Parseval's relation *See* Parseval's equation. { 'pär·sə·vəlz re'lā·shən }

Parseval's theorem A theorem that gives the integral of a product of two functions,

$f(x)$ and $F(x)$, in terms of their respective Fourier coefficients; if the coefficients are defined by

$$a_n = (1/\pi) \int_0^{2\pi} f(x) \cos nx\, dx$$

$$b_n = (1/\pi) \int_0^{2\pi} f(x) \sin nx\, dx$$

and similarly for $F(x)$, the relationship is

$$\int_0^{2\pi} f(x)F(x)dx = \pi \left[\frac{1}{2} a_0 A_0 + \sum_{n=1}^{\infty} (a_n A_n + b_n B_n) \right]$$

{ 'pär·sə·vəlz 'thir·əm }

partial correlation The strength of the linear relationship between two random variables where the effect of other variables is held constant. { 'pär·shəl ¦kär·ə'lā·shən }

partial correlation analysis A technique used to measure the strength of the relationship between the dependent variable and one independent variable in such a way that variations in other independent variables are taken into account. { 'pär·shəl ¦kär·ə'lā·shən ə¦nal·ə·səs }

partial correlation coefficient A measure of the strength of association between a dependent variable and one independent variable when the effect of all other independent variables is removed; equal to the square root of the partial coefficient of determination. { 'pär·shəl ¦kä·rə'lā·shən ¦kō·i¦fish·ənt }

partial derivative A derivative of a function of several variables taken with respect to one variable while holding the others fixed. { 'pär·shəl də'riv·əd·iv }

partial differential equation An equation that involves more than one independent variable and partial derivatives with respect to those variables. { 'pär·shəl ¦dif·ə'ren·chəl i¦kwā·zhən }

partial fractions A collection of fractions which when added are a given fraction whose numerator and denominator are usually polynomials; the partial fractions are usually constants or linear polynomials divided by factors of the denominator of the given fraction. { 'pär·shəl 'frak·shənz }

partially balanced incomplete block design An experimental design in which, while all treatments are not represented in each block, each treatment is tested the same number of times and certain aspects of the design satisfy conditions which simplify the least squares analysis. { ¦pär·shə·lē ¦bal·ənst ¦in·kəm¦plēt 'bläk di¦zīn }

partially ordered set A set on which a partial order is defined. Also known as poset. { 'pär·shə·lē ¦ör·dərd 'set }

partial order *See* ordering. { 'par·shəl ¦örd·ər }

partial ordering *See* ordering. { 'pär·shəl 'ör·də·riŋ }

partial plane In projective geometry, a plane in which at most one line passes through any two points. { 'pär·shəl 'plān }

partial product The product of a multiplicand and one digit of a multiplier that contains more than one digit. { 'pär·shəl 'präd·əkt }

partial recursive function A function that can be computed by using a Turing machine for some inputs but not necessarily for all inputs. { ¦pär·shəl rē¦kər·siv 'fəŋk·shən }

partial regression coefficient Statistics in the population multiple linear regression equation that indicate the effect of each independent variable on the dependent variable with the influence of all the remaining variables held constant; each coefficient is the slope between the dependent variable and each of the independent variables. { 'pär·shəl ri'gresh·ən ¦kō·ə¦fish·ənt }

partial sum A partial sum of an infinite series is the sum of its first n terms for some n. { 'pär·shəl 'səm }

particular integral *See* particular solution. { pər¦tik·yə·lər 'int·ə·grəl }

particular solution A solution to an ordinary differential equation obtained by assigning numerical values to the parameters in the general solution. Also known as particular integral. { pər¦tik·yə·lər sə'lü·shən }

partition 1. For an integer n, any collection of positive integers whose sum equals n. **2.** For a set A, a collection of disjoint sets whose union is A. **3.** For a closed interval I, a finite set of closed subintervals of I that intersect only at their end points and whose union is I. { pär'tish·ən }

partition of unity On a topological space X, this is a covering by open sets U_α with continuous functions f_α from X to $[0,1]$, where each f_α is zero on all but a finite number of the U_α, and the sum of all these f_α at any point equals 1. { pär'tish· ən əv 'yü·nəd·ē }

Pascal distribution *See* negative binomial distribution. { pa¦skal ‚di·strə'byü·shən }

Pascal's identity The equation $C(n,r) = C(n - 1, r) + C(n - 1, r - 1)$ where, in general, $C(n,r)$ is the number of distinct subsets of r elements in a set of n elements (the binomial coefficient). { pa¦skalz ī'den·ə·dē }

Pascal's limaçon *See* limaçon. { pa'skalz ‚lē·mə'son }

Pascal's theorem The theorem that when one inscribes a simple hexagon in a conic, the three pairs of opposite sides meet in collinear points. { pa'skalz ‚thir·əm }

Pascal's triangle A triangular array of the binomial coefficients, bordered by ones, where the sum of two adjacent entries from a row equals the entry in the next row directly below. Also known as binomial array. { pa'skalz 'trī‚aŋ·gəl }

path 1. In a topological space, a path is a continuous curve joining two points. **2.** In graph theory, a walk whose vertices are all distinct. Also known as simple path. **3.** *See* walk. { path }

path-connected set *See* arcwise-connected set. { 'path kə‚nek·təd ‚set }

path curve A curve whose equation is given in parametric form. { 'path ‚kərv }

pathwise-connected set *See* arcwise-connected set. { 'path‚wīz kə‚nek·təd ‚set }

pattern An equivalence class of colorings of the elements of a finite set, which are indistinguishable with respect to a group of permutations of the colors. { 'pad·ərn }

payoff matrix A matrix arising from certain two-person games which gives the amount gained by a player. { 'pā‚óf ‚mā·triks }

Peano continuum A compact, connected, and locally connected metric space. { pā'än· ō kən‚tin·yü·əm }

Peano curve 1. A continuous curve that passes through each point of the unit square. **2.** *See* Peano space. { pā'än·ō ‚kərv }

Peano space Any Hausdorff topological space that is the image of the closed unit interval under a continuous mapping. Also known as Peano curve. { pā'än·ō ‚spās }

Peano's postulates The five axioms by which the natural numbers may be formally defined; they state that (1) there is a natural number 1; (2) every natural number n has a successor n^+; (3) no natural number has 1 as its successor; (4) every set of natural numbers which contains 1 and the successor of every member of the set contains all the natural numbers; (5) if $n^+ = m^+$, then $n = m$. { pā'än·ōz 'päs· chə·ləts }

Pearl-Reed curve *See* logistic curve. { ¦pərl 'rēd ‚kərv }

Pearson Type I distribution *See* beta distribution. { 'pir·sən 'tīp 'wən ‚dis·trə'byü· shən }

pedal coordinates The coordinates r and p describing a point P on a plane curve C, where r is the distance from a fixed point O to P, and p is the perpendicular distance from O to the tangent to C at P. { 'ped·əl kō'órd·ən·əts }

pedal curve 1. The pedal curve of a given curve C with respect to a fixed point P is the locus of the foot of the perpendicular from P to a variable tangent to C. Also known as first pedal curve; first positive pedal curve; positive pedal curve. **2.** Any curve that can be derived from a given curve C by repeated application of the procedure specified in the first definition. { 'ped·əl ‚kərv }

pedal equation An equation that characterizes a plane curve in terms of its pedal coordinates. { 'ped·əl i‚kwā·zhən }

pedal point 1. The fixed point with respect to which a pedal curve is defined. **2.** The fixed point with respect to which the pedal coordinates of a curve are defined. { 'ped·əl ‚póint }

pedal triangle 1. The triangle whose vertices are located at the feet of the perpendiculars from some given point to the sides of a specified triangle.   **2.** In particular, the triangle whose vertices are located at the feet of the altitudes of a given triangle. { 'ped·əl 'trī,aŋ·gəl }

Peirce stroke relationship *See* NOR.   { 'pirs ¦strōk ri'lā·shən,ship }

Pell equation The diophantine equation $x^2 - Dy^2 = 1$, with D a positive integer that is not a perfect square.   { 'pel i,kwā·zhən }

penalty function A function used in treating maxima and minima problems subject to constraints.   { 'pen·əl·tē ,faŋk·shən }

pencil 1. In general, a family of geometric objects which share a common property.   **2.** All the lines that lie in a particular plane and pass through a particular point.   **3.** All the lines parallel to a particular line.   **4.** All the circles that pass through two fixed points and lie in a particular plane.   **5.** All the planes that include a particular line.   **6.** All the spheres that include a particular circle.   { 'pen·səl }

pendulum property The property of a cycloid that, if a simple pendulum is hung from a cusp and made to swing between two branches, and if the length of the pendulum equals the length of the cycloid between successive cusps, then the period of the pendulum's oscillation does not depend on its amplitude, and the end of the pendulum traces out another cycloid.   { 'pen·jə·ləm ,präp·ərd·ē }

pentadecagon A polygon with 15 sides.   { ,pen·tə'dek·ə,gän }

pentagon A polygon with five sides.   { 'pen·tə,gän }

pentagonal number The total number, $P(n)$, of dots marking off unit segments of the sides of a set of $n - 1$ nested pentagons, given by the formula $P(n) = n(3n - 1)/2$.   { pen'tag·ən·əl ¦nəm·bər }

pentagonal prism A prism with two pentagonal sides, parallel and congruent.   { pen 'tag·ən·əl 'priz·əm }

pentagonal pyramid A pyramid whose base is a pentagon.   { pen'tag·ən·əl 'pir·ə,mid }

pentahedron A polyhedron with five faces.   { ,pen·tə'hē·drən }

pentomino One of the 12 plane figures that can be formed by joining five unit squares along their sides.   { pent'äm·ə·nō }

percent A quantitative term whereby n percent of a number is n one-hundredths of the number. Symbolized %.   { pər'sent }

percentage The result obtained by taking a given percent of a given quantity.   { pər'sen·tij }

percentage distribution A frequency distribution in which the individual class frequencies are expressed as a percentage of the total frequency equated to 100.   Also known as relative frequency distribution; relative frequency table.   { pər'sen·tij ,dis·trə'byü·shən }

percentile A value in the range of a set of data which separates the range into two groups so that a given percentage of the measures lies below this value.   { pər'sen,tīl }

percolation network A lattice constructed of a random mixture of conducting and nonconducting links.   { ,pər·kə'lā·shən 'net,wərk }

percolation problem The problem of determining the critical threshold concentration of conducting links in a pecolation network at which an infinite cluster of conducting links is formed and the lattice transforms from an insulator to a conductor.   { ,pər·kə'lā·shən ,präb·ləm }

perfect cube A number or polynomial which is the exact cube of another number or polynomial.   { 'pər·fikt 'kyüb }

perfect field A field such that any irreducible polynomial with coefficients in this field is separable; a field whose finite extensions are all separable.   { ¦pər,fikt 'fēld }

perfect group A group that is equal to its commutator subgroup.   { 'pər·fikt 'grüp }

perfectly separable space A topological space whose topology has a countable base.   Also known as completely separable space.   { ¦pər·fikt·lē ¦sep·rə·bəl 'spās }

perfect number An integer which equals the sum of all its factors other than itself.   { 'pər·fikt 'nəm·bər }

perfect power A number or polynomial which equals another number or polynomial raised to an integral power greater than one.   { ¦pər·fikt 'paủ·ər }

perfect set A set in a topological space which equals its set of accumulation points. { 'pər·fikt 'set }

perfect square A number or polynomial which is the exact square of another number or polynomial. { 'pər·fikt 'skwer }

perfect trinomial square A trinomial that is the exact square of a binomial. { ¦pər·fikt trī¦nō·mē·əl 'skwer }

perigon An angle that contains 360° or 2π radians. Also known as round angle. { 'per·ə,gän }

perimeter The total length of a closed curve; for example, the perimeter of a polygon is the total length of its sides. { pə'rim·əd·ər }

period 1. A number T such that $f(x + T) = f(x)$ for all x, where $f(x)$ is a specified function of a real or complex variable. **2.** The period of an element a of a group G is the smallest positive integer n such that a^n is the identity element; if there is no such integer, a is said to be of infinite period. { 'pir·ē·əd }

periodic continued fraction See recurring continued fraction. { ¦pir·ē¦äd·ik kən¦tin·yüd 'frak·shən }

periodic decimal See repeating decimal. { ¦pir·ē,äd·ik 'des·məl }

periodic function A function $f(x)$ of a real or complex variable is periodic with period T if $f(x + T) = f(x)$ for every value of x. { ¦pir·ē¦äd·ik ¦fəŋk·shən }

periodicity The property of periodic functions. { ,pir·ē·ə'dis·əd·ē }

periodic perturbation A perturbation which is periodic as a function. { ¦pir·ē¦äd·ik pər·tər'bā·shən }

periodogram A graph used in harmonic analysis of a series that oscillates, such as a time series consisting potentially of several cycles differing in length; the square of the amplitude or intensity for each curve covering a length of time is plotted against the lengths of the various curves. { ,pir·ē'äd·ə,gram }

period parallelogram For a doubly periodic function $f(z)$ of a complex variable, a parallelogram with vertices at z_0, $z_0 + a$, $z_0 + a + b$, and $z_0 + b$, where z_0 is any complex number, and a and b are periods of $f(z)$ but are not necessarily primitive periods. { ¦pir·ē·əd ,par·ə'lel·ə,gram }

periphery The bounding curve of a surface or the surface of a solid. { pə'rif·ə·rē }

permanent For a matrix with m rows and n columns, with n equal to or greater than m, the sum, over all permutations of m columns, of a product of m terms, where the ith term in the product is the term in the ith row and the permutation of the ith column. { 'pər·mə·nənt }

permanently convergent series A series that is convergent for all values of the variable or variables involved in its terms. { ¦pər·mə·nənt·lē kən¦vər·jənt 'sir,ēz }

permissible value A value of a variable for which a given function is defined. { pər¦mis·ə·bəl 'val·yü }

permutation A function which rearranges a finite number of symbols; more precisely, a one-to-one function of a finite set onto itself. { ,pər·myə'tā·shən }

permutation character The set of all fixed points of a specified permutation. { ,pər,myü'tā·shən ,kar·ik·tər }

permutation group The group whose elements are permutations of some set of symbols where the product of two permutations is the permutation arising from successive application of the two. Also known as substitution group. { ,pər·myə'tā·shən ,grüp }

permutation matrix A square matrix whose elements in any row, or any column, are all zero, except for one element that is equal to unity. { ,pər·myə'tā·shən ,mā,triks }

permutation tensor See determinant tensor. { ,pər·myə'tā·shən ,ten·sər }

perpendicular Geometric objects are perpendicular if they intersect in an angle of 90°. { ¦pər·pən¦dik·yə·lər }

perpendicular bisector For a line segment in a plane or in space, the line or plane that is perpendicular to this segment and passes through its midpoint. { ,pər·pən¦dik·yə·lər 'bī,sek·tər }

Perron-Frobenius theorem If M is a matrix with positive entries, then its largest eigenvalue λ is positive and simple; moreover, there exist vectors v and w with

Perron-Frobenius theory

positive components such that $vM = \lambda v$ and $Mw = \lambda w$, and if the inner product of v with w is 1, then the limit of $\lambda - n$ times the i,jth entry of M^n as n goes to infinity is the product of the ith component of w and the jth component of v. { pe'rōn frō'bā·nē·ùs ,thir·əm }

Perron-Frobenius theory The study of positive matrices and their eigenvalues; in particular, application of the Perron-Frobenius theorem. { pe'rōn frō'bā·nē·ùs ,thē·ə·rē }

personal probability A number between 0 and 1 assigned to an event based upon personal views concerning whether the event will occur or not; it is obtained by deciding whether one would accept a bet on the event at odds given by this number. Also known as subjective probability. { 'pərs·ən·əl ,präb·ə'bil·əd·ē }

perturbation A function which produces a small change in the values of some given function. { ,pər·tər'bā·shən }

perturbation theory The study of the solutions of differential and partial differential equations from the viewpoint of perturbation of solutions. { ,pər·tər'bā·shən ,thē·ə·rē }

Peters' formula An approximate formula for the probable error in the value of a quantity determined from several equally careful, independent measurements of the value of the quantity. { 'pēd·ərz ,fòr·myə·lə }

Pfaffian differential equation The first-order linear total differential equation $P(x,y,z)\,dx + Q(x,y,z)dy + R(x,y,z)dz = 0$, where the functions P, Q, and R are continuously differentiable. { 'faf·ē·ən ,dif·ə¦ren·chəl i'kwā·zhən }

p-form A totally antisymmetric covariant tensor of rank p. { 'pē ,fòrm }

phase An additive constant in the argument of a trigonometric function. { fāz }

phase space In a dynamical system or transformation group, the topological space whose points are being moved about by the given transformations. { 'fāz ,spās }

phi function See Euler's phi function. { fī or 'fē ,fəŋk·shən }

pi The irrational number which is the ratio of the circumference of any circle to its diameter; an approximation is 3.14159. Symbolized π. { pī }

Picard method A method of successive substitution for solving differential equations. { pi'kär ,meth·əd }

Picard's big theorem The image of every neighborhood of an essential singularity of a complex function is dense in the complex plane. Also known as Picard's second theorem. { pi'kärz 'big ¦thir·əm }

Picard's first theorem See Picard's little theorem. { pi'kärz 'fərst ¦thir·əm }

Picard's little theorem A nonconstant entire function of the complex plane assumes every value save at most one. Also known as Picard's first theorem. { pi'kärz 'lid·əl ¦thir·əm }

Picard's second theorem See Picard's big theorem. { pi'kärz 'sek·ənd 'thir·əm }

pico- A prefix meaning 10^{-12}; used with metric units. Abbreviated p. { 'pē·kō }

piecewise-continuous function A function defined on a given region, which can be divided into a finite number of pieces such that the function is continuous on the interior of each piece and its value approaches a finite limit as the argument of the function in the interior approaches a boundary point of the piece. { ¦pēs,wīz kən¦tin·yə·wəs 'fəŋk·shən }

piecewise-linear A continuous curve or function obtained by joining a finite number of linear pieces. { 'pēs,wīz ,lin·ē·ər }

piecewise linear topology See combinatorial topology. { 'pēs,wīz ¦lin·ē·ər tə'päl·ə·jē }

piecewise-smooth curve The range of a function from a closed interval to a Euclidean space such that each of the Cartesian coordinates of the image point is a continuously differentiable function on the closed interval, except at a finite set of points where the function is differentiable on the left and on the right. { ¦pēs,wīz ,smüth 'kərv }

pie chart A circle divided by several radii into sectors whose relative areas represent the relative magnitudes of quantities or the relative frequencies of items in a frequency distribution. Also known as circle graph; sectorgram. { 'pī ,chärt }

pigeonhole principle The principle, that if a very large set of elements is partitioned into a small number of blocks, then at least one block contains a rather large number of elements. Also known as Dirichlet drawer principle. { 'pij·ən,hōl ,prin·sə·pəl }

piriform A plane curve whose equation in Cartesian coordinates x and y is $y^2 = ax^3 - bx^4$, where a and b are constants. { 'pir·ə,fòrm }

pivotal condensation A method of evaluating a determinant that is convenient for determinants of large order, especially when digital computers are used, involving a repeated process in which a determinant of order n is reduced to the product of one of its elements raised to a power and a determinant of order $n - 1$. { ¦piv·əd·əl ,kän·dən'sā·shən }

pivotal method A technique for passing from one set of double inequalities to another in order to find a confidence interval for a parameter. { ¦piv·əd·əl ,meth·əd }

pivoting In the solution of a system of linear equations by elimination, a method of choosing a suitable equation to eliminate at each step so that certain difficulties are avoided. { 'piv·əd·iŋ }

place A position corresponding to a given power of the base in positional notation. Also known as column. { plās }

place value The value given to a digit by virtue of its location in a numeral. { 'plās ,val·yü }

planar Lying in or pertaining to a Euclidean plane. { 'plā·nər }

planar graph A graph that can be drawn in a plane without any lines crossing. { 'plā·nər ,graf }

planar map A plane or sphere divided into connected regions by a topological graph. { 'plān·ər ,map }

planar point A point on a surface at which the curvatures of all the normal sections vanish. { 'plā·nər *or* 'plā·när ,pòint }

plane 1. A surface such that a straight line that joins any two of its points lies entirely in that surface. 2. In projective geometry, a triple of sets (P,L,I) where P denotes the set of points, L the set of lines, and I the incidence relation on points and lines, such that (1) P and L are disjoint sets, (2) the union of P and L is nonnull, and (3) I is a subset of $P \times L$, the Cartesian product of P and L. { plān }

plane angle An angle between lines in the Euclidean plane. { 'plān ,aŋ·gəl }

plane curve Any curve lying entirely within a plane. { 'plān ,kərv }

plane cyclic curve *See* cyclic curve. { 'plān 'sī·klik ,kərv }

plane field *See* field of planes on a manifold. { 'plān ,fēld }

plane geometry The geometric study of the figures in the Euclidean plane such as lines, triangles, and polygons. { 'plān jē'äm·ə·trē }

plane group One of 17 two-dimensional patterns which can be produced by one asymmetric motif that is repeated by symmetry operations to produce a pattern unit which then is repeated by translation to build up an ordered pattern that fills any two-dimensional area. Also known as plane symmetry group. { 'plān ,grüp }

plane of mirror symmetry Also known as mirror plane of symmetry; plane of symmetry; reflection plane; symmetry plane. An imaginary plane which divides an object into two halves, each of which is the mirror image of the other in this plane. { 'plān əv 'mir·ər ,sim·ə·trē }

plane of reflection *See* plane of mirror symmetry. { 'plān əv ri'flek·shən }

plane of support Relative to a convex body in a three-dimensional space, a plane that contains at least one point of the body but is such that the half-space on one side of the plane contains no points of the body. { 'plān əv sə'pòrt }

plane of symmetry *See* plane of mirror symmetry. { 'plān əv 'sim·ə·trē }

plane polygon A polygon lying in the Euclidean plane. { 'plān 'päl·ə,gän }

plane quadrilateral A four-sided polygon lying in the Euclidean plane. { 'plān ,kwä·drə'lad·ə·rəl }

plane section The intersection of a plane with a surface or a solid. Also known as section. { 'plān ,sek·shən }

plane triangulation The process of adding arcs between pairs of verticles of a planar

plane trigonometry

graph to producer another planar graph, each of whose regions is bounded by three sides. { 'plān trī,aŋ·gyə'lā·shən }

plane trigonometry The study of triangles in the Euclidean plane with the use of functions defined by the ratios of sides of right triangles. { 'plān trig·ə'näm·ə·trē }

plateau problem The problem of finding a minimal surface having as boundary a given curve. { pla'tō 'präb·ləm }

platonic solid *See* regular polyhedron. { plə,tn·ik 'säl·əd }

platykurtic distribution A distribution of a data set which is relatively flat. { ,plad·ə'kərd·ik ,dis·trə'byü·shən }

Plemelj formulas Formulas for the limits of the Cauchy integrals of an arc with respect to a point *z* as *z* approaches the arc from either side. { 'plā·mə·lē ,fór·myə·ləz }

plus A mathematical symbol; *A* plus *B*, where *A* and *B* are mathematical quantities, denotes the quantity obtained by taking their sum in an appropriate context. { pləs }

plus sign *See* addition sign. { 'pləs ,sīn }

p.m.f. *See* probability mass function.

Pockels equation A partial differential equation which states that the Laplacian of an unknown function, plus the product of the value of the function with a constant, is equal to 0; it arises in finding solutions of the wave equation that are products of time-independent and space-independent functions. { 'päk·əlz i,kwā·zhən }

Poincaré-Birkhoff fixed-point theorem The theorem that a bijective, continuous, area-preserving mapping of the ring between two concentric circles onto itself that moves one circle in the positive sense and the other in the negative sense has at least two fixed points. { ,pwän·kə'rā 'bərk,hóf 'fikst 'póint 'thir·əm }

Poincaré conjecture The question as to whether a compact, simply connected three-dimensional manifold without boundary must be homeomorphic to the three-dimensional sphere. { ,pwän,kä'rā kən,jek·chər }

Poincaré recurrence theorem 1. A volume preserving homeomorphism *T* of a finite dimensional Euclidean space will have, for almost all points *x*, infinitely many points of the form $T^i(x)$, $i = 1, 2, \ldots$ within any open set containing *x*. 2. A measure preserving transformation on a space with finite measure is recurrent. { ,pwän,kä'rā ri'kə·rəns ,thir·əm }

Poinsot's spiral Either of two plane curves whose equations in polar coordinates (r,θ) are $r \cosh n\theta = a$ and $r \sinh n\theta = a$, where a is a constant and n is an integer. { pwän'sōz 'spī·rəl }

point 1. An element in a topological space. 2. One of the basic undefined elements of geometry possessing position but no nonzero dimension. 3. In positional notation, the character or the location of an implied symbol that separates the integral part of a numerical expression from its fractional part; for example, it is called the binary point in binary notation and the decimal point in decimal notation. { póint }

point at infinity 1. A single point that is adjoined to the complex plane so that it corresponds to the pole of a stereographic projection of the Riemann sphere onto the complex plane, giving the complex plane a compact topology. 2. *See* ideal point. { 'póint at in'fin·əd·ē }

point biserial correlation coefficient A modification of the biserial correlation coefficient in which one variable is dichotomous and the other is continuous; a product moment correlation coefficient. { 'póint ,bī'sir·ē·əl ,kär·ə'lā·shən ,kō·ə,fish·ənt }

point distal flow A transformation group on a compact metric space for which there exists a distal point with a dense orbit. { 'póint 'dis·təl 'flō }

point estimates Estimates which produce a single value of the population. { 'póint ,es·tə·məts }

point-finite family of subsets A family of subsets of a particular set, *S*, such that any member of *S* is a member of at most a finite number of these subsets. { 'póint ,fī,nīt ,fam·lē əv 'səb,sets }

point function A function whose values are points. { 'póint ,fəŋk·shən }

point of division The point that divides the line segment joining two given points in a given ratio. { ‚póint əv di'vizh·ən }

point of inflection A point where a plane curve changes from the concave to the convex relative to some fixed line; equivalently, if the function determining the curve has a second derivative, this derivative changes sign at this point. Also known as inflection point. { 'póint əv in'flek·shən }

point of osculation *See* double cusp. { 'póint əv ‚äs·kyə'lā·shən }

point set A collection of points in a geometrical or topological space. { 'póint ‚set }

point-set topology *See* general topology. { 'póint ‚set tə'päl·ə·jē }

point-slope form The equation of a straight line in the form $y - y_1 = m(x - x_1)$, where m is the slope of the line and (x_1, y_1) are the coordinates of a given point on the line in a Cartesian coordinate system. { 'póint ‚slōp ‚fórm }

point spectrum Those eigenvalues in the spectrum of a linear operator between Banach spaces whose corresponding eigenvectors are nonzero and of finite norm. { 'póint ‚spek·trəm }

pointwise convergence A sequence of functions $f_1, f_2, \ldots$ defined on a set S converges pointwise to a function f if the sequence $f_1(x), f_2(x), \ldots$ converges to $f(x)$ for each x in S. { 'póint‚wīz kən'vər·jəns }

pointwise equicontinuous family of functions A family of functions defined on a common domain D with the property that for any point x in D and for any $\epsilon > 0$ there is a $\delta > 0$ such that, whenever y is in D and $|x - y| < \delta$, $|f(x) - f(y)| < \epsilon$ for every function f in the family. { ‚póint‚wīz ‚ek·wi·kən‚tin·yə·wəs ‚fam·le əv 'fəŋk·shənz }

Poisson binomial trials model *See* generalized binomial trials model. { pwä'sōn̲ bī'nō·mē·əl ‚trīlz ‚mäd·əl }

Poisson density functions Density functions corresponding to Poisson distributions. { pwä'sōn̲ 'den·səd·ē ‚fəŋk·shənz }

Poisson distribution A probability distribution whose mean and variance have a common value k, and whose frequency is $f(x) = k^x e^{-k}/x!$, for $x = 0, 1, 2, \ldots$. { pwä'sōn̲ ‚dis·trə'byü·shən }

Poisson formula If the infinite series of functions $f(2\pi k + t)$, k ranging from $-\infty$ to ∞, converges uniformly to a function of bounded variation, then the infinite series with term $f(2\pi k)$, k ranging from $-\infty$ to ∞, is identical to the series with term the integral of $f(x)e^{-ikx}dx$, k ranging from $-\infty$ to ∞. { pwä'sōn̲ ‚fór·myə·lə }

Poisson index of dispersion An index used for events which follow a Poisson distribution and should have a chi-square distribution. { ‚pwä·sōn̲ ‚in‚deks əv di'spər·zhən }

Poisson integral formula This formula gives a solution function for the Dirichlet problem in terms of integrals; an integral representation for the Bessel functions. { pwä'sōn̲ 'int·ə·grəl ‚fór·myə·lə }

Poisson process A process given by a discrete random variable which has a Poisson distribution. { 'pwä'sōn̲ ‚prä·səs }

Poisson's equation The partial differential equation which states that the Laplacian of an unknown function is equal to a given function. { pwä'sōn̲z i‚kwā·zhən }

Poisson transform An integral transform which transforms the function $f(t)$ to the function

$$F(x) = (2/\pi) \int_0^\infty [t/(x^2 + t^2)]f(t)dt$$

Also known as potential transform. { pwä'sōn̲ 'tranz‚fórm }

polar 1. For a conic section, the polar of a point is the line that passes through the points of contact of the two tangents drawn to the conic from the point. **2.** For a quadric surface, the polar of a point is the plane that passes through the curve which is the locus of the points of contact of the tangents drawn to the surface from the point. **3.** For a quadric surface, the polar of a line is the line of intersection of the planes which are tangent to the surface at its points of intersection with the original line. { 'pō·lər }

polar angle The angular coordinate θ in a polar coordinate system whose value at a given point p is the angle that a line from the origin to p makes with the polar axis. { 'pō·lər 'aŋ·gəl }

polar axis The directed straight line relative to which the angle is measured for a representation of a point in the plane by polar coordinates. { 'pō·lər 'ak·səs }

polar coordinates A point in the plane may be represented by coordinates (r,θ), where θ is the angle between the positive x-axis and the ray from the origin to the point, and r the length of that ray. { 'pō·lər kō'órd·ən·əts }

polar developable The envelope of the normal planes of a space curve. { ¦pō·lər di¦vel·əp·ə·bəl }

polar equation An equation expressed in polar coordinates. { ¦pō·lər i¦kwā·zhən }

polar form A complex number $x + iy$ has as polar form $re^{i\theta}$, where (r,θ) are the polar coordinates corresponding to the point of the plane with rectangular coordinates (x,y), that is, $r = \sqrt{x^2 y^2}$ and $\theta = $ arc tan y/x. { 'pō·lər 'fórm }

polarity Property of a line segment whose two ends are distinguishable. { pə'lar·əd·ē }

polar line For a point on a space curve, the line that is normal to the osculating plane of the curve and passes through the center of curvature at that point. { 'pō·lər ‚līn }

polar normal For a given point on a plane curve, the segment of the normal between the given point and the intersection of the normal with the radial line of a polar coordinate system that is perpendicular to the radial line to the given point. { 'pō·lər 'nór·məl }

polar reciprocal convex bodies Any two convex bodies, each containing the origin in its interior, such that the Minkowski distance function of each is the support function of the other. { ¦pō·lər ri¦sip·rə·kəl ¦kän‚veks 'bäd‚ēz }

polar-reciprocal curves Two curves configured so that the polar of every point of one of them, with respect to a particular conic, is tangent to the other curve. { ¦pō·lər ri‚sip·rə·kəl 'kərvz }

polar-reciprocal triangles Two triangles configured so that the vertices of each triangle are the poles of the sides of the other with respect to some conic. { ¦pō·lər ri‚sip·rə·kəl 'trī‚aŋ·gəlz }

polar subnormal For a given point on a plane curve, the projection of the polar normal on the radial line of the polar coordinate system that is perpendicular to the radial line to the given point. { 'pō·lər səb'nór·məl }

polar subtangent For a given point on a plane curve, the projection of the polar tangent on the radial line of the polar coordinate system that is perpendicular to the radial line to the given point. { 'pō·lər səb'tan·jənt }

polar tangent For a given point on a plane curve, the segment of the tangent between the given point and the intersection of the tangent with the radial line of a polar coordinate system that is perpendicular to the radial line to the given point. { 'pō·lər 'tan·jənt }

polar triangle A triangle associated to a given spherical triangle obtained from three directed lines perpendicular to the planes associated with the sides of the original triangle. { 'pō·lər 'trī‚aŋ·gəl }

pole 1. An isolated singular point z_0 of a complex function whose Laurent series expansion about z_0 will include finitely many terms of form $a_n(z - z_0)^{-n}$. **2.** For a great circle on a sphere, the pole of the circle is a point of intersection of the sphere and a line that passes through the center of the sphere and is perpendicular to the plane of the circle. **3.** For a conic section, the pole of a line is the intersection of the tangents to the conic at the points of intersection of the conic with the line. **4.** For a quadric surface, the pole of a plane is the vertex of the cone which is tangent to the surface along the curve where the plane intersects the surface. **5.** The origin of a system of polar coordinates on a plane. **6.** The origin of a system of geodesic polar coordinates on a surface. { pōl }

Polish space A separable metric space which is homeomorphic to a complete metric space. { 'pōl·ish 'spās }

polyabolo A plane figure formed by joining isosceles right triangles along their edges. { ‚päl·ē'ab·ə‚lō }

Polya counting formula A formula which counts the number of functions from a finite set D to another finite set, with two functions f and g assumed to be the same if some element of a fixed group of complete permutations of D takes f into g. { ¦pōl·yə 'kaún·tiŋ ‚fȯr·myə·lə }

Polya-Eggenberger distribution A discrete frequency distribution that was originally considered in connection with contagious distributions. { ¦pōl·yə 'eg·ən‚bȯrg·ər ‚dis·trə‚byü·shən }

polyalgorithm A set of algorithms together with a strategy for choosing and switching among them. { ¦päl·ē'al·gə‚rith·əm }

polygon A figure in the plane given by points $p_1, p_2, \ldots, p_n$ and line segments $p_1p_2, p_2p_3, \ldots, p_{n-1}p_n, p_np_1$. { 'päl·i‚gän }

polygonal region The union of the interior of a polygon with some, all, or none of the polygon itself. { pə'lig·ən·əl ‚rē·jən }

polygon of vectors A polygon all but one of whose sides represent vectors to be added, directed in the same sense along the perimeter, and whose remaining side represents the sum of these vectors, directed in the opposite sense. { ¦päl·i‚gän əv 'vek·tərz }

polyhedral angle The shape formed by the lateral faces of a polyhedron which have a common vertex. { ¦päl·i¦hē·drəl 'aŋ·gəl }

polyhedron 1. A solid bounded by planar polygons. **2.** The set of points that belong to the simplexes of a simplicial complex. **3.** *See* triangulable space. { ¦päl·i¦hē·drən }

polyhex A plane figure formed by joining a finite number of regular hexagons along their sides. { 'päl·i‚heks }

polyiamond A plane figure formed by joining a finite number of equilateral triangles along their sides. { 'päl·ē·ə‚mänd }

polyking *See* polyplet. { 'päl·i‚kiŋ }

polymodal distribution A frequency distribution characterized by two or more localized modes, each having a higher frequency of occurrence than other immediately adjacent individuals or classes. { ¦päl·i'mōd·əl ‚dis·trə'byü·shən }

polynomial A polynomial in the quantities $x_1, x_2, \ldots, x_n$ is an expression involving a finite sum of terms of form $bx_1{}^{p_1}x_2{}^{p_2} \ldots x_n{}^{p_n}$, where b is some number, and $p_1, \ldots, p_n$ are integers. { ¦päl·ə¦nō·mē·əl }

polynomial equation An equation in which a polynomial in one or more variables is set equal to zero. { ¦päl·ə¦nō·mē·əl i'kwā·zhən }

polynomial function A function whose values can be found by substituting the value (or values) of the independent variable (or variables) in a polynomial. { ¦päl·ə¦nō·mē·əl 'fəŋk·shən }

polynomial trend A trend line which is best approximated by a polynomial function; used in time series analysis. { ¦päl·i¦nō·mē·əl ‚trend }

polyomino A plane figure formed by joining a finite number of unit squares along their sides. { ‚päl·ē'äm·ə·nō }

polyplet A plane figure formed by joining squares either along their sides or at their corners. Also known as polyking. { 'päl·ə·plət }

polytope A finite region in n-dimensional space ($n = 2, 3, 4, \ldots$), enclosed by a finite number of hyperplanes; it is the n-dimensional analog of a polygon ($n = 2$) and a polyhedron ($n = 3$). { 'päl·i‚tōp }

Pontryagin's maximum principle A theorem giving a necessary condition for the solution of optimal control problems: let $\theta(\tau)$, $\tau_0 \leq \tau \leq T$, be a piecewise continuous vector function satisfying certain constraints; in order that the scalar function $S = \Sigma c_i x_i(T)$ be minimum for a process described by the equation $\partial x_i/\partial \tau = (\partial H/\partial z_i)[z(\tau), x(\tau), \theta(\tau)]$ with given initial conditions $x(\tau_0) = x^0$ it is necessary that there exist a nonzero continuous vector function $z(\tau)$ satisfying $dz_i/d\tau = -(\partial H/\partial x_i)[z(\tau), x(\tau), \theta(\tau)]$, $z_i(T) = -c_i$, and that the vector $\theta(\tau)$ be so chosen that $H[z(\tau), x(\tau), \theta(\tau)]$ is maximum for all τ, $\tau_0 \leq \tau \leq T$. { ‚pän·trē'ä·gənz 'mak·sə·məm ‚prin·sə·pəl }

pooling of error A method used in the analysis of variance to secure more degrees of freedom for estimating the error variance; the sums of squares of several sets of data considered to be generated under the same model are added together and divided by the sum of the degrees of freedom in the several sets of data. { ¦pül·iŋ əv 'er·ər }

population A specified set of objects or outcomes to be measured or observed. { ˌpäp·yə'lā·shən }

population correlation coefficient The ratio of the covariance of two random variables to their standard deviations. { ˌpäp·yə'lā·shən ˌkär·ə'lā·shən ˌkō·ə,fish·ənt }

population covariance The number $(1/N)[(v_1 - \bar{v})(w_1 - \bar{w}) + \cdots + (v_N - \bar{v})(w_N - \bar{w})]$, where v_i and w_i, $i = 1, 2, \ldots, N$, are the values obtained from two populations, and $\bar{v}$ and $\bar{w}$ are the respective means. { ˌpäp·yə'lā·shən kō'ver·ē·əns }

population mean The average of the numbers obtained for all members in a population by measuring some quantity associated with each member. { ˌpäp·yə'lā·shən ˌmēn }

population multiple linear regression equation An equation relating the conditional mean of the dependent variable to each one of the independent variables under the assumption that this relationship is linear; for the multivariate, normal distribution linearity always exists. { ˌpäp·yə¦lā·shən ¦məl·tə·pəl ¦lin·ē·ər ri'gresh·ən i,kwā·zhən }

population variance The arithmetic average of the numbers $(v_1 - \bar{v})^2, \ldots, (v_N - \bar{v})^2$, where v_i are numbers obtained from a population with N members, one for each member, and $\bar{v}$ is the population mean. { ˌpäp·yə'lā·shən 'ver·ē·əns }

poset *See* partially ordered set. { 'pō,set }

positional notation Any of several numeration systems in which a number is represented by a sequence of digits in such a way that the significance of each digit depends on its position in the sequence as well as its numeric value. Also known as notation. { pə'zish·ən·əl nō'tā·shən }

position vector The position vector of a point in Euclidean space is a vector whose length is the distance from the origin to the point and whose direction is the direction from the origin to the point. Also known as radius vector. { pə'zish·ən ˌvek·tər }

positive Having value greater than zero. { 'päz·əd·iv }

positive angle The angle swept out by a ray moving in a counterclockwise direction. { 'päz·əd·iv 'aŋ·gəl }

positive axis The segment of an axis arising from a cartesian coordinate system which is realized by positive values of the coordinate variables. { 'päz·əd·iv 'ak·səs }

positive correlation A relation between two quantities such that when one increases the other does also. { 'päz·əd·iv ˌkä·rə'lā·shən }

positive definite 1. A square matrix A of order n is positive definite if

$$\sum_{i,j=1}^{n} A_{ij} x_i \bar{x}_j > 0$$

for every choice of complex numbers $x_1, x_2, \ldots, x_n$, not all equal to 0, where $\bar{x}_j$ is the complex conjugate of x_j. **2.** A linear operator T on an inner product space is positive definite if (Tu, u) is greater than 0 for all nonzero vectors u in the space. { 'päz·əd·iv 'def·ə·nət }

positive integer An integer greater than zero; one of the numbers $1, 2, 3, \ldots$ { ¦päz·əd·iv 'int·i·jər }

positive linear functional A linear functional on some vector space of real-valued functions which takes every nonnegative function into a nonnegative number. { ¦päz·əd·iv ¦lin·ē·ər 'fəŋk·shən·əl }

positive number A real number that is greater than 0. { 'päz·əd·iv 'nəm·bər }

positive part For a real-valued function f, this is the function, denoted f^+, for which $f^+(x) = f(x)$ if $f(x) \geq 0$ and $f^+(x) = 0$ if $f(x) < 0$. { 'päz·əd·iv ˌpärt }

positive pedal curve *See* pedal curve. { 'päz·əd·iv 'ped·əl ˌkərv }

positive real function An analytic function whose value is real when the independent variable is real, and whose real part is positive or zero when the real part of the independent variable is positive or zero. { 'päz·əd·iv 'rēl 'fəŋk·shən }

positive semidefinite Also known as nonnegative semidefinite.   **1.** A square matrix A is positive semidefinite if

$$\sum_{i,j=1}^{n} A_{ij} x_i \bar{x}_j \geq 0$$

for every choice of complex numbers $x_1, x_2, \ldots, x_n$, where $\bar{x}_j$ is the complex conjugate of x_j.   **2.** A linear operator T on an inner product space is positive semidefinite if (Tu, u) is equal to or greater than 0 for all vectors u in the space. { 'päz·əd·iv ¦sem·i'def·ə·nət }

positive series A series whose terms are all positive real numbers.   { 'päz·əd·iv 'sir¸ēz }

positive sign The symbol $+$, used to indicate a positive number.   { 'päz·əd·iv ¸sīn }

positive skewness Property of a unimodal distribution with a longer tail in the direction of higher values of the random variable.   { 'päz·əd·iv 'skü·nəs }

positive with respect to a signed measure A set A is positive with respect to a signed measure m if, for every measurable set B, the intersection of A and B, $A \cap B$, is measurable and $m(A \cap B) \geq 0$.   { ¦päz·əd·iv wi<u>th</u> ri¸spekt tü ə ¦sīnd 'mezh·ər }

posterior distribution A probability distribution on the values of an unknown parameter that combines prior information about the parameter contained in the observed data to give a composite picture of the final judgments about the values of the parameter.   { pä¦stir·ē·ər ¸dis·trə'byü·shən }

posterior probabilities Probabilities of the outcomes of an experiment after it has been performed and a certain event has occurred.   { pä'stir·ē·ər präb·ə'bil·əd·ēz }

postmultiplication In multiplying a matrix or operator **B** by another matrix or operator **A**, the operation that results in the matrix or operator **BA**.   Also known as multiplication on the right.   { ¸pōst·məl·tə·plə'kā·shən }

postulate See axiom.   { 'päs·chə·lət }

potential theory The study of the functions arising from Laplace's equation, especially harmonic functions.   { pə'ten·chəl ¸thē·ə·rē }

potential transform See Poisson transform.   { pə'ten·chəl 'tranz¸fȯrm }

power 1. The value that is assigned to a mathematical expression and its exponent.   **2.** The power of a set is its cardinality.   **3.** For a point, with reference to a circle, the quantity $(x - a)^2 + (y - b)^2 - r^2$, where x and y are the coordinates of the point, a and b are the coordinates of the center of the circle, and r is the radius of the circle.   **4.** For a point, with reference to a sphere, the quantity $(x - a)^2 + (y - b)^2 + (z - c)^2 - r^2$, where x, y, and z are the coordinates of the point; a, b, and c are the coordinates of the center of the sphere; and r is the radius of the sphere.   **5.** One minus the probability that a given test causes the acceptance of the null hypothesis when it is false due to the validity of an alternative hypothesis; this is the same as the probability of rejecting the null hypothesis by the test when the alternative is true.   { 'paủ·ər }

power curve The graph of the power of a test for various alternatives.   { 'paủ·ər ¸kərv }

power efficiency The probability of rejecting a statistical hypothesis when it is false.   { 'paủ·ər i¸fish·ən·sē }

power function A function whose value is the product of a constant and a power of the independent variable. The function that indicates the probability of rejecting the null hypothesis for all possible values of the population parameter for a given critical region.   { 'paủ·ər ¸fəŋk·shən }

power of the continuum The cardinality of the set of real numbers.   { 'paủ·ər əv <u>th</u>ə kən'tin·yə·wəm }

power series An infinite series composed of functions having nth term of the form $a_n(x - x_0)^n$, where x_0 is some point and a_n some constant.   { 'paủ·ər ¸sir·ēz }

power set The set consisting of all subsets of a given set.   { 'paủ·ər ¸set }

precision The number of digits in a decimal fraction to the right of the decimal point.   { prə'sizh·ən }

precompact set A set in a metric space which can always be covered by open balls of any diameter about some finite number of its points.   Also known as totally bounded set.   { prē'käm¸pakt ¸set }

predecessor For a vertex a in a directed graph, any vertex b for which there is an arc between a and b directed from b to a. { 'pred·ə‚ses·ər }

predicate 1. To affirm or deny, in mathematical logic, one or more subjects. Also known as logical function; propositional function. **2.** *See* propositional function. { 'pred·ə‚kāt }

predicate calculus The mathematical study of logical statements relating to arbitrary sets of objects and involving predicates and quantifiers as well as propositional connectives. { 'pred·ə·kət ‚kal·kyə·ləs }

predictor-corrector methods Methods of calculating numerical solutions of differential equations that employ two formulas, the first of which predicts the value of the solution function at a point x in terms of the values and derivatives of the function at previous points where these have already been calculated, enabling approximations to the derivatives at x to be obtained, while the second corrects the value of the function at x by using the newly calculated values. { 'pri¦dik·tər kə'rek·tər ‚meth·ədz }

pre-Hilbert space *See* inner-product space. { prē'hil·bərt ‚spās }

pre-image 1. For a point y in the range of a function f, the set of points x in the domain of f for which $f(x) = y$. **2.** For a subset A of the range of a function f, the set of points x in the domain of f for which $f(x)$ is a member of A. Also known as inverse image. { ¦prē'im·ij }

premultiplication In multiplying a matrix or operator **B** by another matrix or operator **A**, the operation that results in the matrix or operator **AB**. Also known as multiplication on the left. { ‚prē·məl·tə·plə'kā·shən }

price index A statistic used primarily in economics to indicate an average level of prices in a time series; combines several series of price data into one index. { 'prīs ‚in‚deks }

price relative The ratio of the price of certain goods in a specified period to the price of the same goods in the base period. { ¦prīs 'rel·ə·tiv }

primary decomposition A primary decomposition of a submodule N of a module M is an expression of N as a finite intersection of primary submodules of M. { 'prī‚mer·ē dē‚käm·pə'zish·ən }

primary submodule A submodule N of a module M over a commutative ring R such that $M \neq N$ and, for any a in R, the principal homomorphism of the factor module M/N associated with a, $a_{M/N}$, is either injective or nilpotent. { 'prī‚mer·ē səb'mä·jəl }

prime *See* prime element. { prīm }

prime element A member of an integral domain that is not a unit and is not the product of two members that are not units. Also known as prime. { 'prīm 'el·ə·mənt }

prime factor A prime number or prime polynomial that exactly divides a given number or polynomial. { 'prīm 'fak·tər }

prime field For a field K with multiplicative unit element e, the field consisting of elements of the form $(ne)(me)^{-1}$, where $m \neq 0$ and n are integers. { 'prīm 'fēld }

prime ideal A principal ideal of a ring given by a single element that has properties analogous to those of the prime numbers. { 'prīm ī'dēl }

prime number A positive integer having no divisors except itself and the integer 1. { 'prīm 'nəm·bər }

prime number theorem The theorem that the limit of the quantity $[\pi(x)]$ $(\ln x)/x$ as x approaches infinity is 1, where $\pi(x)$ is the number of prime numbers not greater than x and $\ln x$ is the natural logarithm of x. { ¦prīm ¦nəm·bər ‚thir·əm }

prime polynomial A polynomial whose only factors are itself and constants. { 'prīm ‚päl·i'nō·mē·əl }

prime ring For a field K with multiplicative unit element e, the ring consisting of elements of the form ne, where n is an integer. { ¦prīm 'riŋ }

primitive abundant number An integer that is an abundant number and has no proper divisors that are also abundant numbers. { ¦prim·əd·iv ə¦bən·dənt 'nəm·bər }

primitive circle The stereographic projection of the great circle whose plane is perpendicular to the diameter of the projected sphere that passes through the point of projection. { 'prim·əd·iv 'sər·kəl }

primitive element A member of a finite number field from which all the other members can be generated by repeated multiplication. { 'prim·əd·iv 'el·ə·mənt }

primitive period 1. A period a of a simply periodic function $f(x)$ such that any period of $f(x)$ is an integral multiple of a. **2.** Either of two periods a and b of a doubly periodic function $f(x)$ such that any period of $f(x)$ is of the form $ma + nb$, where m and n are integers. { 'prim·əd·iv 'pir·ē·əd }

primitive period parallelogram For a doubly periodic function $f(z)$ of a complex variable, a parallelogram with vertices at z_0, $z_0 + a$, $z_0 + a + b$, and $z_0 + b$, where z_0 is any complex number and a and b are primitive periods of $f(z)$. { 'prim·əd· iv 'pir·ē·əd ,par·ə'lel·ə,gram }

primitive plane A partial plane in which every line passes through at least two points.

primitive polynomial A polynomial with integer coefficients which have 1 as their greatest common divisor. { 'prim·əd·iv ,päl·i'nō·mē·əl }

primitive pseudoperfect number An integer that is a pseudoperfect number and has no proper divisors that are also pseudoperfect numbers. { 'prim·əd·iv 'süd·ō'pər,fikt 'nəm·bər }

primitive root An nth root of unity that is not an mth root of unity for any m less than n. { 'prim·əd·iv 'rüt }

principal axis 1. One of a set of perpendicular axes such that a quadratic function can be written as a sum of squares of coordinates referred to these axes. **2.** For a conic, a straight line that passes through the midpoints of all the chords perpendicular to it. **3.** For a quadric surface, the intersection of two principal planes. { 'prin·sə· pəl 'ak·səs }

principal branch For complex valued functions such as the logarithm which are multiple-valued, a selection of values so as to obtain a genuine single-valued function. { 'prin·sə·pəl 'branch }

principal curvatures For a point on a surface, the absolute maximum and absolute minimum values attained by the normal curvature. { 'prin·sə·pəl 'kər·və·chərz }

principal diagonal For a square matrix, the diagonal extending from the upper left-hand corner to the lower right-hand corner of the matrix, that is, the diagonal containing the elements a_{ij} for which $i = j$. Also known as main diagonal. { 'prin· sə·pəl dī'ag·ən·əl }

principal directions For a point on a surface, the directions in which the normal curvature attains its absolute maximum and absolute minimum values. { 'prin· sə·pəl di'rek·shənz }

principal homomorphism Let a be an element of a ring R and M be a module over R; the principal homomorphism of M associated with a, denoted a_M, is the mapping which takes each element x in M into ax. { 'prin·sə·pəl ,hō·mō'mòr,fiz·əm }

principal ideal The smallest ideal of a ring which contains a given element of the ring. { 'prin·sə·pəl ī'dēl }

principal ideal ring A commutative ring with a unit element in which every ideal is a principal ideal. { 'prin·sə·pəl i'del ,riŋ }

principal normal The line perpendicular to a space curve at some point which also lies in the osculating plane at that point. { 'prin·sə·pəl 'nòr·məl }

principal normal indicatrix For a space curve, all the end points of those radii of a unit sphere that are parallel to the positive directions of the principal normals to the curve. Also known as spherical indicatrix of the principal normal. { 'prin· sə·pəl ,nòr·məl in'dik·ə,triks }

principal normal section For a point on a surface, a normal section in a direction in which the curvature of this section has a maximum or minimum value. { 'prin· sə·pəl 'nòr·məl 'sek·shən }

principle of duality See duality principle. { 'prin·sə·pəl əv dü'al·əd·ē }

principal part 1. The principal part of an analytic function $f(z)$ defined in an annulus about a point z_0 is the sum of terms in its Laurent expansion about z_0 with negative powers of $(z - z_0)$. **2.** A principal part of a triangle is one of its sides or one of its interior angles. { 'prin·sə·pəl 'pärt }

principal plane For a quadric surface, a plane that passes through the midpoints of all the chords perpendicular to it. { 'prin·sə·pəl 'plān }

principal radii The radii of curvature of the normal sections with maximum and minimum curvature at a given point on a surface; the reciprocals of the principal curvatures. { 'prin·sə·pəl 'rād·ē,ī }

principal radii of normal curvature The reciprocals of the principal curvatures of a surface at a point. { ¦prin·sə·pəl ¦rād·ē,ī əv ¦nȯr·məl 'kər·və·chər }

principal root The positive real root of a positive number, or the negative real root in the case of odd roots of negative numbers. { 'prin·sə·pəl 'rüt }

principal section A normal section at a given point on a surface whose curvature has a maximum or minimum value. { 'prin·sə·pəl 'sek·shən }

principal submatrix An $m \times m$ matrix, P, is an $m \times m$ principal submatrix of an $n \times n$ matrix, A, if P is obtained from A by removing any $n - m$ rows and the same $n - m$ columns. { ¦prin·səpəl səb'ma·triks }

principal value 1. The numerically smallest value of the arc sine, arc cosine, or arc tangent of a number, the positive value being chosen when there are values that are numerically equal but opposite in sign. **2.** *See* Cauchy principal value. { 'prin·sə·pəl 'val·yü }

principle of dichotomy law of the excluded middle. { ¦prin·sə·pəl əv dī'käd·ə·mē }

principle of insufficient reason The principle that cases are equally likely to occur unless reasons to the contrary are known. { 'prin·sə·pəl əv ,in·sə,fish·ənt 'rēz·ən }

principle of the maximum The principle that for a nonconstant complex analytic function defined in a domain, the absolute value of the function cannot attain its maximum at any interior point of the domain. { 'prin·sə·pəl əv thə 'mak·sə·məm }

principle of the minimum The principle that for a nonvanishing nonconstant complex analytic function defined in a domain, the absolute value of the function cannot attain its minimum at any interior point of the domain. { 'prin·sə·pəl əv thə 'min·ə·məm }

prior distribution A probability distribution on the set of all possible values of an unknown parameter of a statistical model that describes information available from sources other than a statistical investigation, in particular, expert judgment, past experience, or prior belief. { ¦prī·ər ,dis·trə'byü·shən }

prior probabilities Probabilities of the outcomes of an experiment before the experiment has been performed. { 'prī·ər ,präb·ə'bil·əd·ēz }

prism A polyhedron with two parallel, congruent faces and all other faces parallelograms. { 'priz·əm }

prismatic surface A surface generated by moving a straight line which always meets a broken line lying in a given plane and which is always parallel to some given line not in that plane. { priz'mad·ik 'sər·fəs }

prismatoid A polyhedron whose vertices all are in one or the other of two parallel planes. { 'priz·mə,tȯid }

prismoid A prismatoid whose two parallel faces are polygons having the same number of sides while the other faces are trapezoids or parallelograms. { 'priz,mȯid }

prismoidal formula A formula that gives the volume of a prismatoid as $(1/6)h(A_1 + 4A_m + A_2)$, where h is the altitude, A_1 and A_2 are the areas of the bases, and A_m is the area of a plane section halfway between the bases. { priz¦mȯid·əl 'fȯr·myə·lə }

probabilistic sampling A process in which the laws of probability determine which elements are to be included in a sample. { ,präb·ə'lis·tik 'sam·pliŋ }

probability The probability of an event is the ratio of the number of times it occurs to the large number of trials that take place; the mathematical model of probability is a positive measure which gives the measure of the space the value 1. { ,präb·ə'bil·əd·ē }

probability density function A real-valued function whose integral over any set gives the probability that a random variable has values in this set. Also known as density function; frequency function. { ‚präb·ə'bil·əd·ē ¦den·səd·ē ‚fəŋk·shən }

probability deviation *See* probable error. { ‚präb·ə'bil·əd·ē ‚dē·vē‚ā·shən }

probability distribution *See* distribution. { ‚präb·ə'bil·əd·ē ‚dis·trə‚byü·shən }

probability mass function A function which gives the relative frequency of each possible value of the random variable in an experiment involving a discrete set of outcomes. Abbreviated p.m.f. { ‚präb·ə'bil·əd·ē ¦mas ‚fəŋk·shən }

probability measure The measure on a probability space. { ‚präb·ə'bil·əd·ē ‚mezh·ər }

probability paper Graph paper with one axis specially ruled to transform the distribution function of a specified function to a straight line when it is plotted against the variate as the abscissa. { ‚präb·ə'bil·əd·ē ‚pā·pər }

probability ratio test Testing a simple hypothesis against a simple alternative by using the ratio of the probability of each simple event under the alternative to the probability of the event under the hypothesis. { ‚präb·ə'bil·əd·ē 'rā·shō ‚test }

probability sampling A method of sampling from a finite population where the probability of each set of units being selected is known. { ‚präb·ə'bil·əd·ē ‚sam·pliŋ }

probability space A measure space such that the measure of the entire space equals 1. { ‚präb·ə'bil·əd·ē ‚spās }

probability theory The study of the mathematical structures and constructions used to analyze the probability of a given set of events from a family of outcomes. { ‚präb·ə'bil·əd·ē ‚thē·ə·rē }

probable error The error that is exceeded by a variable with a probability of 1/2. Also known as probability deviation. { 'präb·ə·bəl 'er·ər }

probit A procedure used in dosage-response studies to avoid obtaining negative response values to certain dosages; five is added to the values of the standardized variate which is assumed to be normal; the term is a contraction of probability unit. { 'prō·bət }

problème des ménages The problem of seating a specified number (greater than two) of married couples at a circular table so that the sexes alternate and no husband and wife sit side by side. Also known as ménage problem. { prò·blem de mā'nazh }

problème des recontres The problem of determining the number of derangements of a specified number of distinct objects. { prò·blem derə'kän·trə }

problem of nontaking rooks The problem of determining the number of ways that a specified number of rooks can be placed on a chessboard of specified size so that no rook can capture another rook (that is, no two rooks are in same row or in the same column). Also known as rook problem. { ¦präb·ləm əv ¦nän‚tāk·iŋ 'rùks }

problem of type The problem of determining whether a given simply connected Riemann surface is of hyperbolic, parabolic, or elliptic type. { ¦präb·ləm əv 'tīp }

product **1.** For two integers, m and n, the number of objects in the set formed by combining m sets, each of which has n objects. **2.** For two rational numbers, a/b and c/d, where a, b, c, and d are integers, the number $(ac)/(bd)$. **3.** For any two real numbers, which are the limits of sequences of rational numbers p_n and q_n respectively, the limit of the sequence $p_n q_n$. **4.** The product of two algebraic quantities is the result of their multiplication relative to an operation analogous to multiplication of real numbers. **5.** The product of a collection of sets A_1, A_2, . . ., A_n is the set of all elements of the form $(a_1, a_2, . . ., a_n)$ where each a_i is an element of A_i for each $i = 1, 2, . . ., n$. **6.** For two transformations, the transformation that results from their successive application. **7.** For two fuzzy sets A and B, with membership functions m_A and m_B, the fuzzy set whose membership function $m_{A·B}$ satisfies the equation $m_{A·B}(x) = m_A(x) \cdot m_B(x)$ for every element x. **8.** The product AB of two matrices A and B, where the number n of columns in A equals the number of rows in B, is the matrix whose element c_{ij} in row i and column j is the sum over $k = 1, 2, . . ., n$ of the product of the elements a_{ik} in A and b_{kj} in B. { 'präd·əkt }

product bundle A bundle whose total space is the cartesian product of the base space

product measure

B and a topological space F and whose projection map sends (b,a) to b. { 'präd·
əkt ‚bən·dəl }
product measure A measure on a product of measure spaces constructed from the
measures on each of the individual spaces by taking the measure of the product
of a finite number of measurable sets, one from each of the measure spaces in
the product, to be the product of the measures of these sets. { 'prä·dəkt ‚mezh·ər }
product model A model for independent repetition of an experiment, or independent
performance of several experiments, obtained by taking the cartesian product of
the probability spaces representing the experiments. { 'prä·dəkt ‚mäd·əl }
product-moment coefficient *See* sample correlation coefficient. { 'prä·dəkt 'mō·mənt
‚kō·i'fish·ənt }
product topology A topology on a product of topological spaces whose open sets are
constructed from cartesian products of open sets from the individual spaces.
{ 'prä‚dəkt tə'päl·ə·jē }
progression A sequence or series of mathematical objects or quantities, each entry
determined from its predecessors by some algorithm. { prə'gresh·ən }
projecting cylinder A cylinder whose elements pass through a given curve and are
perpendicular to one of the three coordinate planes. { prə'jekt·iŋ 'sil·ən·dər }
projecting plane A plane that contains a given straight line in space and is perpendicular
to one of the three coordinate planes. { prə'jekt·iŋ 'plān }
projection 1. The continuous map for a fiber bundle. 2. Geometrically, the image of
a geometric object or vector superimposed on some other. 3. A linear map P
from a linear space to itself such that $P \circ P$ is equal to P. { prə'jek·shən }
projective geometry The study of those properties of geometric objects which are
invariant under projection. { prə'jek·tiv jē'äm·ə·trē }
projective group A group of transformations arising in the general theory of projective
geometry. { prə'jek·tiv 'grüp }
projective line The line obtained from the stereographic projection of the circle.
{ prə'jek·tiv 'līn }
projective plain curve The set of all points in the projective plane for which a particular
homogeneous polynomial in the coordinates equals zero. { prə'jek·tiv ‚plän 'kərv }
projective plane 1. The topological space obtained from the two-dimensional sphere
by identifying antipodal points; the space of all lines through the origin in Euclidean
space. 2. More generally, a plane (in the sense of projective geometry) such that
(1) every two points lie on exactly one line, (2) every two lines pass through exactly
one point, and (3) there exists a four-point. { prə'jek·tiv 'plān }
projective point The point from which a projection by rays is performed, as in stereographic projection. { prə'jek·tiv 'póint }
projective space The topological space obtained from the n-dimensional sphere under
identification of antipodal points. { prə'jek·tiv 'spās }
projective topology The finest topology on the tensor product of two locally convex
topological vector spaces such that the function that maps each element of the
Cartesian product of the two spaces to the corresponding element of their tensor
product is a continuous function. { prə‚jek·tiv tə'päl·ə·jē }
projector One of the lines or rays in a central projection. { prə'jek·tər }
prolate cycloid A trochoid in which the distance from the center of the rolling circle
to the point describing the curve is greater than the radius of the circle. { 'prō
‚lāt 'sī‚klóid }
prolate ellipsoid *See* prolate spheroid. { 'prō‚lāt i'lip‚sóid }
prolate spheroid The ellipsoid or surface obtained by revolving an ellipse about one
of its axes so that the equatorial circle has a diameter less than the length of the
axis of revolution. Also known as prolate ellipsoid. { 'prō‚lāt 'sfir‚óid }
prolate spheroidal coordinate system A three-dimensional coordinate system whose
coordinate surfaces are the surfaces generated by rotating a plane containing a
system of confocal ellipses and hyperbolas about the major axis of the ellipses,
together with the planes passing through the axis of rotation. { ‚prō‚lāt sfir‚óid·
əl kō'órd·ən·ət ‚sis·təm }

proof A deductive demonstration of a mathematical statement. { prüf }

proof by contradiction *See* reductio ad absurdum. { ¦prüf bī ¦kän·trə'dik·shən }

proof by descent *See* mathematical induction. { ¦prüf bī di'sent }

proper divisor A proper divisor of a positive integer *n* is any divisor other than 1 and *n*. { ¦präp·ər di'vīz·ər }

proper face 1. For a simplex, a face whose dimension is strictly less than that of the simplex. **2.** For a convex polytope, the intersection of the convex polytope with one of the hyperplanes enclosing it. { ¦präp·ər 'fās }

proper fraction 1. A fraction *a/b* where the absolute value of *a* is less than the absolute value of *b*. **2.** The quotient of two polynomials in which the degree of the numerator is less than the degree of the denominator. { 'präp·ər 'frak·shən }

proper function *See* eigenfunction. { 'präp·ər 'fəŋk·shən }

properly divergent series A series whose partial sums become either arbitrarily large or arbitrarily small (algebraically). { ¦präp·ər·lē də¦vər·jənt 'sir,ēz }

proper orthogonal transformation An orthogonal transformation such that the determinant of its matrix is + 1. { ¦präp·ər ór¦thäg·ən·əl ,tranz·fər'mā·shən }

proper rational function The quotient of a polynomial *P* by a polynomial *Q* whose order is greater than *P*. { 'präp·ər ¦rash·ən·əl ,fəŋk·shən }

proper subset A set *X* is a proper subset of a set *Y* if there is an element of *Y* which is not in *X* while *X* is a subset of *Y*. { 'präp·ər 'səb,set }

proper value *See* eigenvalue. { 'präp·ər 'val·yü }

proportion 1. The proportion of two quantities is their ratio. **2.** The statement that two ratios are equal. { prə'pór·shən }

proportional parts Numbers in the same proportion as a set of given numbers; such numbers are used in an auxiliary interpolation table based on the assumption that the tabulated quantity and entering arguments differ in the same proportion. { prə'pór·shən·əl 'pärts }

proposition 1. Any problem or theorem. **2.** A statement that makes an assertion that is either false or true or has been designated as false or true. { ,präp·ə'zish·ən }

propositional algebra The study of finite configurations of symbols and the interrelationships between them. { ,prä·pə¦zish·ən·əl 'al·jə·brə }

propositional calculus The mathematical study of logical connectives between propositions and deductive inference. Also known as sentential calculus. { ,präp·ə'zish·ən·əl 'kal·kyə·ləs }

propositional connectives The symbols ∼, ∧, ∨, → or ⊃, and ↔ or ≡, denoting logical relations that may be expressed by the phrases "it is not the case that," "and," "or," "if . . . , then," and "if and only if." Also known as sentential connectives. { ,prä·pə¦zish·ənəl kə'nek·tivz }

propositional function An expression that becomes a proposition when the values of certain symbols in the expression are specified, and that is either true or false depending on these values. Also known as logical function; open sentence; open statement; predicate; sentential function; statement function. { ,präp·ə¦zish·ən·əl 'fəŋk·shən }

Prüfer domain An integral domain in which every nonzero finitely generated ideal is invertible. { 'prüf·ər də,mān }

***p* series** The series $1 + (1/2)^p + (1/3)^p + \cdots$, where *p* is a real number. { 'pē ,sir·ēz }

pseudograph A graph with at least one loop. { 'süd·ə,graf }

pseudometric *See* semimetric. { ¦süd·ə¦me·trik }

pseudoperfect number An integer that is equal to the sum of some of its proper divisors. { ¦süd·ō,pər·fikt 'nəm·bər }

pseudosphere The pseudospherical surface generated by revolving a tractrix about its asymptote. { 'süd·ə,sfir }

pseudospherical surface A surface whose total curvature has a constant negative value. { ,süd·ō¦sfer·ə·kəl 'sər·fəs }

psi function The special function of a complex variable which is obtained from differentiating the logarithm of the gamma function. { 'sī ,fəŋk·shən }

Ptolemy's theorem The theorem that a necessary and sufficient condition for a convex quadrilateral to be inscribed in a circle is that the sum of the products of the two pairs of opposite sides equal the product of the diagonals. { 'täl·ə·mēz ,thir·əm }

pure geometry Geometry studied from the standpoint of its axioms and postulates rather than its objects. { 'pyür jē'äm·ə·trē }

pure imaginary number A complex number $z = x + iy$, where $x = 0$. { 'pyür i¦maj·ə,ner·ē ¦nəm·bər }

purely inseparable An element a is said to be purely inseparable over a field F with characteristic p greater than 0 if it is algebraic over F and if there exists a nonnegative integer n such that ap^n lies in F. { ¦pyür·lē in'sep·rə·bəl }

purely inseparable extension A purely inseparable extension E of a field F is an algebraic extension of F whose separable degree over F equals 1 or, equivalently, an algebraic extension of F in which every element is purely inseparable over F. { ¦pyür·lē in¦sep·rə·bəl ik'sten·chən }

pure mathematics The intrinsic study of mathematical structures, with no consideration given as to the utility of the results for practical purposes. { 'pyür ,math·ə'mad·iks }

pure projective geometry The axiomatic study of geometric systems which exhibit invariance relative to a notion of projection. { 'pyür prə¦jek·tiv jē'äm·ə·trē }

pure strategy In game theory, a predetermined plan covering all possible situations in a game and not involving the use of random devices. { 'pyür 'strad·ə·jē }

pure surd A surd, all of whose terms are irrational numbers. { 'pyür 'sərd }

pyramid A polyhedron with one face a polygon and all other faces triangles with a common vertex. { 'pir·ə,mid }

pyramidal numbers The numbers 1, 4, 10, 20, 35, . . ., which are the number of dots in successive pyramidal arrays and are given by $(1/6)n(n + 1)(n + 2)$, where $n = 1, 2, 3, ...$ { ¦pir·ə,mid·əl 'nəm·bərz }

pyramidal surface A surface generated by a line passing through a fixed point and moving along a broken line in a plane not containing that point. { ¦pir·ə¦mid·əl 'sər·fəs }

Pythagorean numbers Positive integers x, y, and z which satisfy the equation $x^2 + y^2 = z^2$. Also known as Pythagorean triple. { pə,thag·ə'rē·ən 'nəm·bərz }

Pythagorean theorem In a right triangle the square of the length of the hypotenuse equals the sum of the squares of the lengths of the other two sides. { pə,thag·ə'rē·ən 'thir·əm }

Pythagorean triple *See* Pythagorean numbers. { pə,thag·ə¦rē·ən 'trip·əl }

196

Q

quadrangle A geometric figure bounded by four straight-line segments called sides, each of which intersects each of two adjacent sides in points called vertices, but fails to intersect the opposite sides. Also known as quadrilateral.   { 'kwä,draŋ·gəl }

quadrangular prism A prism whose bases are quadrangles   { kwə¦draŋ·gyə·lər 'priz·əm }

quadrangular pyramid A pyramid whose base is a quadrangle.   { kwə¦draŋ·gyələr 'pir·ə,mid }

quadrant 1. A quarter of a circle; either an arc of 90° or the area bounded by such an arc and the two radii.   **2.** Any of the four regions into which the plane is divided by a pair of coordinate axes.   { 'kwä·drənt }

quadrantal angle An angle equal to 90° or π/2 radians multiplied by a positive or negative integer or zero.   { kwä¦drant·əl 'aŋ·gəl }

quadrantal spherical triangle A spherical triangle that has one and only one right angle.   { kwä¦drant·əl ¦sfir·ə·kəl 'trī,aŋ·gəl }

quadratic Any second-degree expression.   { kwä 'drad·ik }

quadratic congruence A statement that two polynomials of second degree have the same remainder on division by a given integer.   { kwä¦drad·ik kən'grü·əns }

quadratic equation Any second-degree polynomial equation.   { kwä'drad·ik i'kwā·zhən }

quadratic form Any second-degree, homogeneous polynomial.   { kwä'drad·ik 'fòrm }

quadratic formula A formula giving the roots of a quadratic equation in terms of the coefficients; for the equation $ax^2 + bx + c = 0$, the roots are $x = (-b \pm \sqrt{b^2 - 4ac})/2a$.   { kwä'drad·ik 'fòr·myə·lə }

quadratic function A function whose value is given by a quadratic polynomial in the independent variable.   { kwə,drad·ik 'fəŋk·shən }

quadratic inequality An inequality in which one side is a quadratic polynomial and the other side is zero.   { kwə,drad·ik ,in·i'kwal·əd·ē }

quadratic polynomial A polynomial where the highest degree of any of its terms is 2.   { kwä'drad·ik ,päl·ə'nō·mē·əl }

quadratic programming A body of techniques developed to find extremal points for systems of quadratic inequalities.   { kwä'drad·ik 'prō,gram·iŋ }

quadratic reciprocity law The law that, if p and q are distinct odd primes, then

$$(p \mid q)(q \mid p) = (-1)^{(1/4)(p-1)(q-1)}$$

(the vertical line inside parentheses is Legendre's symbol).   { kwuä¦drad·ik ,res·ə'präs·əd·ē ,lò }

quadratic residue A residue of order 2.   { kwä¦drad·ik 'rez·ə·dü }

quadratic surd A square root of a rational number that is itself an irrational number.   { kwä'drad·ik ,sərd }

quadratrix of Hippias A plane curve whose equation in cartesian coordinates x and y is $y = x \cot [\pi x/(2a)]$, where a is a constant.   { 'kwäd·rə,triks əv 'hip·ē·əs }

quadrature 1. The construction of a square whose area is equal to that of a given surface.   **2.** The process of calculating a definite integral.   { 'kwä·drə·chər }

quadric cone A conical surface whose directrices are conic curves.   { 'kwä·drik ¦kōn }

quadric curve An algebraic curve whose equation is of the second degree. { 'kwäd·rik 'kərv }

quadric quantic A quantic of the second degree. { 'kwä·drik ¦kwän·tik }

quadrics Homogeneous, second-degree expressions. { 'kwä·driks }

quadric surface A surface whose equation is a second-degree algebraic equation. { 'kwä·drik 'sər·fəs }

quadrillion 1. The number 10^{15}. **2.** In British and German usage, the number 10^{24}. { kwə'dril·yən }

quadrinomial distribution A multinomial distribution with four possible outcomes. { ¦kwä·drə'nō·mē·əl ¦di·strə'byü·shən }

quadruple vector product 1. For any four vectors, the dot product of two derived vectors, one of which is the cross product of two of the original vectors, and the other of which is the cross product of the other two. **2.** For any four vectors, the cross product of two derived vectors, one of which is the cross product of two of the original vectors, and the other of which is the cross product of the other two. { kwə'drüp·əl 'vek·tər ¦präd·əkt }

quadrupole A mass distribution that has unequal components of the moment-of-inertia tensor along the three principal directions. { 'kwä·drə,pōl }

quantal response Response to treatment which has only two outcomes, all or none. { 'kwänt·əl ri,späns }

quantic A homogeneous algebraic polynomial with more than one variable. { 'kwän·tik }

quantifier Either of the phrases "for all" and "there exists"; these are symbolized respectively by an inverted A and a backward E. { 'kwän·tə,fī·ər }

quantile A value which divides a set of data into equal proportions; examples are quartile and decile. { 'kwän,tīl }

quantity Any expression which is concerned with value rather than relations. { 'kwän·əd·ē }

quartic *See* biquadratic. { 'kwȯrd·ik }

quartic equation Any fourth-degree polynomial equation. Also known as biquadratic equation. { 'kwȯrd·ik i'kwä·zhən }

quartic quantic A quantic of the fourth degree. { ¦kwȯr·tik 'kwän·tik }

quartic surd A fourth root of a rational number that is itself an irrational number. { 'kwärd·ik ,sərd }

quartile The value of any of the three random variables which separate the frequency of a distribution into four equal parts. { 'kwȯr,tīl }

quartile deviation One-half of the difference between the upper and lower, that is, the third and first, quartiles. Also known as semi-interquartile range. { ¦kwȯr,tīl ,dē·vē'ā·shən }

quasi-F martingale A stochastic process which is the sum of an F martingale and an F process having bounded variation on every finite time interval. { ¦kwä·zē ¦ef 'märt·ən,gāl }

quasi-perfect number An integer that is 1 less than the sum of all its factors other than itself. { ¦kwäz·ē ¦pər·fikt 'nəm·bər }

quaternary quantic A quantic that contains four variables. { 'kwät·ən,er·ē ¦kwän·tik }

quaternion The division algebra over the real numbers generated by elements i, j, k subject to the relations $i^2 = j^2 = k^2 = -1$ and $ij = -ji = k$, $jk = -kj = i$, and $ki = -ik = j$. Also known as hypercomplex number. { kwə'ter·nē·ən }

quatrefoil A multifoil consisting of four congruent arcs of a circle arranged around a square. { 'kwä·trə,fȯil }

queueing theory The area of stochastic processes emphasizing those processes modeled on the situation of individuals lining up for service. { 'kyü·iŋ ,thē·ə·rē }

quintic A fifth-degree expression. { 'kwin·tik }

quintic equation A fifth-degree polynomial equation. { 'kwin·tik i'kwä·zhən }

quintic quantic A quantic of the fifth degree. { 'kwin·tik ¦kwän·tik }

quintic surd A fifth root of a rational number that is itself an irrational number. { 'kwin·tik ,sərd }

quintillion 1. The number 10^{18}. **2.** In British and German usage, the number 10^{30}. { kwin'til·yən }

quotient The result of dividing one quantity by another. { 'kwō·shənt }

quotient field The smallest field containing a given integral domain; obtained by formally introducing all quotients of elements of the integral domain. { 'kwō· shənt ‚fēld }

quotient group A group G/H whose elements are the cosets gH of a given normal subgroup H of a given group G, and the group operation is defined as $g_1 H \cdot g_2 H \equiv (g_1 \cdot g_2)H$. Also known as factor group. { 'kwō·shənt ‚grüp }

quotient ring A ring R/I whose elements are the cosets rI of a given ideal I in a given ring R, where the additive and multiplicative operations have the form: $r_1 I + r_2 I \equiv (r_1 + r_2)I$ and $r_1 I \cdot r_2 I \equiv (r_1 \cdot r_2)I$. Also known as factor ring; residue class ring. { 'kwō·shənt ‚riŋ }

quotient set The set of all the equivalence classes relative to a given equivalence relation on a given set. { 'kwō·shənt ‚set }

quotient space The topological space Y which is the set of equivalence classes relative to some given equivalence relation on a given topological space X; the topology of Y is canonically constructed from that of X. Also known as factor space. { 'kwō·shənt ‚spās }

quotient topology If X is a topological space, X/R the quotient space by some equivalence relation on X, the quotient topology on X/R is the smallest topology which makes the function which assigns to each element of X its equivalence class in X/R a continuous function. { 'kwō·shənt tə'päl·ə·jē }

R

r$_s$ *See* Spearman's rank correlation coefficient.

Raabe's convergence test An infinite series with positive terms a_n, where, for each n, $a_{n+1}/a_n = 1/(1 + b_n)$, will converge if, after a certain term, nb_n always exceeds a fixed number greater than 1 and will diverge if nb_n always is less than a fixed number less than or equal to 1. { 'räb·əz kən'vər·jəns ˌtest }

rabbit sequence A sequence of binary numbers that is recursively generated by the rules 0→1 (young rabbits grow old) and 1 → 10 (old rabbits stay old and beget young ones); beginning with a single 1, successive generations are 1, 10, 101, 10110, 10110101, and so forth. { 'ab·ət 'sē·kwəns }

Rademacher functions The functions f_n, with $n = 1, 2, 3, \ldots$, defined on the closed interval [0,1] by the equation $f_n(x) = \text{sgn}[\sin(2^n \pi x)]$, where sgn represents the signum function and sin represents the sine function.

radial For a plane curve C, the locus of end points of lines, drawn from a fixed point, that are equal and parallel to the radius of curvature of C. { 'rād·ē·əl }

radial distribution function A function $F(r)$ equal to the average of a given function of the three coordinates over a sphere of radius r centered at the origin of the coordinate system. { 'rād·ē·əl ˌdis·trə'byü·shən ˌfəŋk·shən }

radially related figures *See* homothetic figures. { ¦rad·ē·ə·lē ri¸lād·əd 'fig·yərz }

radian The central angle of a circle determined by two radii and an arc joining them, all of the same length. { 'rād·ē·ən }

radical 1. In a ring, the intersection of all maximal ideals. Also known as Jacobson radical. **2.** An indicated root of a quantity. Symbolized $\sqrt{}$. **3.** *See* nilradical. { 'rad·ə·kəl }

radical axis The line passing through the two points of intersection of a pair of circles. { 'rad·ə·kəl 'ak·səs }

radical center 1. For three circles, the point at which the three radical axes of pairs of the circles intersect. **2.** For four spheres, the point at which the six radical planes of pairs of the spheres intersect. { 'rad·ə·kəl 'sen·tər }

radical equation *See* irrational equation. { 'rad·ə·kəl i'kwā·zhən }

radical plane The plane containing the circle of intersection of a pair of spheres. { 'rad·ə·kəl 'plān }

radical sign The symbol $\sqrt{}$, indicating that a root of a quantity is to be taken. { 'rad·ə·kəl ˌsīn }

radicand The quantity that appears under a radical sign. { 'rad·ə¸kand }

radius 1. A line segment joining the center and a point of a circle or sphere. **2.** The length of such a line segment. { 'rād·ē·əs }

radius of convergence The positive real number corresponding to a power series expansion about some number a with the property that if $x - a$ has absolute value less than this number the power series converges at x, and if $x - a$ has absolute value greater than this number the power series diverges at x. { 'rād·ē·əs əv kən'vər·jəns }

radius of curvature The radius of the circle of curvature at a point of a curve. { 'rād·ē·əs əv 'kər·və·chər }

radius of geodesic curvature For a point on a curve lying on a surface, the reciprocal of the geodesic curvature at the point. { ¦rād·ē·əs əv ¸jē·ə¦des·ik 'kər·və·chər }

radius of geodesic torsion

radius of geodesic torsion The reciprocal of the geodesic torsion of a surface at a point in a given direction. { ¦rād·ē·əs əv ¸jē·ə¦des·ik 'tȯr·shən }

radius of gyration The square root of the ratio of the moment of inertia of a plane figure about a given axis to its area. { 'rād·ē·əs əv ji'rā·shən }

radius of normal curvature The reciprocal of the normal curvature of a surface at a point and in a given direction. { ¦rād·ē·əs əv ¦nȯr·məl 'kər·və·chər }

radius of torsion The reciprocal of the torsion of a space curve at a point. { ¦rād·ē·əs əv 'tȯr·shən }

radius of total curvature The quantity $\sqrt{-1/C}$, where C is the total curvature of a surface at a point. { ¦rād·ē·əs əv ¦tōd·əl 'kər·və·chər }

radius vector The coordinate r in a polar coordinate system, which gives the distance of a point from the origin. { 'rād·ē·əs ¸vek·tər }

radix approximation The approximation of a number by a number that can be expressed by a specified finite number of digits in radix notation. { 'rād·iks ə¸präk·sə'mā·shən }

radix complement A numeral in positional notation that can be derived from another by subtracting the original numeral from the numeral of highest value with the same number of digits, and adding 1 to the difference. Also known as complement; true complement. { 'rād·iks 'käm·plə·mənt }

radix fraction A generalization of a decimal fraction given by an expression of the form $(a/r) + (b/r^2) + (c/r^3) + \cdots$, where r is an integer and $a, b, c, \ldots$ are integers that are less than r. { 'rād·iks ¸frak·shən }

radix-minus-one complement A numeral in positional notation of base (or radix) B derived from a given numeral by subtracting the latter from the highest numeral with the same number of digits, that is, from $B - 1$; it is 1 less than the radix complement. { 'rād·iks ¦mī·nəs ¦wən 'käm·plə·mənt }

radix notation A positional notation in which the successive digits are interpreted as coefficients of successive integral powers of a number called the radix or base; the represented number is equal to the sum of this power series. Also known as base notation. { 'rād·iks nō¸tā·shən }

radix point A dot written either on or slightly above the line, used to mark the point at which place values change from positive to negative powers of the radix in a number system; a decimal point is a radix point for radix 10. { 'rād·iks ¸pȯint }

Radon measure *See* regular Borel measure. { 'rä¸dän ¸mezh·ər }

Radon theorem The theorem that a set in an n-dimensional Euclidean space with at least $n + 2$ points is the union of two sets whose convex spans are disjoint. { 'rä¸dän ¸thir·əm }

Radon transform A mathematical operation that is roughly equivalent to finding the projection of a function along a given line; useful in computerized tomography. { 'rä¸dän 'tranz¸fȯrm }

ramphoid cusp A cusp of a curve which has both branches of the curve on the same side of the common tangent. Also known as single cusp of the second kind. { 'ram¸fȯid ¸kəsp }

Ramsey number For any two positive integers, p and q, the smallest integer, $R(p,q)$, that has the (p,q)-Ramsey property. { 'ram·zē ¸nəm·bər }

Ramsey property For any two positive integers, p and q, a positive integer r is said to have the (p,q)-Ramsey property if in any set of r people there is either a subset of p people who are all mutual acquaintances or a set of q people who are all mutual strangers. { 'ram·zē ¸präp·ərd·ē }

Ramsey theorem The theorem that for any two positive integers p and q there is a positive integer r that has the (p,q)-Ramsey property. { ¦ram·zē ¦thir·əm }

Ramsey theory The theory of order that must exist in subsets of sufficiently large sets, as illustrated by Ramsey's theorem and van der Waerden's theorem. { 'ram·zē ¸thē·ə·rē }

random digit Digit taken from a table of random numbers according to some specified probability rule. { 'ran·dəm 'dij·ət }

random error An error that can be predicted only on a statistical basis. { 'ran·dəm 'er·ər }

random experiments Experiments which do not always yield the same result when repeated under the same conditions. { 'ran·dəm ik'sper·ə·məns }

random function A function whose domain is an interval of the extended real numbers and has range in the set of random variables on some probability space; more precisely, a mapping of the cartesian product of an interval in the extended reals with a probability space to the extended reals so that each section is a random variable. { 'ran·dəm 'fəŋk·shən }

randomization Assigning subjects to treatment groups by use of tables of random numbers. { ‚ran·də·mə'zā·shən }

randomized blocks An experimental design in which the various treatments are reproduced in each of the blocks and are randomly assigned to the units within the blocks, permitting unbiased estimates of error to be made. { 'ran·də‚mīzd 'bläks }

randomized test Acceptance or rejection of the null hypothesis by use of a random variable to decide whether an observation causes rejection or acceptance. { 'ran·də‚mīzd 'test }

random matrices Collections of large matrices, chosen at random from some ensemble. { ‚ran·dəm 'mā·tri‚sēz }

random noise A form of random stochastic process arising in control theory. { 'ran·dəm 'nȯiz }

random numbers A listing of numbers which is nonrepetitive and satisfies no algorithm. { 'ran·dəm 'nəm·bərz }

random ordered sample An ordered sample of size s drawn from a population of size N such that the probability of any particular ordered sample is the reciprocal of the number of permutations of N things taken s at a time. { 'ran·dəm ¦ȯr·dərd 'sam·pəl }

random process *See* stochastic process. { 'ran·dəm 'prä·səs }

random sampling A sampling from some population where each entry has an equal chance of being drawn. { 'ran·dəm 'sam·pliŋ }

random start In a systematic sample, the random selection of a starting point in the first sample block followed by taking that value in the same position in every succeeding block. { 'ran·dəm 'stärt }

random variable A measurable function on a probability space; usually real valued, but possibly with values in a general measurable space. Also known as chance variable; stochastic variable; variate. { 'ran·dəm 'ver·ē·ə·bəl }

random walk A succession of movements along line segments where the direction and possibly the length of each move is randomly determined. { 'ran·dəm 'wȯk }

range 1. The range of a function f from a set X to a set Y consists of those elements y in Y for which there is an x in X with $f(x) = y$. **2.** The difference between the maximums and minimums of a variable quantity. { rānj }

rank 1. The rank of a matrix is its maximum number of linearly independent rows. **2.** The rank of a system of homogeneous linear equations equals the rank of the matrix of its coefficients. **3.** A tensor in an n-dimensional space is of rank r if it has n^r components. **4.** The rank of a group G is the number of elements in the basis of the quotient group of G over the subgroup consisting of all elements of G having finite period. **5.** The rank of a place or valuation is equal to the number of proper prime ideals in its valuation ring. **6.** The rank of a prime ideal P is the largest number n for which there exists a sequence $P_0 = P, P_1, P_2, \ldots , P_n$ of prime ideals such that P_i is a subset of P_{n-1}. **7.** The number assigned to an observation if a collection of observations is ordered from smallest to largest and each observation is given the number corresponding to its place in the order. { raŋk }

rank correlation A nonparametric test of statistical dependence for a random sample of paired observations. { 'raŋk ‚kä·rə'lā·shən }

ranked p₀ set A partially ordered set for which there is a function, r, defined on the elements of the set, such that $r(x) = 0$ if x is a minimal element, and $r(x) = r(y) + 1$ if x covers y. { ¦raŋkt ‚pē'zir·ō *or* 'pē‚səb¦zir·ō ‚set }

rank of an observation The number assigned to an observation if a collection of observations is ordered from smallest to largest and each observation is given the number corresponding to its place in the order. { 'raŋk əv ən 'äb·zər'vā·shən }

rank-order statistics Statistics computed from rankings of the observations rather than from the observations themselves. { ¦raŋk 'òr·dər stə'tis·tiks }

rank tests Tests which use the ranks of observations with respect to one another rather than the observations themselves. { 'raŋk ‚tests }

rare set *See* nowhere dense set. { ¦rär 'set }

ratio A ratio of two quantities or mathematical objects A and B is their quotient or fraction A/B. { 'rā·shō }

ratio estimator A ratio of two random variables that is used as an estimator. { 'rā·shō ‚es·tə‚mād·ər }

rational algebraic expression An algebraic expression that equals a quotient of polynomials. { ¦rash·ən·əl ‚al·jə‚brā·ik ik'spresh·ən }

rational fraction 1. A fraction whose numerator and denominator are both rational numbers. 2. A fraction whose numerator and denominator are both polynomials. { 'rash·ən·əl 'frak·shən }

rational function A function which is a quotient of polynomials. { 'rash·ən·əl 'fəŋk·shən }

rationalize 1. To carry out operations on an algebraic equation that remove radicals containing the varible. 2. To multiply the numerator and denominator of a fraction by a quantity that removes the radicals in the denominator. 3. To make a substitution in an integral that removes the radicals in the integrand. { 'rash·ən·əl‚īz }

rational number A number which is the quotient of two integers. { 'rash·ən·əl 'nəm·bər }

rational root theorem The theorem that, if a rational number p/q, where p and q have no common factors, is a root of a polynomial equation with integral coefficients, then the coefficient of the term of highest order is divisible by q and the coefficient of the term of lowest order is divisible by p. { 'rash·ən·əl ‚rüt ‚thir·əm }

ratio of similitude The ratio of the lengths of corresponding line segments of similar figures.   Also known as homothetic ratio. { ¦rā·shō əv sə'mil·ə‚tüd }

ratio paper *See* semilogarithmic coordinate paper. { 'rā·shō ‚pā·pər }

ratio scale A rule or system for assigning numbers to objects which has all the properties of an interval scale and, in addition, has a natural origin, so that ratios of numbers assigned to different objects have meaning. { 'rā·shō ‚skāl }

ratio test *See* Cauchy ratio test. { 'rā·shō ‚test }

raw score Any number as it originally appears in an experiment; for example, in evaluating test results the raw scores express the number of correct answers, uncorrected for position in the reference population. { 'rò 'skòr }

ray A straight-line segment emanating from a point.   Also known as half line. { rā }

ray center *See* homothetic center. { 'rā ‚sen·tər }

Rayleigh distribution A normal distribution of two uncorrelated variates with the same variance. { 'rā·lē ‚dis·trə‚byü·shən }

Rayleigh-Ritz method An approximation method for finding solutions of functional equations in terms of finite systems of equations. { 'rā·lē 'rits ‚meth·əd }

r-combination An r-combination of a set is an selection of r elements of the set. { 'är ‚käm·bə‚nā·shən }

real axis The horizontal axis of the Cartesian coordinate system for the Euclidean or complex plane. { 'rēl 'ak·səs }

real closed field A real field which has no algebraic extensions other than itself. { ¦rēl ¦klōzd 'fēld }

real closure A real closure of a real field F is a real closed field which is an algebraic extension of F. { ¦rēl 'klō·zhər }

real continuum *See* real number system. { ¦rēl kən'tin·yə·wəm }

realization For a stochastic process, a probability space whose points are sample paths of the stochastic process and whose probability is obtained from the joint probability distributions of the random variables in the process. { ‚rē·ə·lə'zā·shən }

real line A straight line of infinite extent upon which the real numbers are plotted according to their distance in a positive or negative direction from a point arbitrarily chosen as zero. Also known as number line. { 'rēl 'līn }

real linear group The group of all nonsingular linear transformations of a real vector space whose group operation is composition. { ¦rēl ¦lin·ē·ər 'grüp }

real number Any member of the real number system. { 'rēl 'nəm·bər }

real number system The unique (to within isomorphism) complete ordered field; the field of real numbers. Also known as real continuum. { ¦rēl 'nəm·bər ‚sis·təm }

real orthogonal group The group composed of orthogonal matrices having real number entries. { 'rēl ȯr'thäg·ən·əl ‚grüp }

real part The real part of a complex number $z = x + iy$ is the real number x. { 'rēl ¦pärt }

real plane A plane whose points are assigned ordered pairs of real numbers for coordinates. { 'rēl ‚plān }

real unimodular group The group of all square $n \times n$ matrices with real number entries and of determinant 1. { 'rēl ‚yün·i'mäj·ə·lər ‚grüp }

real-valued function A function whose values are real numbers. { ¦rēl ¦val·yüd 'fəŋk·shən }

real variable A variable that assumes real numbers for its values. { 'rēl 'ver·ē·ə·bəl }

reciprocal The reciprocal of a number A is the number $1/A$. { ri'sip·rə·kəl }

reciprocal curve For a particular curve, C, a curve that includes a point with Cartesian coordinates (x, y) if C has a point with cartesian coordinates $(x, 1/y)$. { ri‚sip·rə·kəl 'kərv }

reciprocal differences An interpolation technique using successive quotients of a function with its values so as to obtain a continued fraction expansion approximating the given function by a rational function. { ri'sip·rə·kəl 'dif·rən·səz }

reciprocal equation An algebraic equation in one variable whose roots are unchanged when the unknown is replaced by its reciprocal. { rə'sip·rə·kəl i'kwā·zhən }

reciprocal matrix *See* inverse matrix. { ri¦sip·rə·kəl 'mā·triks }

reciprocal polar figures Two plane figures consisting of lines and their points of intersection such that the points of each of them are the poles of the lines of the other with respect to a given conic. { ri¦sip·rə·kəl ¦pō·lər 'fig·yərz }

reciprocal ratio *See* inverse ratio. { ri¦sip·rə·kəl 'rā·shō }

reciprocal series A series whose terms are reciprocals of the corresponding terms of a given series. { rə'sip·rə·kəl 'sir‚ēz }

reciprocal spiral *See* hyperbolic spiral. { ri'sip·rə·kəl ‚spī·rəl }

reciprocal substitution The substitution of a new variable for the reciprocal of the original variable. { ri¦sip·rə·kəl ‚səb·stə'tü·shən }

reciprocal theorem **1.** In plane geometry, a theorem (which may be true or false) that is obtained from a given theorem by exchanging points and lines, angles and sides, and so forth. **2.** *See* dual theorem. { ri'sip·rə·kəl 'thir·əm }

reciprocal vectors For a set of three linearly independent vectors, a second set of three vectors, each of which is perpendicular to two of the original vectors and has a scalar product of unity with the third. { ri'sip·rə·kəl 'vek·tərz }

rectangle A plane quadrilateral having four interior right angles and opposite sides of equal length. { 'rek‚taŋ·gəl }

rectangular Cartesian coordinate system *See* Cartesian coordinate system. { rek'taŋ· gyə·lər kär'tē·zhən kō'ȯrd·ən·ət ‚sis·təm }

rectangular coordinates *See* Cartesian coordinates. { rek'taŋ·gyə·lər kō'ȯrd·ən·əts }

rectangular distribution *See* uniform distribution. { rek'taŋ·gyə·lər ‚dis·trə'byü·shən }

rectangular game *See* matrix game. { rek'taŋ·gyə·lər 'gām }

rectangular graph *See* bar graph. { rek¦taŋ·gyə·lər 'graf }

rectangular hyperbola A hyperbola whose major and minor axes are equal. { rek'taŋ· gyə·lər hī'pər·bə·lə }

rectangular parallelepiped A parallelepiped with bases as rectangles all perpendicular to its lateral faces. Also known as cuboid; rectangular solid. { rek'taŋ·gyə·lər ,par·ə,lel·ə'pī,ped }

rectangular solid *See* rectangular parallelepiped. { rek'taŋ·gyə·lər 'säl·əd }

rectifiable curve A curve whose length can be computed and is finite. { 'rek·tə,fī·ə· bəl 'kərv }

rectifying developable The envelope of the rectifying planes of a space curve. { 'rek· tə,fī·iŋ də'vel·əp·ə·bəl }

rectifying plane The plane that contains the tangent and binormal to a curve at a given point on the curve. { 'rek·tə,fī·iŋ ,plān }

rectilinear Consisting of or bounded by lines. { ¦rek·tə'lin·ē·ər }

rectilinear generators Straight lines which generate ruled surfaces. { ¦rek·tə'lin·ē·ər 'jen·ə,rād·ərz }

recurrence formula methods Methods of calculating numerical solutions of differential equations in which the equation is written in the form of a recurrence relation between values of the solution function at successive points by replacing the derivatives with corresponding finite difference expressions. { ri¦kər·əns 'fȯr· myə·lə ,meth·ədz }

recurrence relation An equation relating a term in a sequence to one or more of its predecessors in the sequence. { ri'kər·əns ri,lā·shən }

recurrent transformation 1. A measurable function from a measure space T to itself such that for every measurable set A in the space and every point x in A there is a positive integer n such that $T^n(x)$ is also in A. **2.** A continuous function from a topological space T to itself such that for every open set A in the space and every point x in A there is a positive integer n such that $T^n(x)$ is also in A. { ri'kər· ənt ,tranz·fər'mā·shən }

recurring continued fraction A continued fraction in which a finite sequence of terms is repeated indefinitely. Also known as periodic continued fraction. { ri¦kər·iŋ kən¦tin·yüd 'frak·shən }

recurring decimal *See* repeating decimal. { ri'kər·iŋ 'des·məl }

recursion formula An algorithm allowing computation of a succession of quantities. Also known as recursion relation. { ri'kər·zhən ,fȯr·myə·lə }

recursion relation *See* recursion formula. { ri'kər·zhən ri,lā·shən }

recursive Pertaining to a process that is inherently repetitive, with the results of each repetition usually depending upon those of the previous repetition. { ri'kər·siv }

recursive functions Functions that can be obtained by a finite number of operations, computations, or algorithms. { ri'kər·siv 'fəŋk·shənz }

reduced characteristic equation The polynomial equation of lowest degree that is satisfied by a given matrix. Also known as minimal equation. { ri'düst ,kär·ik· tə'ris·tik i,kwā·zhən }

reduced cubic equation A cubic equation in a variable x, where the coefficient of x^2 is zero. { ri¦düst ¦kyü·bik i'kwā·zhən }

reduced echelon matrix A matrix in which the first nonzero term in a row is the only nonzero term in its column. { ri¦düst ,esh·ə,län 'mā·triks }

reduced equation *See* auxiliary equation. { ri'düst i'kwā·zhən }

reduced form A lambda expression that has no subexpressions of the form (λxMA), where M and A are lambda expressions, is said to be in reduced form. { ri'düst 'fȯrm }

reduced residue system modulo n A set of integers that includes those members of a complete residue system modulo n that are relatively prime to n. { ri¦düst ¦res· ə,dü ¦sis·təm ¦mäj·ə,lō 'en }

reducible configuration A graph such that the four-colorability of any planar graph containing the configuration can be deduced from the four-colorability of planar graphs with fewer vertices. { ri'dü·sə·bəl kən,fig·yə'rā·shən }

reducible curve A curve that can be shrunk to a point by a continuous deformation without passing outside a given region. { ri¦düs·ə·bəl 'kərv }

reducible polynomial A polynomial relative to some field which can be written as the product of two polynomials of degree at least 1. { ri'dü·sə·bəl ,päl·i'nō·mē·əl }

reducible representation of a group A representation of a group as a family of linear operators on a vector space V such that there is a proper closed subspace of V that is invariant under these operators. { ri¦düs·ə·bəl ,rep·rə·zen'tā·shən əv ə ,grüp }

reducible transformation A linear transformation T on a vector space V that can be completely specified by describing its effect on two subspaces, M and N, that are each transformed into themselves by T and are such that any vector of V can be uniquely represented as the sum of a vector of M and a vector of N. { ri¦düs·ə· bəl ,tranz·fər'mā·shən }

reductio ad absurdum A method of proof in which it is first supposed that the fact to be proved is false, and then it is shown that this supposition leads to the contradiction of accepted facts. Also known as indirect proof; proof by contradiction. { ri¦dək·tē·ō äd ab'sərd·əm }

reduction formula 1. An equation that expresses an integral as the sum of certain functions and a simpler integral. **2.** An identity that expresses the values of a trigonometric function of an angle greater than 90° in terms of a function of an angle less than 90°. { ri'dək·shən ,fȯr·myə·lə }

reduction sequence A sequence of applications of the reduction rule to a lambda expression. { ri'dək·shən ,sē·kwəns }

redundancy A repetitive statement. { ri'dən·dən·sē }

redundant equation An equation with roots that have been introduced in the process of solving another equation but that are not solutions of the equation to be solved. { ri'dən·dənt i'kwā·zhən }

redundant number *See* abundant number. { ri'dən·dənt ¦nəm·bər }

reentrant angle An interior angle of a polygon that is greater than 180°. { rē'en·trənt ,aŋ·gəl }

refinement A tower that can be obtained by inserting a finite number of subsets in a given tower. { ri'fīn·mənt }

reflection 1. The reflection of a configuration in a line, in a plane, or in the origin of a coordinate system is the replacement of each point in the configuration by a point that is symmetric to the given point with respect to the line, plane, or origin. **2.** Two permutations, a and b, of the same objects are reflections of each other if the first object in a is the last object in b, the second object in a is the next-to-last object in b, and so forth, with the last object in a being the first object in b. { ri'flek·shən }

reflection plane *See* plane of mirror symmetry. { ri'flek·shən ,plān }

reflex angle An angle greater than 180° and less than 360°. { 'rē,fleks ,aŋ·gəl }

reflexive Banach space A Banach space B such that, for every continuous linear functional F on the conjugate space B^*, there corresponds a point x_0 of B such that $F(f) = f(x_0)$ for each element f of B^*. Also known as regular Banach space. { ri¦flek·siv 'bä,näk ,spās }

reflexive relation A relation among the elements of a set such that every element stands in that relation to itself. { ri'flek·siv ri,lā·shən }

region 1. The union of an open connected set with a subset of its boundary points (which may be the entire boundary or the empty set). **2.** *See* domain. { 'rē·jən }

regression Given two stochastically dependent random variables, regression functions measure the mean expectation of one relative to the other. { ri'gresh·ən }

regression analysis The description of the nature of the relationship between two or more variables; it is concerned with the problem of describing or estimating the value of the dependent variable on the basis of one or more independent variables. { ri'gresh·ən ə,nal·ə·səs }

regression coefficient The coefficient of the independent variables in a regression equation. { ri'gresh·ən ,kō·ə,fish·ənt }

regression curve A plot of a regression equation; for two variables, the independent variable is plotted as the abscissa and the dependent variable as the ordinate; for three variables, a solid model can be constructed or the representation can be reduced by an isometric chart or stereogram. { ri'gresh·ən ,kərv }

regression estimate An estimate of one variable obtained by substituting the known value of another variable in a regression equation calculated on sample values of the two variables. { ri'gresh·ən ˌes·tə·mət }

regression line A linear regression equation with two or more variables. { ri'gresh· ən ˌlīn }

regret criterion *See* Savage principle. { ri'gret krī,tir·ē·ən }

regula falsi A method of calculating an unknown quantity by first making an estimate and then using this and the properties of the unknown to obtain it. Also known as rule of false position. { 'reg·yə·lə 'fäl·sē }

regular analytic curve An analytic curve for which the derivatives of the analytic functions that represent the coordinates of points on the curve do not all vanish at any point. { ¦reg·yə·lər ¸an·ə¸lid·ik 'kərv }

regular Baire measure A Baire measure such that the measure of any Baire set *E* is equal to both the greatest lower bound of measures of open Baire sets containing *E*, and to the least upper bound of closed, compact sets contained in *E*. { 'reg· yə·lər 'bār ¸mezh·ər }

regular Banach space *See* reflexive Banach space. { 'reg·yə·lər 'bä¸näk ¸späs }

regular Borel measure A Borel measure such that the measure of any Borel set *E* is equal to both the greatest lower bound of measures of open Borel sets containing *E*, and to the least upper bound of measures of compact sets contained in *E*. Also known as Radon measure. { 'reg·yə·lər bə'rel ¸mezh·ər }

regular curve A curve that has no singular points. { 'reg·yə·lər 'kərv }

regular definition A definition of the sum of a divergent series which yields the ordinary sum when applied to a convergent series. { ¦reg·yə·lər ¸def·ə'nish·ən }

regular dodecahedron A regular polyhedron of 12 faces. { 'reg·yə·lər dō¸dek·ə'hē· drən }

regular extension An extension field *K* of a field *F* such that *F* is algebraically closed in *K* and *K* is separable over *F*; equivalently, an extension field *K* of a field *F* such that *K* and F̄ are linearly disjoint over *F*, where F̄ is the algebraic closure of *F*. { 'reg·yə·lər ik'sten·chən }

regular function An analytic function of one or more complex variables. { 'reg·yə· lər 'fəŋk·shən }

regular graph A graph whose vertices all have the same degree. { 'reg·gə·lər ¦graf }

regular icosahedron A 20-sided regular polyhedron, having five equilateral triangles meeting at each face. { 'reg·yə·lər ī¸käs·ə'hē·drən }

regular map *See* normal map. { 'reg·yə·lər 'map }

regular octahedron A regular polyhedron of eight faces. { 'reg·yə·lər ¸äk·tə'hē·drən }

regular parameter The independent variable that parametrizes a regular analytic curve. { ¦reg·yə·lər pə'ram·əd·ər }

regular permutation group A permutation group of order *n* on *n* objects, where *n* is a positive integer. { 'reg·yə·lər ¸pər·myə'tā·shən ¸grüp }

regular point 1. Any point or a surface that is not a singular point. **2.** *See* ordinary point. { ¦reg·yə·lər 'pȯint }

regular polygon A polygon with congruent sides and congruent interior angles. { 'reg· yə·lər 'päl·i¸gän }

regular polyhedron A polyhedron all of whose faces are regular polygons, and whose polyhedral angles are congruent. Also known as platonic solid. { 'reg·yə·lər ¸päl· i'hē·drən }

regular polytope A geometric object in multidimensional Euclidean space that is analogous to the regular polygons (in two-dimensional space) and the regular polyhedra (in three-dimensional space). { ¦reg·yə·lər 'päl·ə¸tōp }

regular prism A right prism whose bases are regular polygons. { ¦reg·yə·lər 'priz·əm }

regular pyramid A pyramid whose base is a regular polygon and whose lateral faces are inclined at equal angles to the base. { ¦reg·yə·lər 'pir·ə¸mid }

regular representation A regular representation of a finite group is an isomorphism of it with a group of permutations. { 'reg·yə·lər ¸rep·rə·zən'tā·shən }

regular sequence *See* Cauchy sequence. { ¦reg·yə·lər 'sē·kwəns }

regular singular point A regular singular point of a differential equation is a singular point of the equation at which none of the solutions has an essential singularity. { 'reg·yə·lər ¦siŋ·gyə·lər 'póint }

regular space A topological space such that any neighborhood of a point in the space contains the closure of another neighborhood of the same point. { ¦reg·yə·lər 'spās }

regular tetrahedron A regular polyhedron of four faces. { 'reg·yə·lər ¸te·trə'hē·drən }

regular topological space A topological space where any point and a closed set not containing it can be enclosed in disjoint open sets. { 'reg·yə·lər ¦täp·ə¦läj·ə·kəl 'spās }

related angle The acute angle at which trigonometric functions have the same absolute values as at a given angle outside the first quadrant. { ri'lād·əd ¸aŋ·gəl }

relation A set of ordered pairs. Also known as correspondence. { ri'lā·shən }

relative coordinates Coordinates given as offsets from some point whose location can be adjusted. { 'rel·əd·iv kō'órd·ən·əts }

relative efficiency 1. Of an estimator, the comparative efficiency of the two estimators of the same parameter. **2.** For experimental design, the number of replications each design requires to reach the same precision. { 'rel·ə·tiv ə'fish·ən·sē }

relative error The absolute error in estimating a quantity divided by its true value. { 'rel·əd·iv 'er·ər }

relative frequency The ratio of the number of occurrences of a given type of event or the number of members of a population in a given class to the total number of events or the total number of members of the population. { 'rel·əd·iv frē·kwən·sē }

relative frequency distribution *See* percentage distribution. { ¦rel·ə·tiv 'frē·kwən·sē ¸dis·trə¸byü·shən }

relative frequency table *See* percentage distribution. { 'rel·ə·tiv 'frē·kwən·sē ¸tā·bəl }

relatively closed set A subset of a topological space is relatively closed if it is a closed set in some relative topology of a subset. { 'rel·ə¸tiv·lē ¦klōzd ¸set }

relatively compact set *See* conditionally compact set. { 'rel·ə¸tiv·lē ¦käm¸pakt ¸set }

relatively open set A subset of a topological space is relatively open if it is an open set in some relative topology of a subset. { 'rel·ə¸tiv·lē ¦ō·pən ¸set }

relatively prime Integers m and n are relatively prime if there are integers p and q so that $pm + qn = 1$; equivalently, if they have no common factors other than 1. { 'rel·ə¸tiv·lē 'prīm }

relative maximum A value of a function at a point x_0 which is equal to or greater than the values of the function at all points in some neighborhood of x_0. { 'rel·ə·tiv 'mak·sə·məm }

relative minimum A value of a function at a point x_0 which is equal to or less than the values of the function at all points in some neighborhood of x_0. { 'rel·ə·tiv 'min·ə·məm }

relative primes Two positive integers with no common positive divisor other than 1. { 'rel·ə·tiv 'prīmz }

relative topology In a topological space X any subset A has a topology on it relative to the given one by intersecting the open sets of X with A to obtain open sets in A. { 'rel·əd·iv tə'päl·ə·jē }

relaxation *See* relaxation method. { ¸rē¸lak'sā·shən }

relaxation method A successive approximation method for solving systems of equations where the errors from an initial approximation are viewed as constraints to be minimized or relaxed within a toleration limit. Also known as relaxation. { ¸rē¸lak'sā·shən ¸meth·əd }

reliability 1. The amount of credence placed in a result. **2.** The precision of a measurement, as measured by the variance of repeated measurements of the same object. { ri¸lī·ə'bil·əd·ē }

remainder 1. The remaining integer when a division of an integer by another is performed; if $l = m \cdot p + r$, where l, m, p, and r are integers and r is less than p, then r is the remainder when l is divided by p. **2.** The remaining polynomial when division of a polynomial is performed; if $l = m \cdot p + r$, where l, m, p, and

r are polynomials, and the degree of r is less than that of p, then r is the remainder when l is divided by p.   **3.** The remaining part of a convergent infinite series after a computation, for some n, of the first n terms.   { ri'mān·dər }

remainder formula A formula by which the remainder resulting from an approximation of a function by a partial sum of a power series can be computed or analyzed.   { ri'mān·dər ‚fȯr·myə·lə }

remainder theorem Dividing a polynomial $p(x)$ by $(x - a)$ gives a remainder equaling the number $p(a)$.   { ri'mān·dər ‚thir·əm }

removable discontinuity A point where a function is discontinuous, but it is possible to redefine the function at this point so that it will be continuous there.   { ri'müv·ə·bəl ¦dis‚känt·ən'ü·əd·ē }

renaming rule A transformation rule in the lambda calculus that allows conflicts of variables to be eliminated; it states that a bound variable x in a lambda expression M may be uniformly replaced by some other bound variable y, provided y does not occur in M.   Also known as alpha rule.   { rē'nām·iŋ ‚rül }

renormalization transformation A transformation of a mathematical function involving a change of scale.   { rē‚nȯr·mə·lə'zā·shən ‚tranz·fər‚mā·shən }

repeated measurements model A product model in which each factor is the same.   { ri'pēd·əd ¦mezh·ər·məns ‚mäd·əl }

repeated root *See* multiple root.   { ri'pēd·əd 'rüt }

repeating decimal A decimal that is either finite or infinite with a finite block of digits repeating indefinitely.   Also known as periodic decimal; recurring decimal.   { ri'pēd·iŋ 'des·məl }

replication In experimental design, the repetition of an experiment or parts of an experiment to secure more data as an aid to determining the experimental error and to arrive at better estimates of the effects of various treatments with smaller standard errors.   { ‚rep·lə'kā·shən }

representation A representation of a group is given by a homomorphism of it onto some group either of matrices or unitary operators of a Hilbert space.   { ‚rep·ri‚zen'tā·shən }

representation theory 1. The study of groups by the use of their representations. **2.** The determination of representations of specific groups.   { ‚rep·ri‚zen'tā·shən ‚thē·ə·rē }

representative sample A sample whose characteristics reflect those of the population from which it is drawn.   { ¦rep·ri¦zen·təd·iv 'sam·pəl }

residual set In a topological space, the complement of a set which is a countable union of nowhere dense sets.   { rə'zij·ə·wəl 'set }

residual spectrum Those members λ of the spectrum of a linear operator A on a Banach space X for which $(A - \lambda I)^{-1}$, I being the identity operator, is unbounded with domain not dense in X.   { rə'zij·ə·wəl 'spek·trəm }

residual sum of squares *See* error sum of squares.   { rə¦zij·ə·wəl ¦səm əv 'skwerz }

residual variance In analysis of variance and regression analysis, that part of the variance which cannot be attributed to specific causes.   { rə'zij·ə·wəl 'ver·ē·əns }

residue 1. The residue of a complex function $f(z)$ at an isolated singularity z_0 is given by $(1/2\pi i)\int f(z)dz$ along a simple closed curve interior to an annulus about z_0; equivalently, the coefficient of the term $(z - z_0)^{-1}$ in the Laurent series expansion of $f(z)$ about z_0. **2.** In general, a coset of an ideal in a ring. **3.** A residue of m of order n, where m and n are integers, is a remainder that results from raising some integer to the nth power and dividing by m.   { 'rez·ə‚dü }

residue class A set of numbers satisfying a congruency relation.   { 'rez·ə‚dü ‚klas }

residue class ring *See* quotient ring.   { 'rez·ə‚dü ¦klas ‚riŋ }

residue theorem The value of the integral of a complex function, taken along a simple closed curve enclosing at most a finite number of isolated singularities, is given by $2\pi i$ times the sum of the residues of the function at each of the singularities.   { 'rez·ə‚dü ‚thir·əm }

resolution For a vector, the determination of vectors parallel to specified (usually perpendicular) axes such that their sum equals the given vector.   { ‚rez·ə'lü·shən }

resolution of the identity A family of linear projection operators on a Banach space used in studying the spectra of linear operators. { ‚rez·ə'lü·shən əv the̱ ə'den·əd·ē }

resolvable balanced incomplete block design A balanced incomplete block design such that the blocks themselves are partitioned into r families of v/k blocks, such that every element occurs in exactly one block of each of these families. { ri¦zäl· və·bəl ¦bal·ənst ‚iŋ·kəm‚plēt 'bläk di‚zīn }

resolvent For a linear operator T on a Banach space, the function, defined on the complement of the spectrum of T given by $(T - \lambda I)^{-1}$ for each λ in this complement, where I is the identity operator; this enables a study of T relative to its eigenvalues. { ri'zäl·vənt }

resolvent kernel A function appearing as an integrand in an integral representation for a solution of a linear integral equation which often completely determines the solutions. { ri'zäl·vənt 'kər·nəl }

resolvent set Those scalars λ for which the operator $T - \lambda I$ has a bounded inverse, where T is some linear operator on a Banach space, and I is the identity operator. { ri'zäl·vənt 'set }

response The value of some measurable quantity after a treatment has been applied. { ri'späns }

restricted limit See limit inferior. { ri¦strik·təd 'lim·ət }

resultant 1. For a set of polynomial equations, a function of the coefficients which equals zero if the equations have at least one solution. Also known as eliminant. **2.** See vector sum. { ri'zəl·tənt }

reticular density The number of points per unit area in a two-dimensional lattice, such as the plane of a crystal lattice. { re'tik·yə·lər 'den·səd·ē }

retract A subset R of a topological space X is a retract of X if there is a continuous map f from X to R, with $f(r) = r$ for all points r of R. { 're‚trakt }

Reuleaux triangle A closed plane curve, not actually a triangle, that consists of three arcs, each of which joins two vertices of an equilateral triangle and is part of a circle centered at the remaining vertex. { ‚re‚lō 'trī‚aŋ·gəl }

reverse curve An S-shaped curve, that is, one having two arcs with their centers on opposite sides of the curve. Also known as S curve. { ri'vərs 'kərv }

reversion For a series, the process of constructing a new series in which the dependent and independent variables of the original series are interchanged. { ri'vər·zhən }

rhomb See rhombus. { räm }

rhombohedron A prism with six parallelogram faces. { ¦räm·bō¦hē·drən }

rhomboid A parallelogram whose adjacent sides are not equal. { 'räm‚bȯid }

rhombus A parallelogram with all sides equal. Also known as rhomb. { 'räm·bəs }

ribbon The plane figure generated by a straight line which moves so that it is always perpendicular to the path traced by its middle point. { 'rib·ən }

Riccati-Bessel functions Solutions of a second-order differential equation in a complex variable which have the form $z f(z)$, where $f(z)$ is a function in terms of polynomials and cos (z), sin (z). { ri'käd·ē 'bes·əl ‚fəŋk·shənz }

Riccati equation 1. A first-order differential equation having the form $y' = A_0(x) + A_1(x)y + A_2(x)y^2$; every second-order linear differential equation can be transformed into an equation of this form. **2.** A matrix equation of the form $dP(t)/dt + P(t)F(t) + F^T(t)P(t) - P(t)G(t)R^{-1}(t)G^T(t)P(t) + Q(t) = 0$, which frequently arises in control and estimation theory. { ri'käd·ē i‚kwä·zhən }

Ricci equations Equations relating the components of the Ricci tensor, the curvature tensor, and an arbitrary tensor of a Riemann space. Also known as Ricci identities. { 'rē‚chē i‚kwä·zhənz }

Ricci identities See Ricci equations. { 'rē‚chē ī‚den·ə·dēz }

Ricci tensor See contracted curvature tensor. { 'rē‚chē ‚ten·sər }

Ricci theorem The covariant derivative vanishes for either of the fundamental tensors of a Riemann space. { 'rē‚chē ‚thir·əm }

ridge regression analysis A form of regression analysis in which damping factors are added to the diagonal of the correlation matrix prior to inversion, a procedure

Riemann-Christoffel tensor

which tends to orthogonalize interrelated variables; study of the robustness of the regression coefficients with changes in the damping factors is then used to determine sets of variables that should be removed. Also known as damped regression analysis. { 'rij ri'gresh·ən ə,nal·ə·səs }

Riemann-Christoffel tensor The basic tensor used for the study of curvature of a Riemann space; it is a fourth-rank tensor, formed from Christoffel symbols and their derivatives, and its vanishing is a necessary condition for the space to be flat. Also known as curvature tensor. { 'rē,män 'kris·tə·fəl ,ten·sər }

Riemann function A type of Green's function used in solving the Cauchy problem for a real hyperbolic partial differential equation. { 'rē,män ,fəŋk·shən }

Riemann hypothesis The conjecture that the only zeros of the Riemann zeta function with positive real part must have their real part equal to 1/2. { 'rē,män hī,päth·ə·səs }

Riemannian curvature A general notion of space curvature at a point of a Riemann space which is directly obtained from orthonormal tangent vectors there. { rē'män·ē·ən 'kər·və·chər }

Riemannian geometry See elliptic geometry. { rē'män·ē·ən jē'äm·ə·trē }

Riemannian manifold A differentiable manifold where the tangent vectors about each point have an inner product so defined as to allow a generalized study of distance and orthogonality. { rē'män·ē·ən 'man·ə,fōld }

Riemann integral The Riemann integral of a real function $f(x)$ on an interval (a,b) is the unique limit (when it exists) of the sum of $f(a_i)(x_i - x_{i-1})$, $i = 1, \ldots, n$, taken over all partitions of (a,b), $a = x_0 < a_1 < x_1 < \cdots < a_n < x_n = b$, as the maximum distance between x_i and x_{i-1} tends to zero. { 'rē,män ,int·ə·grəl }

Riemann-Lebesgue lemma If the absolute value of a function is integrable over the interval where it has a Fourier expansion, then its Fourier coefficients a_n tend to zero as n goes to infinity. { 'rē,män lə'beg ,lem·ə }

Riemann mapping theorem Any simply connected domain in the plane with boundary containing more than one point can be conformally mapped onto the interior of the unit disk. { 'rē,män 'map·iŋ ,thir·əm }

Riemann method A method of solving the Cauchy problem for hyperbolic partial differential equations. { 'rē,män ,meth·əd }

Riemann P function A scheme for exhibiting the singular points of a second-order ordinary differential equation, and the orders at these points of solutions of the equation. { 'rē·män 'pē ,fəŋk·shən }

Riemann space A Riemannian manifold or subset of a euclidean space where tensors can be defined to allow a general study of distance, angle, and curvature. { 'rē,män ,spās }

Riemann sphere The two-sphere whose points are identified with all complex numbers by a stereographic projection. Also known as complex sphere. { 'rē,män ,sfir }

Riemann-Stieltjes integral See Stieltjes integral. { ¦rē,män 'stēl·tyəs ,int·i·grəl }

Riemann surfaces Sheets or surfaces obtained by analyzing multiple-valued complex functions and the various choices of principal branches. { 'rē,män ,sər·fə·səz }

Riemann tensors Various types of tensors used in the study of curvature for a Riemann space. { 'rē,män ,ten·sərz }

Riemann zeta function The complex function $\zeta(z)$ defined by an infinite series with nth term $e^{-z \log n}$. Also known as zeta function. { 'rē,män 'zād·ə ,fəŋk·shən }

Riesz-Fischer theorem The vector space of all real- or complex-valued functions whose absolute value squared has a finite integral constitutes a complete inner product space. { 'rēsh 'fish·ər ,thir·əm }

right angle An angle of 90°. { 'rīt 'aŋ·gəl }

right circular cone A circular cone whose axis is perpendicular to its base. { 'rīt 'sər·kyə·lər 'kōn }

right circular cylinder A solid bounded by two parallel planes and by a cylindrical surface consisting of the straight lines perpendicular to the planes and passing through a circle in one of them. { 'rīt 'sər·kyə·lər 'sil·ən·dər }

right coset A right coset of a subgroup H of a group G is a subset of G consisting of

Rodrigues formula

all elements of the form ha, where a is a fixed element of G and h is any element of H. { 'rīt 'kō,set }

right-handed coordinate system 1. A three-dimensional rectangular coordinate system such that when the thumb of the right hand extends in the positive direction of the first (or x) axis the fingers fold in the direction in which the second (or y) axis could be rotated about the first axis to coincide with the third (or z) axis. **2.** A Riemann space which has negative scalar density function. { 'rīt ¦han·dəd kō'ȯrd·ən,ət ,sis·təm }

right-handed curve A space curve whose torsion is negative at a given point. Also known as dextrorse curve; dextrorsum. { ¦rīt ¦hand·əd 'kərv }

right-hand limit See limit on the right. { 'rīt ¦hand 'lim·ət }

right helicoid The surface that is swept out by a ray that originates at an axis and remains perpendicular to this axis while the ray is rotated about the axis and is translated in the direction of the axis, both at a constant rate. { 'rīt 'hel·ə,kȯid }

right identity In a set on which a binary operation ∘ is defined, an element e with the property that $a \circ e = a$ for every element a in the set. { 'rīt ī'den·əd·ē }

right inverse For a set S with a binary operation $x \circ y$ that has an identity element e, the right inverse of a member, x of S is another member, $\bar{x}$, of S for which $x \circ \bar{x} = e$. { ¦rīt 'in,vers }

right-invertible element An element x of a groupoid with a unit element e for which there is an element x such that $x \circ \bar{x} = e$. { ¦rīt in,vərd·ə·bə 'el·ə·mənt }

right module A module over a ring in which the product of a member x of the module and a member a of the ring is written xa. { 'rīt ,mäj·əl }

right parallelepiped A parallelepiped whose lateral faces are perpendicular to its bases. { 'rīt ,par·ə,lel·ə'pī·pəd }

right prism A prism whose lateral edges are perpendicular to the bases. { 'rīt ,priz·əm }

right section A plane section by a plane perpendicular to the elements of a given cylinder, or to the lateral faces of a given prism. { 'rīt 'sek·shən }

right spherical triangle A spherical triangle that has at least one right angle. { ¦rīt ¦sfir·ə·kəl 'trī,aŋ·gəl }

right strophoid A plane curve derived from a straight line L and a point called the pole, consisting of the locus of points on a rotating line L' passing through the pole whose distance from the intersection of L and L' is equal to the distance of this intersection from the foot of the perpendicular from the pole to L. { 'rīt 'strä,fȯid }

right triangle A triangle one of whose angles is a right angle. { 'rīt 'trī,aŋ·gəl }

right truncated prism A truncated prism, in which one of the cutting planes is perpendicular to the lateral edges. { ¦rīt ¦trən·kād·əd 'priz·əm }

ring 1. An algebraic system with two operations called multiplication and addition; the system is a commutative group relative to addition, and multiplication is associative, and is distributive with respect to addition. **2.** A ring of sets is a collection of sets where the union and difference of any two members is also a member. { riŋ }

ring isomorphism An isomorphism between rings. { 'riŋ ,ī·sō'mȯr,fiz·əm }

ring of operators See von Neumann algebra. { ¦riŋ əv 'äp·ə,rād·ərz }

ring permutation An arrangement of objects about a ring whose orientation is unspecified. { 'riŋ ,pər·myə¦tā·shən }

ring theory The study of the structure of rings in algebra. { 'riŋ ,thē·ə·rē }

rising factorial polynomials The polynomials $[x]^n = x(x + 1)(x + 2) \cdots (x + n - 1)$. { ,rīz·iŋ fak¦tȯr·ē·əl ,päl·ə'nō·mē·əlz }

Ritz method A method of solving boundary value problems based upon reformulating the given problem as a minimization problem. { 'ritz ,meth·əd }

robustness The property of statistical procedures that are insensitive to small departures from the assumptions on which they depend, such as the assumption that certain distributions are normal. { rō'bəst·nəs }

Rodrigues formula 1. The equation giving the nth function in a class of special functions in terms of the nth derivatives of some polynomial. **2.** The formula $d\mathbf{n} + k\, d\mathbf{r} = 0$, expressing the difference $d\mathbf{n}$ in the unit normals to a surface at two neighboring points on a line of curvature, in terms of the difference $d\mathbf{r}$ in the position vectors

213

Rolle's theorem

of the two points and the principal curvature k. **3.** The formula for a matrix that is used to transform the Cartesian coordinates of a vector in three-space under a rotation through a specified angle about an axis with specified direction cosines. { rə'drē·gəs ˌfȯr·myə·lə }

Rolle's theorem If a function $f(x)$ is continuous on the closed interval $[a,b]$ and differentiable on the open interval (a,b) and if $f(a) = f(b)$, then there exists x_0, $a < x_0 < b$, such that $f'(x_0) = 0$. { 'rȯlz ˌthir·əm }

rook polynomial A polynomial in which the coefficient of xk is the number of ways the k rooks can be placed on a chessboard of specified size so that no rook can capture another rook (that is, so that no two rooks are in the same row or the same column). { 'rük ˌpäl·ə'nō·mē·əl }

rook problem *See* problem of nontaking rooks. { 'rük ˌpräb·ləm }

root 1. A root of a given real or complex number is a number which when raised to some exponent equals that number. Also known as radix. **2.** A root of a polynomial $p(x)$ is a number a such that $p(a) = 0$. **3.** A root of an equation is a number or quantity that satisfies that equation. { rüt }

rooted ordered tree A rooted tree in which the order of the subtrees formed by deleting the root vertex is significant. { 'rüd·əd 'ȯr·dərd 'trē }

rooted tree A directed tree graph in which one vertex has no predecessor, and each of the remaining vertices has a unique predecessor. { 'rüd·əd 'trē }

root field *See* Galois field. { 'rüt ˌfēld }

root-mean-square deviation The square root of the sum of squared deviations from the mean divided by the number of observations in the sample. { ¦rüt ¦mēn ¦skwer ˌder·ə'vā·shən }

root-mean-square error The square root of the second moment corresponding to the frequency function of a random variable. { 'rüt ˌmēn 'skwer 'er·ər }

root-mean-square value Abbreviated rms value. The square root of the average of the squares of a series of related values. { 'rüt ˌmēn 'skwer 'val·yü }

root of unity A root of unity in a field F is an element a in F such that $a^n = 1$ for some positive integer n. { ¦rüt əv 'yü·nəd·ē }

root-squaring methods Methods of solving algebraic equations which involve calculating the coefficients in a sequence of equations, each of which has roots which are the squares of the roots in the previous equation. { 'rüt ˌskwer·iŋ ˌmeth·ədz }

root test An infinite series of nonnegative terms a_n converges if, after some term, the ith root of a_i is less than a fixed number smaller than 1. { 'rüt ˌtest }

root vertex The vertex of a rooted tree that has no predecessor. { 'rüt 'vər,teks }

rose A graph consisting of loops shaped like rose petals arising from the equations in polar coordinates $r = a \sin n\theta$ or $r = a \cos n\theta$. Also known as rhodonea. { rōz }

rotation *See* curl. { rō'tā·shən }

rotation angle A directed angle together with a signed measure of the angle. { rō'tā·shən ˌaŋ·gəl }

rotation group The group consisting of all orthogonal matrices or linear transformations having determinant 1. { rō'tā·shən ˌgrüp }

rotund space *See* strictly convex space. { rō¦tənd 'spās }

Rouche's theorem If analytic functions $f(z)$ and $g(z)$ in a simply connected domain satisfy on the boundary $|g(z)| < |f(z)|$, then $f(z)$ and $f(z) + g(z)$ have the same number of zeros in the domain. { 'rüsh·əz ˌthir·əm }

roulette The curve traced out by a point attached to a given curve that rolls without slipping along another given curve that remains fixed. { rü'let }

round angle *See* perigon. { 'raund ˌaŋ·gəl }

rounding Dropping or neglecting decimals after some significant place. Also known as truncation. { 'raund·iŋ }

rounding error The computational error due to always rounding numbers in a calculation. Also known as round-off error. { 'raund·iŋ ˌer·ər }

round off To truncate the least significant digit or digits of a numeral, and adjust the remaining numeral to be as close as possible to the original number. { 'raun,dȯf }

round-off error *See* rounding error. { 'raun,dȯf ˌer·ər }

Routh's rule The number of roots with positive real parts of an algebraic equation is equal to the number of changes of algebraic sign of a sequence whose terms are formed from coefficients of the equation in a specified manner. Also known as Routh test. { 'raüths ˌrül }

Routh table An array of numbers each of which is formed from coefficients of an algebraic equation in a specified manner; the first row of this array constitutes the sequence used in Routh's rule. { 'raüth ˌtā·bəl }

Routh test *See* Routh's rule. { 'raüth ˌtest }

row matrix *See* row vector. { 'rō ˌmā,triks }

row vector A matrix consisting of only one row. Also known as row matrix. { 'rō ˌvek·tər }

r-permutation An *r*-permutation of a set is an ordered selection of *r* elements of the set. { 'är ˌpər·myə,tā·shən }

rule An antecedent condition and a consequent proposition that can support deductive processes. { rül }

ruled surface A surface that can be generated by the motion of a straight line. { 'rüld 'sər·fəs }

rule of detachment The rule that if an implication is true and its antecedent is true, then the consequent is true. { 'rül əv di,tach·mənt }

rule of false position *See* regula falsi. { 'rül əv 'fòls pə'zish·ən }

ruling One of the positions of the straight line that generates a ruled surface. { 'rül·iŋ }

run The occurrence of the same characteristic in a series of observations; can be used to test whether or not two random samples come from populations having the same frequency distribution. { rən }

Runge-Kutta method A numerical approximation technique for solving differential equations. { 'rəŋ·ə 'küd·ə ˌmeth·əd }

Russell's paradox The paradox concerning the concept of all sets which are not members of themselves which forces distinctions in set theory between sets and classes. { 'rəs·əlz 'par·ə,däks }

S

saddle point 1. A point where all the first partial derivatives of a function vanish but which is not a local maximum or minimum. **2.** For a matrix of real numbers, an element that is both the smallest element of its row and the largest element of its column, or vice versa. **3.** For a two-person, zero-sum game, an element of the payoff matrix that is the smallest element of its row and the largest element of its column, so that the corresponding strategies are optimal for each player, given the strategy chosen by the other player. { 'sad·əl ˌpȯint }

saddle-point method *See* steepest-descent method. { 'sad·əl ¦pȯint ˌmeth·əd }

saddle-point theory The study of differentiable functions and their derivatives from the viewpoint of saddle points, especially applicable to the calculus of variations. { 'sad·əl ¦pȯint ˌthē·ə·rē }

sagitta The distance between the midpoint of an arc and the midpoint of its chord. { sə'jid·ə }

salient angle An interior angle of a polygon that is less than 180°. { ¦sāl·yənt ¦aŋ·gəl }

salient point A point at which two branches of a curve with different tangents meet and terminate. { ¦sāl·yənt ¦pȯint }

salinon A plane figure bounded by a semicircle, two smaller semicircles which lie inside the larger semicircle with diameters along the diameter of the larger semicircle and are tangent to the larger semicircle, and another semicircle which is outside the larger. { 'sal·əˌnän }

saltus *See* oscillation. { 'sal·təs }

sample A selection of a certain collection from a larger collection. { 'sam·pəl }

sample correlation coefficient The ratio of the sample covariance of x and y to the standard deviation of x times the standard deviation of y. Also known as product-moment coefficient. { ¦sam·pəl ˌkä·rə'lā·shən ˌkō·əˌfish·ənt }

sampled data Data that are obtained at discrete rather than continuous intervals. { 'sam·pəld 'dad·ə }

sample design A procedure or plan drawn up before any data are collected to obtain a sample from a given population. Also known as sampling plan; survey design. { 'sam·pəl diˌzīn }

sample function A function or procedure which, when applied repeatedly to a given population, produces a collection of samples. { 'sam·pəl ˌfəŋk·shən }

sample path If $\{X_t: t$ in $T\}$ is a stochastic process, a sample path for the process is the function on T to the range of the process which assigns to each t the value $X_t(w)$, where w is a previously given fixed point in the domain of the process. { 'sam·pəl ˌpath }

sample size The number of objects in the sample. { 'sam·pəl ˌsīz }

sample space A concept in probability theory which considers all possible outcomes of an experiment, game, and so on, as points in a space. { 'sam·pəl ˌspās }

sample survey A survey of a population made by using only a portion of the population. { 'sam·pəl ˌsər·vā }

sampling A drawing of a collection from a given population. { 'sam·pliŋ }

sampling distribution A distribution of the estimates that can be made by each of all possible samples of a fixed size that could be taken from a universe. { 'samp·liŋ ˌdis·trəˌbyü·shən }

sampling error That portion of the difference between the value of a statistic derived from observations and the value that it is supposed to estimate; attributed to the fact that samples represent only a portion of a population. { 'samp·liŋ ,er·ər }

sampling fraction The ratio of the sample size to the population size. { 'sam·pliŋ ,frak·shən }

sampling plan *See* sample design. { 'sam·pliŋ ,plan }

sampling techniques The methods used in drawing samples from a population usually in such a manner that the sample will facilitate determination of some hypothesis concerning the population. { 'sam·pliŋ tek,nēks }

sampling theory The mathematical study of sampling techniques. { 'sam·pliŋ ,thē·ə·rē }

Savage principle A technique used in decision theory; a criterion is used to construct a regret matrix in which each outcome entry represents a regret defined as the difference between best possible outcome and the given outcome; the matrix is then used as in decision making under risk with expected regret as the decision-determining quality. Also known as regret criterion. { 'sav·ij ,prin·sə·pəl }

scalar One of the algebraic quantities which form a field, usually the real or complex numbers, by which the vectors of a vector space are multiplied. { 'skā·lər }

scalar field 1. The field consisting of the scalars of a vector space. **2.** A function on a vector space into the scalars of the vector space. { 'skā·lər 'fēld }

scalar function A function from a vector space to its scalar field. { 'skā·lər 'faŋk·shən }

scalar gradient The gradient of a function. { 'skā·lər 'grā·dē·ənt }

scalar matrix A diagonal matrix whose diagonal elements are all equal. { ¦skā·lər 'mā,triks }

scalar multiplication The multiplication of a vector from a vector space by a scalar from the associated field; this usually contracts or expands the length of a vector. { 'skā·lər ,məl·tə·pli'kā·shən }

scalar product 1. A symmetric, alternating, or Hermitian form. **2.** *See* inner product. { 'skā·lər 'präd·əkt }

scalar triple product The scalar triple product of vectors v_1, v_2, and v_3 from Euclidean three-dimensional space determines the volume of the parallelepiped with these vectors as edges; it is given by the determinant of the 3×3 matrix whose rows are the components of v_1, v_2, and v_3. Also known as triple scalar product. { 'skā·lər 'trip·əl ,präd·əkt }

scalene spherical triangle A spherical triangle no two of whose sides are equal. { ¦skā,lēn ¦sfir·ə·kəl 'trī,aŋ·gəl }

scalene triangle A triangle where no two angles are equal. { 'skā,lēn 'trī,aŋ·gəl }

scaling symmetry The property of an object each part of which is identical to the whole seen at a different magnification; the property that characterizes a fractal. { 'skāl·iŋ ,sim·ə·trē }

scatter diagram A plot of the pairs of values of two variates in rectangular coordinates. Also known as scatter gram. { 'skad·ər ,dī·ə,gram }

Schauder's fixed-point theorem A continuous mapping from a closed, compact, convex set in a Banach space into itself has at least one fixed point. { ¦shaùd·ərz ¦fikst ,pȯint 'thir·əm }

schedule A group or list of questions used by an interviewer to obtain information directly from a subject. { 'skej·əl }

schlicht function *See* simple function. { 'shlikt ,fəŋ·shən }

Schroeder-Bernstein theorem If a set A has at least as many elements as another set B and B has at least as many elements as A, then A and B have the same number of elements. { 'shrād·ər 'bərn,stīn ,thir·əm }

Schur-Cohn test A test to determine whether all the coefficients of a polynomial have magnitude less than one; the polynomial has this property only if each of a series of determinants formed from the coefficients of the polynomial in a specified manner is positive for determinants of even degree and negative for determinants of odd degree. { 'shür 'kōn ,test }

Schur's lemma For certain types of modules M, the ring consisting of all homomorphisms of M to itself will be a division ring. { 'shürz ˌlem·ə }

Schwartz's theory of distributions A theory that treats distributions as continuous linear functionals on a vector space of continuous functions which have continuous derivatives of all orders and vanish appropriately at infinity. { ¦shwórts ¦thē·ə·rē əv ˌdis·trə'byü·shənz }

Schwarz-Christoffel transformations Those complex transformations which conformally map the interior of a given polygon onto the portion of the complex plane above the real axis. { 'shvärts 'kris·tə·fel ˌtranz·fər'mā·shənz }

Schwarz' inequality *See* Cauchy-Schwarz inequality. { 'shvärts ˌin·i‚kwäl·əd·ē }

Schwarz reflection principle To obtain the analytic continuation of a given function $f(z)$ analytic in a region R, whose boundary contains a segment of the real axis, into a region reflected from R through this segment, one takes the complex conjugate function $f(\bar{z})$. { 'shvärts ri'flek·shən ˌprin·sə·pəl }

Schwarz's lemma If an analytic function of the unit disk to itself sends the origin to the origin then it must be distance-decreasing. { 'shvärt·səz ˌlem·ə }

S curve *See* reverse curve. { 'es ˌkərv }

sec *See* secant.

secant 1. The function given by the reciprocal of the cosine function. Abreviated sec. 2. The secant of an angle A is $1/\cos A$. 3. A line of unlimited length that intersects a given curve. { 'sē‚kant }

sech *See* hyperbolic secant. { sek }

second A unit of plane angle, equal to 1/60 minute, or 1/3,600 degree, or $\pi/648,000$ radian. { 'sek·ənd }

secondary diagonal The elements of a square matrix that lie on the straight line extending from the lower left-hand corner to the upper right-hand corner of the matrix. { ¦sek·ən‚der·ē dī'ag·ən·əl }

second category A set is of second category if it cannot be expressed as a countable union of nowhere dense sets. { 'sek·ənd 'kad·ə‚gór·ē }

second countable topological space A topological space that has a countable base. { ¦sek·ənd ¦kaún·tə·bəl ˌtäp·ə¦läj·ə·kəl 'spās }

second curvature *See* torsion. { 'sek·ənd ¦kər·və·chər }

second derivative The derivative of the first derivative of a function. { 'sek·ənd də¦riv·əd·iv }

second mean-value theorem The theorem that for two functions $f(x)$ and $g(x)$ that are continuous on a closed interval $[a,b]$ and differentiable on the open interval (a,b), such that $g(b) \neq g(a)$, there exists a number x_1 in (a,b) such that either $[f(b) - f(a)]/[g(b) - g(a)] = f'(x_1)/g'(x_1)$ or $f'(x_1) = g'(x_1) = 0$. Also known as Cauchy's mean-value theorem; double law of the mean; extended mean-value theorem; generalized mean-value theorem. { ¦sek·ənd ¦mēn ¦val·yü ˌthir·əm }

second moment of area *See* geometric moment of inertia. { ¦sek·ənd ¦mō·mənt əv 'er·ē·ə }

second-order difference One of the first-order differences of the sequence formed by taking the first-order differences of a given sequence. { 'sek·ənd ¦ór·dər 'dif·rəns }

second-order equation A differential equation where some term includes the second derivative of the unknown function and no derivative of higher order is present. { 'sek·ənd ¦ór·dər i'kwä·zhən }

second quadrant 1. The range of angles from 90 to 180°. 2. In a plane with a system of cartesian coordinates, the region in which the x coordinate is negative and the y coordinate is positive. { 'sek·ənd 'kwä·drənt }

second species The class of sets G_0 such that all the sets G_n are nonempty, where, in general, G_n is the derived set of G_{n-1}. { ¦sek·ənd 'spē‚shēz }

section 1. For a polyhedral angle, the polygon formed by the intersection of the faces of the angle with a plane that does not pass through the vertex. 2. *See* plane section. { 'sek·shən }

sector A portion of a circle bounded by two radii and an arc joining their end points. { 'sek·tər }

sectoral harmonic A spherical harmonic which is 0 on a set of equally spaced meridians of a sphere with center at the origin of spherical coordinates, dividing the sphere into sectors. { ¦sek·tə·rəl här'män·ik }

sectorgram *See* pie chart. { 'sek·tər‚gram }

secular determinant For a square matrix *A*, the determinant of the matrix whose off-diagonal components are equal to those of *A*, and whose diagonal components are equal to the difference between those of *A* and a parameter λ; it is equal to the characteristic polynomial in λ of the linear transformation represented by *A*. { 'sek·yə·lər di'tər·mən·ənt }

secular trend A concept in time series analysis that refers to a movement or trend in a series over very long periods of time. Also known as long-time trend. { 'sek·yə·lər 'trend }

segment 1. A segment of a line or curve is any connected piece. **2.** A segment of a circle is a portion of the circle bounded by a chord and an arc subtended by the chord. **3.** A segment of a totally ordered Abelian group *G* is a subset *D* of *G* such that if *a* is in *D* then so are all elements *b* satisfying $-a \leq b \leq a$. { 'seg·mənt }

Segrè characteristic A set of integers that are the orders of the Jordan submatrices of a classical canonical matrix, with integers that correspond to submatrices containing the same characteristic root being bracketed together. { 'se‚grä ‚kär·ik·tə‚ris·tik }

Seidel method A basic iterative procedure for solving a system of linear equations by reducing it to triangular form. Also known as Gauss-Seidel method. { 'zīd·əl ‚meth·əd }

selection bias A bias built into an experiment by the method used to select the subjects which are to undergo treatment. { si'lek·shən ‚bī·əs }

self-adjoint operator A linear operator which is identical with its adjoint operator. { ¦self ə¦jóint 'äp·ə‚räd·ər }

self-complementary graph A simple graph that is isomorphic to its complement. { ¦self ‚käm·plə‚men·trē 'graf }

self-conjugate partition A partition that is its own conjugate. { ¦self ‚kän·jə·gət pär 'tish·ən }

self-dual switching function A switching function whose value remains unchanged when the digits 0 and 1 are interchanged in each element of the domain of the function. { ¦self ‚dül 'swich·iŋ ‚fəŋk·shən }

self-selection bias Bias introduced into an experiment by having the subjects decide themselves whether or not they will receive treatment. { ¦self si¦lek·shən 'bī·əs }

self-similarity The property whereby an object or mathematical function preserves its structure when multiplied by a certain scale factor. { ¦self ‚sim·ə'lar·əd·ē }

semiaxis A line segment that forms half of the axis of a geometric figure (such as an ellipse), having one end point at the center of symmetry of the figure. { ¦sem·ē'ak·səs }

semicircle One of the two parts of a circle that extend from one end of a diameter to the other, and whose length is one-half that of the circle. { 'sem·i‚sər·kəl }

semicircumference One-half the circumference of a circle. { ‚sem·i·sər'kəm·frəns }

semiconjugate axis Either of the equal line segments into which the conjugate axis of a hyperbola is divided by the center of symmetry. { ‚sem·ē¦kän·jə·gət 'ak·səs }

semicubical parabola A plane curve whose equation in Cartesian coordinates *x* and *y* is $y^2 = ax^3$, where *a* is some constant. Also known as isochrone. { ¦sem·i'kyü·bə·kəl pə'rab·ə·lə }

semigroup A set which is closed with respect to a given associative binary operation. { 'sem·i‚grüp }

semigroup theory The formal algebraic study of the structure of semigroups. { 'sem·i‚grüp ‚thē·ə·rē }

semi-interquartile range *See* quartile deviation. { ¦sem·ē‚in·tər'kwȯr‚tīl ‚ränj }

semi-invariants *See* cumulants. { ¦sem·ē·in'ver·ē·əns }

semilogarithmic coordinate paper Paper ruled with two sets of mutually perpendicular, parallel lines, one set being spaced according to the logarithms of consecutive numbers, and the other set uniformly spaced. Also known as ratio paper. { ¦sem·i¦läg·ə¦ri<u>th</u>·mik kō¦órd·ən·ət ¸pā·pər }

semimagic square *See* magic square. { 'sem·i¸maj·ik ¸skwer }

semimajor axis Either of the equal line segments into which the major axis of an ellipse is divided by the center of symmetry. { ¸sem·ē¦mā·jər 'ak·səs }

semimetric A real valued function $d(x,y)$ on pairs of points from a topological space which has all the same properties as a metric save that $d(x,y)$ may be zero even if x and y are distinct points. Also known as pseudometric. { ¦sem·i'me·trik }

semiminor axis Either of the equal line segments into which the minor axis of an ellipse is divided by the center of symmetry. { ¸sem·i¦mīn·ər 'ak·səs }

seminorm A scalar-valued function on a real or complex vector space satisfying the axioms of a norm, except that the seminorm of a nonzero vector may equal zero. { 'sem·ē¸nórm }

semiperfect number A number which is the sum of some set of its own proper divisors. { 'sem·i¸pər·fekt ¸nəm·bər }

semiperimeter One-half the length of a closed curve. { ¸sem·i·pə'rim·əd·ər }

semiprime A positive integer that is the product of exactly two primes. { 'sem·i¸prīm }

semiregular solid *See* Archimedean solid. { ¦sem·i¸reg·yə·lər 'säl·əd }

semiring of sets A collection S of sets that includes the empty set and the intersection of any two of its members, and is such that if A and B are members of S and A is a subset of B, then $B − A$ is the union of a finite number of disjoint members of S. { ¸sem·ē'riŋ əv 'setz }

semisecant *See* transversal. { 'sem·i¸sē¸kant }

semisimple module A module which is the sum of a family of simple modules. { ¦sem·i¸sim·pəl 'mä·jəl }

semisimple representation *See* completely reducible representation. { ¦sem·i¸sim·pəl ¸rep·ri·zen'ta·shən }

semisimple ring A ring in which 1 does not equal 0, and which is semisimple as a left module over itself. { ¦sem·i¸sim·pəl 'riŋ }

semitransverse axis Either of the equal line segments into which the transverse axis of a hyperbola is divided by the center of symmetry. { ¦sem·i¦tranz¸vərs 'ak·səs }

semivariogram A mathematical function used to quantify the dissimilarity between groups of values. { ¸sem·i'ver·ē·ə¸gram }

sentential calculus *See* propositional calculus. { sen'ten·chəl 'kal·kyə·ləs }

sentential connectives *See* propositional connectives. { sen¦ten·chəl kə'nek·tivz }

sentential function *See* propositional function. { sen¦ten·chəl 'fəŋk·shən }

separable degree Let E be an algebraic extension of a field F, and let f be any embedding of F in a field L such that L is the algebraic closure of the image of F under f; the separable degree of E over F is the number of distinct embeddings of E in L which are extensions of f. { 'sep·rə·bəl di'grē }

separable element An element a is said to be separable over a field F if it is algebraic over F and if the extension field of F generated by a is a separable extension of F. { 'sep·rə·bəl 'el·ə·mənt }

separable extension A field extension K of a field F is separable if every element of K is a root of a separable polynomial whose coefficients are elements of F. { 'sep·rə·bəl ik'sten·chən }

separable polynomial A polynomial with no multiple roots. { 'sep·rə·bəl ¸päl·i'nō·mē·əl }

separable space A topological space which has a countable subset that is dense. { 'sep·rə·bəl 'spās }

separated sets Sets A and B in a topological space are separated if both the closure of A intersected with B and the closure of B intersected with A are disjoint. { 'sep·ə¸rād·əd 'sets }

separating transcendence base A transcendence base of a field E over a field F such that E is algebraic and separable over the field generated by F and the transcendence base. { ¦sep·ə¸rād·iŋ tran'sen·dəns ¸bās }

separation axioms Properties of topological spaces such as Hansdorff, regular, and normal which reflect how points and closed sets may be enclosed in disjoint neighborhoods. { ˌsep·ə'rā·shən 'ak·sē·əmz }

separation of the first kind A division of an ordered set into two classes in which each member of one class is greater than every member of the other, and there exists a separating element that belongs to one class or the other. { ˌsep·ə¦rā·shən əv thə 'first ˌkīnd }

separation of the second kind A division of an ordered set into two classes in which each member of one class is greater than every member of the other, and there is no least member in the class of greater elements and no greatest member in the class of lesser elements. { ˌsep·ə¦rā·shən əv thə 'sek·ənd kīnd }

separation of variables 1. A technique where certain differential equations are rewritten in the form $f(x)dx = g(y)dy$ which is then solvable by integrating both sides of the equation. **2.** A method of solving partial differential equations in which the solution is written in the form of a product of functions, each of which depends on only one of the independent variables; the equation is then arranged so that each of the terms involves only one of the variables and its corresponding function, and each of these terms is then set equal to a constant, resulting in ordinary differential equations. Also known as product-solution method. { ˌsep·ə'rā·shən əv 'ver·ē·ə·bəlz }

septillion 1. The number 10^{24}. **2.** In British and German usage, the number 10^{42}. { sep'til·yən }

septinary number A number in which the quantity represented by each figure is based on a radix of 7. { 'sep·tə,ner·ē ¦nəm·bər }

sequence A listing of mathematical entities $x_1, x_2 \ldots$ which is indexed by the positive integers; more precisely, a function whose domain is an infinite subset of the positive integers. Also known as infinite sequence. { 'sē·kwəns }

sequential analysis The continuous analysis of data, obtained via sampling, performed as the amount of sampling increases. { si'kwen·chəl ə'nal·ə·səs }

sequentially compact space A topological space with the property that every sequence formed from its points has a subsequence that converges to a point in the space. { si¦kwen·chə·lē ˌkäm,pakt 'spās }

sequential trials The outcome of each trial is known before the next trial is performed. { si'kwen·chəl 'trīlz }

serial correlation The correlation between values of events in a time series and those values ahead or behind by a fixed amount in time or space or between parts of two different time series. { 'sir·ē·əl ˌkär·ə'lā·shən }

serially ordered set *See* linearly ordered set. { ¦sir·ē·ə·lē ¦órd·ərd 'set }

serial order *See* linear order. { 'sir·ē·əl ˌórd·ər }

serial sampling A method of gathering samples by a set pattern, such as a grid, to ensure randomness. { 'sir·ē·əl 'sam·pliŋ }

series An expression of the form $x_1 + x_2 + x_3 + \cdots$, where x_i are real or complex numbers. { 'sir·ēz }

serpentine curve The curve given by the equation $x^2y + b^2y - a^2x = 0$, passing through and having symmetry about the origin while being asymptotic to the x axis in both directions. { 'sər·pən,tēn 'kərv }

Serret-Frenet formulas *See* Frenet-Serret formulas. { sə'rā frə'nā ˌfór·myə·ləz }

sesquillinear form A mapping $f(x,y)$ from $E \times F$ into R, where R is a commutative ring with an automorphism with period 2 and $E \times F$ is the Cartesian product of two modules E and F over R, such that for each x in E the function which takes y into $f(x,y)$ is antilinear, and for each y in F the function which takes x into $f(x,y)$ is linear. { ¦ses·kwə,lin·ē·ər 'fórm }

set A collection of objects which has the property that, given any thing, it can be determined whether or not the thing is in the collection. { set }

set function A relation that assigns a value to each member of a collection of sets. { 'set ˌfəŋk·shən }

set of first category *See* meager set. { ¦set əv 'ərst ˌkat·ə,gór·ē }

set of second category Any set that is not a meager set. { ¦set əv 'sek·ənd ˌkat·ə· gȯr·ē }

set theory The study of the structure and size of sets from the viewpoint of the axioms imposed. { 'set ˌthē·ə·rē }

sexadecimal *See* hexadecimal. { ¦sek·sə'des·məl }

sexadecimal number system *See* hexadecimal number system. { ¦sek·sə'des·məl 'nəm·bər ˌsis·təm }

sexagesimal Pertaining to a multiplicity of 60 distinct alternative states or conditions or, simply, a positional numeration system to radix (or base) 60. { ¦sek·sə¦jez· ə·məl }

sexagesimal counting table A table for converting numbers using the 60 system into decimals, for example, minutes and seconds. { ¦sek·sə¦jez·ə·məl 'kaùnt·iŋ ˌtā·bəl }

sexagesimal measure of angles A system of angular units in which a complete revolution is divided into 360 degrees, a degree into 60 minutes, and a minute into 60 seconds. { ¦sek·sə¦jez·ə·məl 'mezh·ər əv 'aŋ·gəlz }

sextant A unit of plane angle, equal to 60° or $\pi/3$ radians. { 'sek·stənt }

sextic Having the sixth degree or order. { 'sek·stik }

sextillion 1. The number 10^{21}. 2. In British and German usage, the number 10^{36}. { sek'stil·yən }

Shannon-McMillan-Breiman theorem Given an ergodic measure preserving transformation T on a probability space and a finite partition ζ of that space, the limit as $n \to \infty$ of $1/n$ times the information function of the common refinement of ζ, $T^{-1}\zeta, \ldots, T^{-n+1}\zeta$ converges almost everywhere and in the L_1 metric to the entropy of T given ζ. { 'shan·ən mik'mil·ən 'brī·mən ˌthir·əm }

Shannon's theorems These results are foundational to the mathematical study of information; mathematically they link the concept of entropy with the amount of efficient transmittal and reception of information. { 'shan·ənz ˌthir·əmz }

sheaf A fiber bundle with algebraic and topological structure usually associated to a differentiable manifold M which reflects the local behavior of differentiable functions on M. { shēf }

sheaf of planes All the planes passing through a given point. Also known as bundle of planes. { 'shēf əv 'plānz }

sheet 1. A portion of a surface such that it is possible to travel continuously between any two points on it without leaving the surface. 2. A part of a Riemann surface such that any extension results in a multiple covering of some part of the complex plane over which the surface lies. { shēt }

sheffer stroke *See* NAND. { 'shef·ər ˌstrōk }

shell method A method of computing the volume of a solid of revolution by integrating over the volumes of infinitesimal shell-shaped sections bounded by cylinders with the same axis of revolution as the solid. { 'shel ˌmeth·əd }

Sheppard's correction A correction for moments computed for a frequency distribution of grouped data to adjust for the error that is introduced by the assumption that all the data within a class are at the midpoint of the class; the adjustment is made by subtracting one-twelfth of the square of a grouping unit from the estimated variance. { 'shep·ərdz kəˌrek·shənz }

shifting theorem 1. If the Fourier transform of $f(t)$ is $F(x)$, then the Fourier transform of $f(t - a)$ is exp $(iax)F(x)$. 2. If the Laplace transform of $f(x)$ is $F(y)$, then the Laplace transform of $f(x - a)$ is exp $(-ay)F(y)$. { 'shift·iŋ ˌthir·əm }

shoemaker's knife *See* arbilos. { ¦shü·māk·ərz 'nīf }

short arc The shorter of the two arcs between the points of intersection of a chord with a circle. { ¦shȯrt 'ärk }

short division 1. Division of numbers in which the divisor contains only one digit 2. Division of algebraic quantities in which the divisor contains only one term. { 'shȯrt dəˌvizh·ən }

short radius *See* apothem. { ¦shȯrt ¦rād·ē·əs }

shrinking of the plane A homothetic transformation in which the ratio of similitude is less than 1. Also known as shrinking transformation. { ¦shriŋk·iŋ əvthə 'plān }

shrinking space The conjugate space of a Banach space with basis $x_1, x_2, \ldots$, which satisfies the property that, for any continuous linear functional f, the norm of f with domain restricted to the linear span of $x_{n+1}, x_{n+2}, \ldots$ approaches zero as n approaches infinity. { |shriŋk·iŋ 'spās }

shrinking transformation See shrinking of the plane. { 'shriŋk·iŋ,tranz·fər,mā·shən }

side 1. One of the line segments that bound a polygon. **2.** One of the two rays that extend from the vertex of an angle. { sīd }

Sierpinski gasket A fractal which can be constructed by a recursive procedure; at each step a triangle is divided into four new triangles, only three of which are kept for further iterations. { sir'pin·skē ,gas·kət }

Sierpinski set 1. A set of points S on a line such that both S and its complement contain at least one point in each uncountable set on the line that is a countable intersection of open sets. **2.** A set of points in a plane that includes at least one point of each closed set of nonzero measure and does not include any subsets consisting of three collinear points. { sər'pin·skē ,set }

sieve of Eratosthenes An iterative procedure which determines all the primes less than a given number. { 'siv əv ,er·ə'täs·thə,nēz }

sigma algebra A collection of subsets of a given set which contains the empty set and is closed under countable union and complementation of sets. Also known as sigma field. { 'sig·mə 'al·jə·brə }

sigma field See sigma algebra. { 'sig·mə ,fēld }

sigma finite A measure is sigma finite on a space X if X is a countable disjoint union of sets each of which is measurable and has finite measure. { 'sig·mə 'fī,nīt }

sigma ring A ring of sets where any countable union of its members is also a member. { 'sig·mə ,riŋ }

sign 1. A symbol which indicates whether a quantity is greater than zero or less than zero; the signs are often the marks $+$ and $-$ respectively, but other arbitrarily selected symbols are used, especially in automatic data processing. **2.** A unit of plane angle, equal to 30° or $\pi/6$ radians. { sīn }

signature 1. For a quadratic or Hermitian form, the number of positive coefficients minus the number of negative coefficients when the form is reduced by a linear transformation to a sum of squares of absolute values. **2.** For a symmetric or Hermitian matrix, the number of positive entries minus the number of negative entries when the matrix is transformed to diagonal form. { 'sig·nə·chər }

signed measure An extended real-valued function m defined on a sigma algebra of subsets of a set S such that (1) the value of m on the empty set is 0, (2) the value of m on a countable union of disjoint sets is the sum of its values on each set, and (3) m assumes at most one of the values $+\infty$ and $-\infty$. { |sīnd 'mezh·ər }

significance The arbitrary rank, priority, or order of relative magnitude assigned to a given position in a number. { sig'nif·i·kəns }

significance level See level of significance. { sig'nif·i·kəns ,lev·əl }

significance probability The probability of observing a value of a test statistic as significant as, or even more significant than, the value actually observed. { sig'nif·i·kəns ,präb·ə,bil·əd·ē }

significant digit See significant figure. { sig'nif·i·kənt ,dij·ət }

significant figure A prescribed decimal place which determines the amount of rounding off to be done; this is usually based upon the degree of accuracy in measurement. Also known as significant digit. { sig'nif·i·kənt ,fig·yər }

signless Stirling number The absolute value of a Stirling number of the first kind. { |sīn·ləs 'stər·liŋ ,nəm·bər }

sign of aggregation One of a pair of parentheses, braces, brackets, or bars which signify that the terms they enclose are to be treated as a single term. { |sīn əv ,ag·rə'gā·shən }

sign test A test which can be used whenever an experiment is conducted to compare a treatment with a control on a number of matched pairs, provided the two treatments are assigned to the members of each pair at random. { 'sīn ,test }

signum The function sgn(x), defined for all real values of x, where sgn(x) = 1 if $x > 0$, sgn(x) = −1 if $x < 0$, and sgn(0) = 0. { 'sig·nəm }

similar decimals Decimals that have the same number of decimal places. { ¦sim·ə·lər ¦des·məlz }

similar figures Two figures or bodies that are identical except for size; similar figures can be placed in perspective, so that straight lines joining corresponding parts of the two figures will pass through a common point. { 'sim·ə·lər 'fig·yərz }

similar fractions Two or more common fractions that have the same denominator. { ¦sim·ə·lər ¦frak·shənz }

similarity transformation 1. A transformation of a euclidean space obtained from such transformations as translations, rotations, and those which either shrink or expand the length of vectors. **2.** A mapping that associates with each linear transformation P on a vector space the linear transformation $R^{-1}PR$ that results when the coordinates of the space are subjected to a nonsingular linear transformation R. **3.** A mapping that associates with each square matrix P the matrix $Q = R^{-1}PR$, where R is a nonsingular matrix and R^{-1} is the inverse matrix of R; if P is the matrix representation of a linear transformation, then this definition is equivalent to the second definition. { ˌsim·ə'lar·əd·ē ˌtranz·fər ˌmā·shən }

similarly placed conics Conics of the same type (both ellipses, both parabolas, or both hyperbolas) whose corresponding axes are parallel. { ¦sim·ə·lər·lē ¦plāst 'kän·iks }

similar matrices Two square matrices A and B related by the transformation $B = SAT$, where S and T are nonsingular matrices and T is the inverse matrix of S. { ¦sim· i·lər 'mā·tri ˌsēz }

similar terms Terms that contain the same unknown factors and the same powers of these factors. Also known as like terms. { ¦sim·ə·lər ¦tərms }

similar triangles Triangles whose corresponding angles are equal; the corresponding sides are then proportional in length. { ¦sim·ə·lər ¦trī ˌaŋ·gəlz }

simple aggregative index A statistic computed for a collection of items by taking the ratio of the sum of their given-year values or amounts to the sum of their base-year values or amounts and usually multiplying by 100 to express the figure as a percentage. { ¦sim·pəl ˌag·rə ˌgād·iv 'in ˌdeks }

simple algebra An algebra over a field that is also a simple ring. { 'sim·pəl ¦al·jə·brə }

simple alternative An alternative to the null hypothesis which completely specifies the distribution of the observed random variables. { 'sim·pəl ȯl'tər·nəd·iv }

simple arc The image of a closed interval under a continuous, injective mapping from the interval into a plane. Also known as Jordan arc. { 'sim·pəl 'ärk }

simple character The character of an irreducible representation of a group. { 'sim· pəl 'kar·ik·tər }

simple closed curve A closed curve which never crosses itself. { 'sim·pəl 'klōzd 'kərv }

simple compression A transformation that compresses a configuration in a given direction, given by $x' = kx$, $y' = y$, $z' = z$, with $k < 1$, when the direction is that of the x axis. { ¦sim·pəl kəm¦presh·ən }

simple curve A curve that does not cross itself or touch itself. { 'sim·pəl ¦kərv }

simple cusp See cusp of the first kind. { 'sim·pəl 'kəsp }

simple dipath A directed path in which no two vertices are the same (except that the initial and final vertices may be the same). { 'sim·pəl ¦dī ˌpath }

simple elongation A transformation that elongates a configuration in a given direction, given by $x' = kx$, $y' = y$, $z' = z$, with $k > 1$, when the direction is that of the x axis. { 'sim·pəl ˌē ˌloŋ'gā·shən }

simple event See elementary event. { 'sim·pəl i¦vent }

simple extension An extension of a field that has an element c such that the extension field consists of the set of all quotients (with nonzero denominator) of polynomials in c with coefficients in the original field. { 'sim·pəl ik¦sten·shən }

simple fraction See common fraction. { ¦sim·pəl 'frak·shən }

simple function 1. For a region D of the complex plane, an analytic, injective function on D. Also known as schlicht function. **2.** Any measurable function whose range is a finite set. **3.** See step function. { 'sim·pəl 'fəŋk·shən }

simple graph A graph with no loops and no parallel edges. { 'sim·pəl ¦graf }

simple group A group G that is nontrivial and contains no normal subgroups other than the identity element and G itself. { ¦sim·pəl 'grüp }

simple hypothesis A hypothesis which completely specifies the distribution of the observed random variables. { 'sim·pəl hī'päth·ə·səs }

simple integral An integral over only one variable. { 'sim·pəl 'int·ə·grəl }

simple order *See* linear order. { 'sim·pəl 'ȯr·dər }

simple point *See* ordinary point. { 'sim·pəl ¦pȯint }

simple polyhedron A polyhedron with no holes inside it; technically, a polyhedron that is topologically equivalent to a solid sphere. { ¦sim·pəl ¸päl·ə'hē·drən }

simple random samples Samples in which every possible sample of size n, that is, every combination of n items from the number in the population, is equally likely to be part of the sample. { ¦sim·pəl ¦ran·dəm 'sam·pəlz }

simple results Results of observations such that on each trial of an experiment one and only one of these results will occur. { 'sim·pəl ri'zəls }

simple ring A semisimple ring R such that for any two left ideals in R there is an isomorphism of R which maps one onto the other. { 'sim·pəl 'riŋ }

simple root A polynomial $f(x)$ has c as a simple root if $(x - c)$ is a factor but $(x - c)^2$ is not.

simple shear A transformation that corresponds to a shearing motion in which a coordinate axis in the plane or a coordinate plane in space does not move, having the form $x' = x$, $y' = ax + y$, $z' = z$, where a is a constant, for a suitable choice of axes. { 'sim·pəl 'shir }

simple strain A one-dimensional strain or a simple shear. { ¦sim·pəl 'strān }

simplex An n-dimensional simplex in a euclidean space consists of $n + 1$ linearly independent points $p_0, p_1, \ldots, p_n$ together with all the points given by $a_0 p_0 + a_1 p_1 + \cdots + a_n p_n$ where the $a_i \geq 0$ and $a_0 + a_1 + \cdots + a_n = 1$; a triangle with its interior and a tetrahedron with its interior are examples. { 'sim¸pleks }

simplex method A finite iterative algorithm used in linear programming whereby successive solutions are obtained and tested for optimality. { 'sim¸pleks ¦meth·əd }

simplicial complex A set consisting of finitely many simplices where either two simplices are disjoint or intersect in a simplex which is a face common to each. Also known as geometric complex. { sim'plish·əl 'käm¸pleks }

simplicial graph A graph in which no line starts and ends at the same point, and in which no two lines have the same pair of end points. { sim'plish·əl 'graf }

simplicial homology A homology for a topological space where the nth group reflects how the space may be filled out by n-dimensional simplicial complexes and detects the presence of analogs of n-dimensional holes. { sim'plish·əl hə'mäl·ə·jē }

simplicial mapping A mapping of one simplicial complex into another in which the images of the simplexes of one complex are simplexes of the other complex. { sim'plish·əl 'map·iŋ }

simplicial subdivision A decomposition of the simplices composing a simplicial complex which results in a simplicial complex with a larger number of simplices. { sim'plish·əl 'səb·di¸vizh·ən }

simply connected region A region having no holes; all closed curves can be shrunk to a point without passing through points in the complement of the region. { 'sim·plē kə¦nek·təd 'rē·jən }

simply connected space A topological space whose fundamental group consists of only one element; equivalently, all closed curves can be shrunk to a point. { 'sim·plē kə¦nek·təd 'spās }

simply normal number A number whose expansion with respect to a given base (not necessarily 10) is such that all the digits occur with equal frequency. { ¦sim·plē ¦nȯr·məl 'nəm·bər }

simply ordered set *See* linearly ordered set. { ¦sim·plē ¦ȯrd·ərd 'set }

simply periodic function A periodic function $f(x)$ for which there is a period a such that every period of $f(x)$ is an integral multiple of a. Also known as singly periodic function. { ¦sim·plē ¦pir·ē¸äd·ik 'fəŋk·shən }

Simpson's rule Also known as parabolic rule.   **1.** A basic approximation formula for definite integrals which states that the integral of a real-valued function f on an interval $[a,b]$ is approximated by $h[f(a) + 4f(a + h) + f(b)]/3$, where $h = (b - a)/2$; this is the area under a parabola which coincides with the graph of f at the abscissas a, $a + h$, and b.   **2.** A method of approximating a definite integral over an interval which is equivalent to dividing the interval into equal subintervals and applying the formula in the first definition to each subinterval.   { 'sim·sənz ‚rül }

simson *See* Simson line.   { 'sim·sən }

Simson line The Simson line of a point P on the circumcircle of a triangle ABC is the line passing through the collinear points L, M, and N, where L, M, and N are the projections of P upon the sides BC, CA, and AB, respectively.   Also known as simson.   { 'sim·sən ‚līn }

simultaneous equations A collection of equations considered to be a set of joint conditions imposed on the variables involved.   { ‚sī·məl'tā·nē·əs i'kwā·zhənz }

sin A *See* sine.   { 'sīn 'ā }

sine The sine of an angle A in a right triangle with hypotenuse of length c given by the ratio a/c, where a is the length of the side opposite A; more generally, the sine function assigns to any real number A the ordinate of the point on the unit circle obtained by moving from $(1,0)$ counterclockwise A units along the circle, or clockwise $|A|$ units if A is less than 0.   Denoted sin A.   { sīn }

sine curve The graph of $y = \sin x$, where x and y are Cartesian coordinates.   Also known as sinusoid.   { 'sīn ‚kərv }

sine series A Fourier series containing only terms that are odd in the independent variable, that is, terms involving the sine function.   { 'sīn ‚sir‚ēz }

single-blind technique An experimental procedure in which the experimenters but not the subjects know the makeup of the test and control groups during the actual course of the experiments.   { 'siŋ·gəl ¦blīnd tek'nēk }

single cusp of the first kind *See* keratoid cusp.   { ¦siŋ·gəl ‚kəsp əv thə 'fərst ‚kīnd }

single cusp of the second kind *See* ramphoid cusp.   { ¦siŋ·gəl ‚kəsp əv thə 'sek·ənd ‚kīnd }

singleton A set that has only one element.   { 'siŋ·gəl·tən }

single-valued function A function for which exactly one point in the range corresponds to each point in the domain; a function that associates to each value of the independent variable exactly one value of the dependent variable.   Also known as one-valued function.   { ¦siŋ·gəl ‚val·yüd 'fəŋk·shən }

singly periodic function *See* simply periodic function.   { ¦siŋ·glē ‚pir·ē‚äd·ik 'fəŋk·shən }

singular integral *See* singular solution.   { 'siŋ·gyə·lər ¦int·ə·grəl əv ə ‚dif·ə'ren·chəl i'kwā·zhən }

singular integral equation An integral equation where the integral appearing either has infinite limits of integration or the kernel function has points where it is infinite.   { 'siŋ·gyə·lər ¦int·ə·grəl i'kwā·zhən }

singularity A point where a function of real or complex variables is not differentiable or analytic.   Also known as singular point of a function.   { ‚siŋ·gyə'lar·əd·ē }

singular matrix A matrix which has no inverse; equivalently, its determinant is zero.   { 'siŋ·gyə·lər 'mā·triks }

singular point 1. For a differential equation, a point that is a singularity for at least one of the known functions appearing in the equation.   **2.** A point on a curve at which the curve possesses no smoothly turning tangent, or crosses or touches itself, or has a cusp or isolated point.   **3.** A point on a surface whose coordinates, x, y, and z, depend on the parameters, u and v, at which the Jacobians $D(x,y)/D(u,v)$, $D(y,z)/D(u,v)$, and $D(z,x)/D(u,v)$ all vanish.   **4.** *See* singularity.   { 'siŋ·gyə·lər 'pöint }

singular solution For a differential equation, a solution that is not generic, that is, not obtainable from the general solution.   Also known as singular integral.   { 'siŋ·gyə·lər sə'lü·shən }

singular transformation A linear transformation which has no corresponding inverse transformation. { 'siŋ·gyə·lər tranz·fər'mā·shən }

singular values For a matrix A these are the positive square roots of the eigenvalues of A^*A, where A^* denotes the adjoint matrix of A. { 'siŋ·gyə·lər 'val·yüz }

sinh *See* hyperbolic sine. { 'sīn'äch }

sinistrorse curve *See* left-handed curve. { ¦sin·ə¦stròrs 'kərv }

sinistrorsum *See* left-handed curve. { ‚sin·ə'stròrs·əm }

sinusoid *See* sine curve. { 'sī·nə‚sòid }

sinusoidal function The real or complex function $\sin(u)$ or any function with analogous continuous periodic behavior. { ‚sī·nə'sòid·əl 'fəŋk·shən }

sinusoidal spiral A plane curve whose equation in polar coordinates (r,θ) is $r^n = a^n \cos n\theta$, where a is a constant and n is a rational number. { ‚sī·nə'sòid·əl 'spī·rəl }

size 1. The number of edges of a graph. **2.** For a test of a hypothesis, the probability of a type I error. { sīz }

size of a critical region For statistical hypotheses, the probability of committing a type I error, that is, rejecting the hypothesis tested when it is true. { ¦sīz əv ə ¦krid·ə·kəl 're·jən }

skeleton 1. For a simplex, the set of all the vertices. **2.** For a simplicial complex, the class of all simplexes which belong to the simplicial complex and have dimension less than that of the simplicial complex. { 'skel·ət·ən }

skewed density function A density function which is not symmetrical, and which depends not only on the magnitude of the difference between the average value and the variate, but also on the sign of this difference. { 'skyüd 'den·səd·ē ‚fəŋk·shən }

Skewes number The first integer n for which the number of primes not greater than n is greater than the Cauchy principal value of the integral over x from 0 to n of the reciprocal of the natural logarithm of x. { 'skyüz ‚nəm·bər }

skew field A ring whose nonzero elements form a non-Abelian group with respect to the multiplicative operation. { 'skyü ‚fēld }

skew Hermitian matrix A square matrix which equals the negative of its adjoint. { 'skyü hər'mish·ən 'mā·triks }

skew lines Lines which do not lie in the same plane in Euclidean three-dimensional space. { 'skyü ‚līnz }

skew matrix *See* antisymmetric matrix. { 'skyü ‚mā‚triks }

skewness The degree to which a distribution departs from symmetry about its mean value. { 'skyü·nəs }

skew product A multiplicative operation or structure induced upon a Cartesian product of sets, where each has some algebraic structure. { 'skyü ‚präd·əkt }

skew quadrilateral A quadrilateral all four of whose vertices do not lie in a single plane. { 'skyü ‚kwäd·rə'lad·ə·rəl }

skew surface A ruled surface that is not a developable surface. { 'skyü ‚sər·fəs }

skew-symmetric determinant *See* antisymmetric determinant. { ¦skyü sə¦me·trik də'tər·mə·nənt }

skew-symmetric matrix *See* antisymmetric matrix. { 'skyü si¦me·trik 'mā·triks }

skew-symmetric tensor A tensor where interchanging two indices will only change the sign of the corresponding component. { 'skyü si¦me·trik 'ten·sər }

slant height 1. The common length of the elements of a right circular cone. **2.** The common altitude of the lateral faces of a regular pyramid. { 'slant hīt }

slide rule A mechanical device, composed of a ruler with sliding insert, marked with various number scales, which facilitates such calculations as division, multiplication, finding roots, and finding logarithms. { 'slīd ‚rül }

slope 1. The slope of a line through the points (x_1,y_1) and (x_2,y_2) is the number $(y_2 - y_1)/(x_2 - x_1)$. **2.** The slope of a curve at a point p is the slope of the tangent line to the curve at p. { slōp }

slope angle The angle of inclination of a line in the plane, where this angle is measured from the positive x axis to the line in the counterclockwise direction. { 'slōp ‚aŋ·gəl }

slope-intercept form In a cartesian coordinate system, the equation of a straight line in the form $y = mx + b$, where m is the slope of the line and b is its intercept on the y axis. { ¦slōp 'in·tər,sept ,förm }

small circle The intersection of a sphere with a plane that does not pass through the center of the sphere. { 'smȯl 'sər·kəl }

Smarandache function A function η defined on the integers with the property that η(n) is the smallest integer m such that $m!$ is divisible by n. { ,smär·ən'dä·chē ,fənk·shən }

smooth To modify a sequential set of numerical data items in a manner designed to reduce the differences in value between adjacent items. { smü̲t̲h̲ }

smooth curve The range of a function from a closed interval to a Euclidean space such that each of the Cartesian coordinates of the image point is a continuously differentiable function on the closed interval. { ¦smü̲t̲h̲ 'kərv }

smoothed data Information that has been averaged or processed with a curve-fitting algorithm so that curves that are free from singularities result when the data are plotted on a graph. { 'smü̲t̲h̲d 'dad·ə }

smoothing 1. Approximating or perturbing a function by one which has a higher degree of differentiability. **2.** A process that uses either freehand methods, moving averages, or fitting a curve by least squares method to remove fluctuations in the data in a time series. { 'smü̲t̲h̲·iŋ }

smooth manifold A differentiable manifold whose local coordinate systems depend upon those of Euclidean space in an infinitely differentiable manner. { 'smü̲t̲h̲ 'man·ə,fōld }

smooth map An infinitely differentiable function. { 'smü̲t̲h̲ 'map }

smooth surface A surface which has a tangent plane at each point, and for which the direction of the normal to this plane is a continuous function of the point of tangency. { ¦smü̲t̲h̲ 'sər·fəs }

solenoidal A vector field has this property in a region if its divergence vanishes at every point of the region. { ¦säl·ə¦nȯid·əl }

solenoid group A compact Abelian, topological group that is one-dimensional and connected. { 'sä·lə,nȯid ,grüp }

solid angle A surface formed by all rays joining a point to a closed curve. { 'säl·əd 'aŋ·gəl }

solid geometry The geometric study of objects in three-dimensional space, such as spheres and polyhedrons. { ¦säl·əd jē'äm·ə·trē }

solid of revolution A solid that can be generated by rotating a plane area about a line. { ¦säl·əd əv ,rev·ə'lü·shən }

solid sphere The union of a sphere with its interior. { 'säl·əd ,sfir }

solidus A sloping line that indicates division in a fraction. { 'säl·əd·əs }

soliton A solution of a nonlinear differential equation that propogates with a characteristic constant shape. { 'säl·ə,tän }

solution set The set of values that satisfy a given equation. { sə'lü·shən ,set }

solvable extension A finite extension E of a field F such that the Galois group of the smallest Galois extension of F containing E is a solvable group. { 'säl·və·bəl ik'sten·chən }

solvable group A group G which has subgroups $G_0, G_1, \ldots, G_n$, where $G_0 = G$, $G_n =$ the identity element alone, and each G_i is a normal subgroup of G_{i-1} with the quotient group G_{i-1}/G_i Abelian. { 'säl·və·bəl 'grüp }

solvmanifold A homogeneous space obtained by factoring a connected solvable Lie group by a closed subgroup. { sälv'man·ə,fōld }

Sommerfeld-Watson transformation *See* Watson-Sommerfeld transformation. { 'zȯm·ər,felt 'wät·sən ,tranz·fər,mā·shən }

source The vertex with indegree 0 that is specified in the definition of an s-t network. { sȯrs }

Souslin's conjecture The conjecture that a topological space is homeomorphic to the real line if (1) it is totally ordered with no first or last element, (2) the open intervals are a base for its topology, (3) it is connected, and (4) any collection of disjoint open intervals. { 'sü,slan̲z kən,jek·chər }

Souslin's line A topological space that satisfies the four conditions of Souslin's conjecture but is not separable, and thus not homeomorphic to the real line, in contradiction of Souslin's conjecture. { 'sü,slanz ,līn }

Souslin's theorem The theorem that, if both a subset of a separable, complete metric space and its complement in this space are continuous images of Borel sets in this space, then the subset is itself a Borel set. { 'sü,slanz thir·əm }

space In context, usually a set with a topology on it or some other type of structure. { spās }

space coordinates A three-dimensional system of Cartesian coordinates by which a point is located by three magnitudes indicating distance from three planes which intersect at a point. { 'spās kó'órd·ən·əts }

space curve A curve in three-dimensional Euclidean space; it may be a twisted curve or a plane curve. { 'spās ,kərv }

space-filling curve See Peano curve. { ¦spās ,fil·iŋ ,kərv }

space polar coordinates A system of coordinates by which a point is located in space by its distance from a fixed point called the pole, the colatitude or angle between the polar axis (a reference line through the pole) and the radius vector (a straight line connecting the pole and the point), and the longitude or angle between a reference plane containing the polar axis and a plane through the radius vector and polar axis. { 'spās 'pō·lər kō'órd·ən·əts }

span 1. For a set A, the intersection of all sets that contain A and have some specified property. Also known as hull. **2.** For a set of vectors, the set of all possible linear combinations of those vectors. Also known as linear span. **3.** The difference between the highest value and the lowest value in a range of values. { span }

spanning subgraph With reference to a graph G, a subgraph of G that contains all the vertices of G. { 'span·iŋ 'səb,graf }

spanning tree A spanning tree of a graph G is a subgraph of G which is a tree and which includes all the vertices in G. { 'span·iŋ ,trē }

sparse matrix A matrix most of whose entries are zeros. { 'spärs 'mā·triks }

sparseness The property of a nonlinear programming problem which has many variables, but whose objective and constraint functions each involve only relatively few variables. { 'spärs·nəs }

Spearman-Brown formula A formula to estimate the reliability of a test n times as long as one for which reliability is known; the tests must be comparable in all aspects other than size. { ¦spir·mən 'braùn ,fòr·myə·lə }

Spearman's rank correlation coefficient A statistic used as a measure of correlation in nonparametric statistics when the data are in ordinal form; a product moment correlation coefficient. Also known as Spearman's rho. { ¦spir·mənz ¦raŋk ,kär·ə'lā·shən ,kō·ə,fish·ənt }

Spearman's rho See Spearman's rank correlation coefficient. { ¦spir·mənz 'rō }

special functions The various families of solution functions corresponding to cases of the hypergeometric equation or functions used in the equation's study, such as the gamma function. { 'spesh·əl 'fəŋk·shənz }

special Jordan algebra A Jordan algebra that can be written as a symmetrized product over a matrix algebra. { 'spesh·əl 'jórd·ən ,al·jə·brə }

special orthogonal group of dimension n The Lie group of special orthogonal transformations on an n-dimensional real inner product space. Symbolized SO_n; $SO(n)$. { ¦spesh·əl ór¦thäg·ən·əl ,grüp əv di¦men·chən 'en }

special orthogonal transformation An orthogonal transformation whose matrix representation has determinant equal to 1. { ¦spesh·əl ór¦thäg·ən·əl ,tranz·fər'mā·shən }

special unitary group of dimension n The Lie group of special unitary transformations on an n-dimensional inner product space over the complex numbers. Symbolized $SU(n)$. { ¦spesh·əl ¦yü·nə,ter·ē ,grüp əv di¦men·chən 'en }

special unitary transformation A unitary transformation whose matrix representation has determinant equal to 1. { ¦spesh·əl ¦yü·nə,ter·ē ,tranz·fər'mā·shən }

spectral approximation A numerical approximation of a function of two or more variables that involves the expansion of the function into a generalized Fourier series, followed by computation of the Fourier coefficients. { ˌspek·trəl ə͟ˌpȯk·sə'mā·shən }

spectral density The density function for the spectral measure of a linear transformation on a Hilbert space. { 'spek·trəl 'den·səd·ē }

spectral factorization A process sometimes used in the study of control systems, in which a given rational function of the complex variable s is factored into the product of two functions, $F_R(s)$ and $F_L(s)$, each of which has all of its poles and zeros in the right and left half of the complex plane, respectively. { 'spek·trəl ˌfak·tə·rə'zā·shən }

spectral function In the theory of stationary stochastic processes, the function

$$F(y) = (2/\pi) \int_0^\infty \rho(x)[(\sin xy)/x] \, dx, \, 0 \le y \le \infty$$

where $\rho(x)$ is the autocorrelation function of a stationary time series. { 'spek·trəl 'fəŋk·shən }

spectral measure A measure on the spectrum of an operator on a Hilbert space whose values are projection operators there; spectral theorems concerning linear operators often give an integral representation of the operator in terms of these projection valued measures. { 'spek·trəl 'mezh·ər }

spectral radius For the spectrum of an operator, this is the least upper bound of the set of all $|\lambda|$, where λ is in the spectrum. { 'spek·trəl 'rād·ē·əs }

spectral theorems Spectral theorems enable detailed study of various types of operators on Banach spaces by giving an integral or series representation of the operator in terms of its spectrum, eigenspaces, and simple projectionlike operators. { 'spek·trəl 'thir·əmz }

spectrum If T is a linear operator of a normed space X to itself and I is the identity transformation ($I(x) \equiv x$), the spectrum of T consists of all scalars λ for which either $T - \lambda I$ has no inverse or the range of $T - \lambda I$ is not dense in X. { 'spek·trəm }

speed-up theorem There is a computable function f with the property that for any algorithm A there is another algorithm B which computes f much faster than A. { 'spēd͟ˌəp ˌthir·əm }

Sperner set A set S of subsets of a given set such that if A and B are in S, and A does not equal B, then neither A nor B is a subset of the other. Also known as antichain; clutter. { 'spər·nər ˌset }

Sperner's theorem A theorem which gives the maximum possible cardinality of an antichain in a finite set. { 'spər·nərz ˌthir·əm }

sphere 1. The set of all points in a euclidean space which are a fixed common distance from some given point; in Euclidean three-dimensional space the Riemann sphere consists of all points (x,y,z) which satisfy the equation $x^2 + y^2 + z^2 = 1$. **2.** The set of points in a metric space whose distance from a fixed point is constant. { sfir }

spherical angle The figure formed by the intersection of two great circles on a sphere, and equal in size to the angle formed by the tangents to the great circles at the point of intersection. { ͟ˈsfir·ə·kəl 'aŋ·gəl }

spherical Bessel functions Bessel functions whose order is half of an odd integer; they arise as the radial functions that result from solving Pockel's equation (or, equivalently, the time-independent Schrödinger equation for a free particle) by separation of variables in spherical coordinates. { ͟ˈsfir·i·kəl 'bes·əl ˌfəŋk·shənz }

spherical cone 1. A solid consisting of the cap and cone formed by the intersection of a plane with a sphere, the cone extending from the plane to the center of the sphere and the cap extending from the plane to the surface of the sphere. **2.** The surface of this solid. { ͟ˈsfir·ə·kəl 'kōn }

spherical coordinates A system of curvilinear coordinates in which the position of a point in space is designated by its distance r from the origin or pole, called the radius vector, the angle ϕ between the radius vector and a vertically directed polar axis, called the cone angle or colatitude, and the angle θ between the plane of ϕ and a fixed meridian plane through the polar axis, called the polar angle or longitude. Also known as spherical polar coordinates. { 'sfir·ə·kəl kō'ȯrd·ən·əts }

spherical curve A curve that lies entirely on the surface of a sphere.   { 'sfir·ə·kəl 'kərv }

spherical cyclic curve *See* cyclic curve.   { 'sfir·ə·kəl 'sī·klik 'kərv }

spherical degree A solid angle equal to one-ninetieth of a spherical right angle.   { 'sfir·ə·kəl di'grē }

spherical distance The length of a great circle arc between two points on a sphere.   { 'sfir·i·kəl 'dis·təns }

spherical excess The sum of the angles of a spherical triangle, minus 180°.   { 'sfir·ə·kəl ek'ses }

spherical geometry The geometry of points on a sphere.   { 'sfir·ə·kəl je'äm·ə·trē }

spherical harmonics Solutions of Laplace's equation in spherical coordinates.   { 'sfir·ə·kəl här'män·iks }

spherical image Also known as spherical representation.   **1.** For a point on a surface, the end point of the radius of a unit sphere parallel to the positive direction of the normal to the surface at the point.   **2.** For a surface, a portion of a unit sphere consisting of all the end points of those radii of the sphere that are parallel to the positive directions of normals to the surface.   Also known as Gaussian representation.   **3.** *See* spherical indicatrix.   { ¦sfir·ə·kəl 'im·ij }

spherical indicatrix For a space curve, those points on the unit sphere traced out by a radius moving from point to point always parallel with the tangent to the curve. Also known as spherical image; spherical indicatrix of the tangent; spherical representation; tangent indicatrix.   { ¦sfir·ə·kəl in'dik·ə,triks }

spherical indicatrix of the binormal *See* binormal indicatrix.   { ¦sfir·ə·kəl in¦dik·ə,triks əv <u>th</u>ə 'bī,nór·məl }

spherical indicatrix of the principal normal *See* principal normal indicatrix.   { ¦sfir·ə·kəl in¦dik·ə,triks əv <u>th</u>ə 'prin·sə·pəl ,nór·məl }

spherical indicatrix of the tangent *See* spherical indicatrix.   { ¦sfir·ə·kəl in¦dik·ə,triks əv <u>th</u>ə 'tan·jənt }

spherical polar coordinates *See* spherical coordinates.   { 'sfir·ə·kəl 'pō·lər kö'órd·ən·əts }

spherical polygon A part of a sphere that is bounded by arcs of great circles.   { 'sfir·ə·kəl 'päl·ə,gän }

spherical pyramid A solid bounded by a spherical polygon and portions of planes passing through the sides of the polygon and the center of the sphere.   { 'sfir·ə·kəl 'pir·ə,mid }

spherical radius For a circle on a sphere, the smaller of the spherical distances from one of the two poles of the circle to any point on the circle.   { 'sfir·ə·kəl 'rād·ē·əs }

spherical right angle The solid angle subtended at the center of a sphere by a portion of the surface of the sphere bounded by a trirectangular spherical triangle's equal to π/2 steradians.   { ¦sfir·ə·kəl ,rīt 'aŋ·gəl }

spherical representation *See* spherical image; spherical indicatrix.   { 'sfir·ə·kəl ,rep·ri·zen¦tā·shən }

spherical sector A solid formed by rotating a sector of a circle about a diameter of the circle; the diameter may contain one of the radii bounding the circular sector or it may lie outside the circular sector.   { 'sfir·ə·kəl 'sek·tər }

spherical segment A solid that is bounded by a sphere and two parallel planes which intersect the sphere or are tangent to it.   { 'sfir·ə·kəl 'seg·mənt }

spherical surface A surface whose total curvature has a constant positive value but that is not necessarily a sphere.   { 'sfir·ə·kəl 'sər·fəs }

spherical surface harmonics Functions of the two angular coordinates of a spherical coordinate system which are solutions of the partial differential equation obtained by separation of variables of Laplace's equation in spherical coordinates.   Also known as surface harmonics.   { ¦sfir·ə·kəl ¦sər·fəs här'män· iks }

spherical triangle A three-sided surface on a sphere the sides of which are arcs of great circles.   { 'sfir·ə·kəl 'trī,aŋ·gəl }

spherical trigonometry The study of spherical triangles from the viewpoint of angle, length, and area.   { 'sfir·ə·kəl ,trig·ə'näm·ə·trē }

spherical wedge The portion of a sphere bounded by two semicircles and a lune (the surface of the sphere between the semicircles). { 'sfir·ə·kəl 'wej }

spheroid *See* ellipsoid of revolution. { 'sfir‚óid }

spheroidal excess The amount by which the sum of the three angles of a triangle on the surface of a spheroid exceeds 180°. { sfir'óid·əl ek‚ses }

spheroidal harmonics Solutions to Laplace's equation when phrased in ellipsoidal coordinates. { sfir'óid·əl här'män·iks }

spheroidal triangle The figure formed by three geodesic lines joining three points on a spheroid. Also known as geodetic triangle. { sfir'óid·əl 'trī‚aŋ·gəl }

spinor 1. A vector with two complex components, which undergoes a unitary unimodular transformation when the three-dimensional coordinate system is rotated; it can represent the spin state of a particle of spin 1/2. **2.** More generally, a spinor of order (or rank) n is an object with 2^n components which transform as products of components of n spinors of rank one. **3.** A quantity with four complex components which transforms linearly under a Lorentz transformation in such a way that if it is a solution of the Dirac equation in the original Lorentz frame it remains a solution of the Dirac equation in the transformed frame; it is formed from two spinors (definition 1). Also known as Dirac spinor. { 'spin·ər }

spin space The two-dimensional vector space over the complex numbers, whose unitary unimodular transformations are a two-dimensional double-valued representation of the three-dimensional rotation group; its vectors can represent the various spin states of a particle with spin 1/2, and its unitary unimodular transformations can represent rotations of this particle. { 'spin ‚spās }

spiral A simple curve in the plane which continuously winds about itself either into some point or out from some point. { 'spī·rəl }

spiral of Archimedes The curve spiraling into the origin which in polar coordinates is given by the equation $r = a\theta$. Also known as Archimedes' spiral. { ‚spī·rəl əv ‚är·kə'mē·dēz }

spline A function used to approximate a specified function on an interval, consisting of pieces which are defined uniquely on a set of subintervals, usually as polynomials or some other simple form, and which match up with each other and the prescribed function at the end points of the subintervals with a sufficiently high degree of accuracy. { splīn }

split-half method A method used to gage the reliability of a test; two sets of scores are obtained from the same test, one set from odd items and one set from even items, and the scores of the two sets are correlated. { 'split ‚haf ‚meth·əd }

splitting field *See* Galois field. { 'splid·iŋ ‚fēld }

split-plot design An experimental design that enables an additional factor or treatment to be included at more than one level; each plot is split into two or more parts. { 'split ‚plät di‚zīn }

sporadic simple group A simple group which cannot be classified in any known infinite family of simple groups. { spə‚rad·ik ‚sim·pəl ‚grüp }

spread The range within which the values of a variable quantity occur. { spred }

spur *See* trace. { spər }

spurious correlation The value of the coefficient of correlation when it is computed correctly but its relationship implications are nonsensical or unreasonable. { 'spyùr·ē·əs ‚kä·rə'lā·shən }

square 1. The square of a number r is the number r^2, that is, r times r. **2.** The plane figure with four equal sides and four interior right angles. { skwer }

square degree A unit of a solid angle equal to $(\pi/180)^2$ steradian, or approximately 3.04617×10^{-4} steradian. { 'skwer di'grē }

squarefree number A positive integer which is not a multiple of the square of any integer other than 1. { ‚skwer‚frē 'nəm·bər }

square grade A unit of solid angle equal to $(\pi/200)^2$ steradian, or approximately 2.46740×10^{-4} steradian. { 'skwer 'grād }

square matrix A matrix with the same number of rows and columns. { 'skwer 'mā·triks }

square number A number that is derived by squaring an integer.   { 'skwer 'nəm·ber }

square root A square root of a real or complex number s is a number t for which $t^2 = s$.   { 'skwer 'rüt }

square-root law The standard deviation of the ratio of the number of successes to number of trials is inversely proportional to the square root of the number of trials.   { 'skwer ¦rüt ‚lȯ }

square-root transformation A conversion or transformation of data having a Poisson distribution where sample means are approximately proportional to the variances of the respective samples; replacing each measurement by its square root will often result in homogeneous variances.   { 'skwer ‚rüt ‚tranz·fər'mā·shən }

squaring the circle For a circle with a specified radius, the problem of constructing a square that has the same area as the circle.   { 'skwer·iŋ th̲ə 'sər·kəl }

sr *See* steradian.

stability Stability theory of systems of differential equations deals with those solution functions possessing some particular property that still maintain the property after a perturbation.   { stə'bil·əd·ē }

stabilizer The stabilizer of a point x in a Riemann surface X, relative to a group G of conformal mappings of X onto itself, is the subgroup G_x of G consisting of elements g such that $g(x) = x$. Also known as stability subgroup.   { 'stā·bə‚līz·ər }

stable bundle The bundle E^s of a hyperbolic structure.   { 'stā·bəl 'bən·dəl }

stable graph A graph from which an edge can be deleted to produce a subgraph whose group of automorphisms is a subgroup of the group of automorphisms of the original graph.   { 'stā·bəl 'graf }

stable homeomorphism conjecture For dimension n, the assertion that each orientation-preserving homeomorphism of the real n space, R^n, into itself can be expressed as a composition of homeomorphisms, each of which is the identity on some nonempty open set in R^n.   { 'stā·bəl ¦hō·mē·ō'mȯr‚fiz·əm kən‚jek·chər }

standard deviate For a variable x, the quantity $(x - \bar{x})/\sigma$, where $\bar{x}$ is the mean value of x and σ is the standard deviation of x.   { ¦stan·dərd 'dē·vē·ət }

standard deviation The positive square root of the expected value of the square of the difference between a random variable and its mean.   { 'stan·dərd ‚dē·vē'ā·shən }

standard error A measure of the variability any statistical constant would be expected to show in taking repeated random samples of a given size from the same universe of observations.   { 'stan·dərd 'er·ər }

standard error of the estimate Standard deviation of observed values about the regression line; computed by dividing the unexplained variation or the error sum of squares by its degrees of freedom.   { ¦stan·dərd ¦er·ər əv th̲ə 'es·tə·mət }

standard error of the forecast Standard deviation of the estimate (point or interval) of a dependent variable for a given value of an independent variable.   { ¦stan·dərd ¦er·ər əv th̲ə 'fȯr‚kast }

standard error of the regression coefficient The standard deviation of an estimated regression coefficient; depends on sample size and model assumptions.   { ¦stan·dərd ¦er·ər əv th̲ə ri'gresh·ən ‚kō·ə‚fish·ənt }

standardized test statistic A test statistic which has been reduced to standardized units.   { 'stan·dər‚dīzd 'test stə‚tis·tik }

standardized units A random variable Z has been reduced to standardized units when it has zero expected value and standard deviation 1; this is accomplished by dividing the difference of Z and the expected value of Z by the standard deviation of Z.   { 'stan·dər‚dīzd 'yü·nəts }

standard measure *See* standard score.   { 'stan·dərd 'mezh·ər }

standard score A test score converted or transformed into a common scale, such as standard units, to effect a more reasonable scale of measurement in order to make comparisons between different tests. Also known as standard measure.   { 'stan·dərd 'skȯr }

star For a member S of a family of sets, the collection of all sets in the family that contain S as a subset.   { stär }

steepest descent method

star algebra A real or complex algebra on which an involution is defined. { 'stär ˌal·jə·brə }

star diagram A graphical way of representing a partition of a positive integer using asterisks that are arranged in rows corresponding to the parts. { 'stär ˌdī·ə,gram }

starlike region A region in the complex number plane such that the line segment joining any of its points to the origin lies entirely in the region. { 'stär,līk ˌrē·jən }

star-shaped set With respect to a point P of a Euclidean space or vector space, a set such that if Q is a member of the set, then so is any point on the line segment PQ. { 'stär ˌshāpt ˌset }

star subalgebra A subalgebra of a star algebra which is mapped onto itself by the involution operation. { 'stär ˌsəb,al·jə·brə }

statement function *See* propositional function. { 'stāt·mənt ˌfəŋk·shən }

static error Error independent of the time-varying nature of a variable. { 'stad·ik 'er·ər }

stationary phase A method used to find approximations to the integral of a rapidly oscillating function, based on the principle that this integral depends chiefly on that part of the range of integration near points at which the derivative of the trigonometric function involved vanishes. { 'stā·shə,ner·ē 'fāz }

stationary point 1. A point on a curve at which the tangent is horizontal. **2.** For a function of several variables, a point at which all partial derivatives are 0. { 'stā·shə,ner·ē 'pöint }

stationary stochastic process A stochastic process $x(t)$ is stationary if each of the joint probability distributions is unaffected by a change in the time parameter t. { 'stā·shə,ner·ē stō'kas·tik 'prä·səs }

stationary time series A time series which as a stochastic process is unchanged by a uniform increment in the time parameter defining it. { 'stā·shə,ner·ē 'tīm ,sir·ēz }

statistic An estimate or piece of data, concerning some parameter, obtained from a sampling. { stə'tis·tik }

statistical analysis The body of techniques used in statistical inference concerning a population. { stə'tis·tə·kəl ə'nal·ə·səs }

statistical computing *See* computational statistics. { stə'tis·tə·kəl kəm'pyüd·iŋ }

statistical distribution *See* distribution. { stə'tis·tə·kəl ,dis·trə'byü·shən }

statistical hypothesis A statement about the way a random variable is distributed. { stə'tis·tə·kəl hī'päth·ə·səs }

statistical independence Two events are statistically independent if the probability of their occurring jointly equals the product of their respective probabilities. Also known as stochastic independence. { stə'tis·tə·kəl ,in·də'pen·dəns }

statistical inference The process of reaching conclusions concerning a population upon the basis of random samplings. { stə'tis·tə·kəl 'in·frəns }

statistical weight A number assigned to each value or range of values of a given quantity, giving the number of times this value or range of values is found to be observed. { stə'tis·tə·kəl 'wāt }

statistics A discipline dealing with methods of obtaining data, analyzing and summarizing it, and drawing inferences from data samples by the use of probability theory. { stə'tis·tiks }

s-t cut The set of all the arcs in an *s-t* network that originates in X and terminate in the complement of X, where X is a set of vertices in the *s-t* network that contains the source but not the terminal. { ¦es¦tē 'kət }

Steenrod algebra The cohomology groups of a topological space have additive operations on them, which can be added and multiplied so as to form the Steenrod algebra. { 'sten,räd 'al·jə·brə }

Steenrod squares Operations which associate elements from different cohomology groups of a topological space and produce an element in another of the groups; these operations can be so added and multiplied as to produce the Steenrod algebra. { 'sten,räd 'skwerz }

steepest descent method 1. Certain functions can be approximated for large values by an asymptotic formula derived from a Taylor series expansion about a saddle

Steiner triple system

point. Also known as saddle point method.   **2.** A method of approximating extreme values of the functions of two or more variables, in which the gradient of the function is used to obtain a sequence of approximations of the point at which the extreme value occurs.   { 'stēp·əst di'sent ,meth·əd }

Steiner triple system A balanced incomplete block design in which the number k of distinct elements in each block equals 3, and the number λ of blocks in which each combination of elements occurs together equals 1.   { ¦stīn·ər ¦trip·əl 'sis·təm }

Steinitz theorem The theorem that an interior point of the convex span of a set in n-dimensional Euclidean space is also an interior point of the convex span of a subset of that set which has at most two n points.   { 'shtī,nits ,thir·əm }

step function 1. A function f defined on an interval $[a,b]$ so that $[a,b]$ can be partitioned into a finite number of subintervals on each of which f is a constant.   Also known as simple function.   **2.** More generally, a real function with finite range.   { 'step ,fəŋk·shən }

sterad *See* steradian.   { 'sti,rad }

steradian The unit of measurement for solid angles; it is equal to the solid angle subtended at the center of a sphere by a portion of the surface of the sphere whose area equals the square of the sphere's radius.   Abbreviated sr; sterad.   { stə'rād·ē·ən }

steregon The entire solid angle bounded by a sphere; equal to 4π steradians.   { 'ster·ə,gän }

stereographic projection The projection of the Riemann sphere onto the Euclidean plane performed by emanating rays from the north pole of the sphere through a point on the sphere.   { ¦ster·ē·ə¦graf·ik prə'jek·shən }

Stieltjes integral The Stieltjes integral of a real function $f(x)$ relative to a real function $g(x)$ of bounded variation on an interval $[a,b]$ is defined, analogously to the Riemann integral, as a limit of a sum of terms $f(a_i)\,[g(x_i) - g(x_{i-1})]$ taken as partitions of the interval shrink. Denoted

$$\int_a^b f(x)\,dg(x)$$

Also known as Riemann-Stieltjes integral.   { 'stēlt·yəs ,int·ə·grəl }

Stieltjes transform A form of the Laplace transform of a function where the usual Riemann integral is replaced by a Stieltjes integral.   { 'stēlt·yəs ,tranz·form }

Stirling numbers The coefficients which occur in the Stirling interpolation formula for a difference operator.   { 'stir·liŋ ,nəm·bərz }

Stirling numbers of the first kind The numbers $s(n,r)$ giving the coefficient of x^r in the falling factorial polynomial $x(x - 1)(x - 2) \cdots (x - n + 1)$.   { ¦stər·liŋ ,num·bərz əv thə 'fərst ,kind }

Stirling numbers of the second kind The numbers $S(n,r)$ giving the numbers of ways that n elements can be distributed among r indistinguishable cells so that no cell remains empty.   { ¦stər·liŋ ,nəm·bərz əv thə 'sek·ənd ,kīnd }

Stirling's formula The expression $(n/e)^n \sqrt{2\pi n}$ is asymptotic to factorial n; that is, the limit as n goes to ∞ of their ratio is 1.   { 'stir·liŋz ,för·myə·lə }

Stirling's series An asymptotic expansion for the logarithm of the gamma function, or an equivalent asymptotic expansion for the gamma function itself, from which Stirling's formula may be derived.   { 'stər·liŋz ,sir,ēz }

stochastic Pertaining to random variables.   { stō'kas·tik }

stochastic calculus The mathematical theory of stochastic integrals and differentials, and its application to the study of stochastic processes.   { stō'kas·tik 'kal·kyə·ləs }

stochastic chain rule A generalization of the ordinary chain rule to stochastic processes; it states that the process $U_t = u(X_t^1, X_t^2, \ldots, X_t^n)$ satisfies

$$dU = \sum_i \partial_i u\,dX^i + \frac{1}{2}\sum_{i,j} \partial_i \partial_j u\,dX^i dX^j$$

with the conventions $(dt)^2 = 0$ and $dW^\alpha dW^\beta = \partial_{\alpha\beta}dt$, where the X^i are processes satisfying

$$dX_t^i = a_t^i dt + \sum_{\alpha=1}^{m} b_t^{i\alpha} \, dW_t^{\alpha}, \; i = 1, 2, \ldots, n;$$

$\{W_t^{\alpha}, t \geq 0\}$, $\alpha = 1, 2, \ldots, m$, are independent Wiener processes; the dW_t^{α} are the corresponding random disturbances occurring in the infinitesimal time interval dt; the a_t^i and $b_t^{i\alpha}$ are independent of future disturbances, and $u(x_1, x_2, \ldots, x_n)$ is a function whose derivatives $\partial_i u$ and $\partial_i \partial_j u$ are continuous. Also known as Itô's formula. { stō'kas·tik 'chān ‚rül }

stochastic differential An expression representing the random disturbances occurring in an infinitesimal time interval; it has the form dW_t, where $\{W_t, t \geq 0\}$ is a Wiener process. { stō'kas·tik ‚dif·ə'ren·chəl }

stochastic independence *See* statistical independence. { stō¦kas·tik in·də'pen·dəns }

stochastic integral An integral used to construct the sample functions of a general diffusion process from those of a Wiener process; it has the form

$$\int_{W_0}^{W_s} a_t dW_t$$

where $\{W_t, t \geq 0\}$ is a Wiener process, dW_t represents the random disturbances occurring in an infinitesimal time interval dt, and a_t is independent of future disturbances. Also known as Itô's integral. { stō'kas·tik 'int·ə·grəl }

stochastic matrix A square matrix with nonnegative real entries such that the sum of the entries of each row is equal to 1. { stō'kas·tik 'mā·triks }

stochastic process A family of random variables, dependent upon a parameter which usually denotes time. Also known as random process. { stō'kas·tik 'prä·səs }

stochastic variable *See* random variable. { stō'kas·tik 'ver·ē·ə·bəl }

Stokes' integral theorem The analog of Green's theorem in n-dimensional Euclidean space; that is, a line integral of $F_1(x_1, x_2, \ldots, x_n)dx_1 + \cdots + F_n(x_1, x_2, \ldots, x_n)dx_n$ over a closed curve equals an integral of an expression containing various partial derivatives of $F_1, \ldots, F_n$ over a surface bounded by the curve. { ¦stōks 'int·ə·grəl ‚thir·əm }

Stokes phenomenon A change in the asymptotic representation of certain analytic functions that occurs in passing from one section of the complex plane to another. { 'stōks fə‚näm·ə‚nän }

Stone-Čech compactification The Stone-Čech compactification of a completely regular space is a Hausdorff space in which the original space forms a dense subset, such that any continuous function from the original space to a compact space has a unique continuous extension to the Hausdorff space. { ¦stōn ¦chek kəm‚pak·tə·fə'kā·shən }

Stone's representation theorem This theorem determines the nature of all unitary representations of locally compact Abelian groups. { 'stōnz ‚rep·rə·zən'tā·shən ‚thir·əm }

Stone's theorem Every Boolean ring is isomorphic to a ring of subsets of some set. { 'stōnz ‚thir·əm }

Stone-Weierstrass theorem If S is a collection of continuous real-valued functions on a compact space E, which contains the constant functions, and if for any pair of distinct points x and y in E there is a function f in S such that $f(x)$ is not equal to $f(y)$, then for any continuous real-valued function g on E there is a sequence of functions, each of which can be expressed as a polynomial in the functions of S with real coefficients, that converges uniformly to g. { 'stōn 'vī·ər‚sträs ‚thir·əm }

stopping rule A rule which specifies when observation is to be discontinued in sequential trials. { 'stäp·iŋ ‚rül }

straight angle An angle of measure one-half revolution or 180°, whose sides lie on the same straight line but extend in opposite directions from the vertex. { 'strāt ‚aŋ·gəl }

strategy In game theory a strategy is a specified collection of moves, which cover all possible situations, for the complete play of a given game. { 'strad·ə·jē }

strategy vector A vector characterizing a mixed strategy, whose components are the probability weights of the strategy. { 'strad·ə·jē ˌvek·tər }

stratified sampling A random sample of specified size is drawn from each stratum of a population. { 'strad·ə‚fīd 'sam·pliŋ }

stratum *See* subpopulation. { 'strad·əm }

stretching transformation A homothetic transformation in which the ratio of similitude is less than 1. { 'strech·iŋ ‚tranz·fər‚mā·shən }

strictly convex space A normal linear space such that, for any two vectors x and y, if $\| x + y \| = \| x \| + \| y \|$, then either $y = 0$ or $x = cy$, where c is a number. Also known as rotund space. { 'strik·lē ¦kän‚veks 'spās }

strictly decreasing function *See* decreasing function. { ¦strik·lē di¦krēs·iŋ 'fəŋk·shən }

strictly dominant strategy Relative to a given pure strategy for one player of a game, a second pure strategy for that player that has a greater payoff than the given strategy for any pure strategy of the opposing player. { ¦strik·lē ¦däm·ə·nənt 'strad·ə·jē }

strictly Hurwitz polynomial A polynomial whose roots all have strictly negative real parts. { 'strik·lē 'hər‚vits ‚päl·i'nō·mē·əl }

strictly increasing function *See* increasing function. { ¦strik·lē in‚krēs·iŋ 'fəŋk·shən }

string One of the space curves that form a braid. { striŋ }

strongly connected digraph A directed graph in which there is a directed path from every vertex to every other vertex. { ¦stróŋ·lē kə¦nek·təd 'dī‚graf }

strongly continuous semigroup A semigroup of bounded linear operators on a Banach space B, together with a bijective mapping T from the positive real numbers onto the semigroup, such that $T(0)$ is the identity operator on B, $T(s + t) = T(s)T(t)$ for any two positive numbers s and t and, for each element x of B, $T(t)x$ is a continuous function of t. { ¦stróŋ·lē kən¦tin·yə·wəs 'sem·i‚grüp }

strong topology The topology on a normed space obtained from the given norm; the basic open neighborhoods of a vector x are sets consisting of all those vectors y where the norm of $x - y$ is less than some number. { 'stróŋ tə'päl·ə·jē }

strophoid 1. A curve derived from a given curve C and two points, called the pole and the fixed point, consisting of the locus of points on a rotating line L passing through the pole whose distance from the intersection of L and C is equal to the distance of this intersection from the fixed point. **2.** The special case of the first definition in which C is a straight line and the fixed point lies on C. { 'strä‚fȯid }

structural stability Property of a differentiable flow on a compact manifold whose orbit structure is insensitive to small perturbations in the equations governing the flow or in the vector field generating the flow. { 'strək·chər·əl stə'bil·əd·ē }

structure constants A set of numbers that serve as coefficients in expressing the commutators of the elements of a Lie algebra. { 'strək·chər ‚kän·stəns }

structured grid In the discretization of partial differential equations, an organized set of points formed by the intersections of the lines of a boundary-conforming curvilinear coordinate system, at which the equations are expressed in discrete form. { ¦strək·chərd 'grid }

Student's distribution The probability distribution used to test the hypothesis that a random sample of n observations comes from a normal population with a given mean. { 'stüd·əns ‚dis·trə'byü·shən }

Student's t-statistic A one-sample test statistic computed by $T = \sqrt{n}(\bar{X} - \mu_H)/S$, where $\bar{X}$ is the mean of a collection of n observations, S is the square root of the mean square deviation, and μ_H is the hypothesized mean. { 'stüd·əns 'tē stə‚tis·tik }

Student's t-test A test in a one-sample problem which uses Student's t-statistic. { 'stüd·əns 'tē ‚test }

Sturges rule A rule for determining the desirable number of groups into which a distribution of observations should be classified; the number of groups or classes is $1 + 3.3 \log n$, where n is the number of observations. { 'stər·jəs ‚rül }

Sturm-Liouville problem The general problem of solving a given linear differential equation of order $2n$ together with $2n$-boundary conditions. Also known as eigenvalue problem. { 'stərm lyü'vil ‚präb·ləm }

Sturm-Liouville system A given differential equation together with its boundary conditions having Sturm-Liouville problem form. { 'stərm lyü'vil ˌsis·təm }

Sturm separation theorem The theorem that if u and v are real, linearly independent solutions of a second-order linear homogenous differential equation in which the coefficient of the second derivative is unity and the other two coefficients are continuous functions, then there is exactly one zero of u between any two zeros of v. { ¦stərm ˌsep·ə'rä·shən ˌthir·əm }

Sturm sequence For a polynomial $p(x)$, this is the sequence of functions $f_0(x), f_1(x),$ ..., where $f_0(x) = p(x)$, $f_1(x) = p'(x)$, and $f_n(x)$ is the negative remainder that occurs by finding the greatest common divisor of $f_{n-2}(x)$ and $f_{n-1}(x)$ via the Euclidean algorithm. { 'stərm ˌsēkwəns }

Sturm's theorem This gives a method to determine the number of real roots of a polynomial $p(x)$ which lie between two given values of x; the Sturm sequence of $p(x)$ provides the necessary information. { 'stərmz ˌthir·əm }

subadditive function A function f such that $f(x + y)$ is less than or equal to $f(x) + f(y)$ for all x and y in its domain. { ¦səb'ad·əd·iv ˌfəŋk·shən }

subadditive set function A set function with the property that the union of any finite or countable collection of sets in the range of the function is also in this range, and the value of the function at this union is equal to or less than the sum of its values at each set of the collection. { səbˌad·əd·iv 'set ˌfəŋk·shən }

subalgebra 1. A subset of an algebra which itself forms an algebra relative to the same operations. **2.** A subalgebra (of sets) is any algebra (of sets) contained in some given algebra. { ¦səb'al·jə·brə }

subbase for a topology A family S of subsets of a topological space X where by taking all finite intersections of sets from S and all unions of such intersections the entire topology of open sets of X is obtained. { 'səbˌbās fȯr ə tə'päl·ə·jē }

subdivision graph A graph which can be obtained from a given graph by breaking up each edge into one or more segments by inserting intermediate vertices between its two ends. { 'səb·diˌvizh·ən ˌgraf }

subfactorial For an integer n, the number that is expressed as $n![(1/2!) - (1/3!) + (1/4!) - \cdots + [(-1)^n/n!]]$. { ˌsəbˌfak'tȯr·ē·əl }

subfield 1. A subset of a field which itself forms a field relative to the same operations. **2.** A subfield (of sets) is any field (of sets) contained in some given field of sets. { 'səbˌfēld }

subgraph A graph contained in a given graph which has as its vertices some subset of the vertices of the original. { 'səbˌgraf }

subgroup A subset N of a group G which is itself a group relative to the same operation. { 'səbˌgrüp }

subharmonic function A continuous function is subharmonic in a region R of the plane if its value at any point z_0 of R is less than or equal to its integral along a circle centered at z_0. { ¦səb·här'män·ik 'fəŋk·shən }

subjective probability See personal probability. { səb'jek·tiv ˌpräb·ə'bil·əd·ē }

submodule A subset N of a module M over a ring R such that, if x and y are in N and a is in R, then $x + y$ and ax are in N, so that N is also a module over R. { 'səbˌmä·jəl }

submultiple A number or quantity divided by an integer. { səb'məl·tə·pəl }

subnormal For a given point on a plane curve, the projection on the x axis of a rectangular coordinate system of the segment of the normal between the given point and the intersection of the normal with the x axis. { səb'nȯr·məl }

subnormal operator An operator A on a Hilbert space **H** is said to be subnormal if there exists a normal operator B on a Hilbert space **K** such that **H** is a subspace of **K**, the subspace **H** is invariant under the operator B, and the restriction of B to **H** coincides with A. { 'səbˌnȯr·məl 'äp·ə,rād·ər }

subpopulation A subset of population. Also known as stratum. { ¦səbˌpäp·yə'lā·shən }

subrange A subset of the range of values that a function may assume. { 'səbˌränj }

subring 1. A subset I of a ring R where I is also a ring relative to the operations of R. **2.** A subring (of sets) is any ring (of sets) contained in some given ring (of sets). { 'səb,riŋ }

subsampling Taking samples from a sample of a population. { 'səb,sam·pliŋ }

subscripted variable A symbolic name for an array of variables whose elements are identified by subscripts. { səb'skrip·təd 'ver·ē·ə·bəl }

subsequence A subsequence of a given sequence is any sequence all of whose entries appear in the original sequence and in the same manner of succession. { 'səb·sə·kwəns }

subset 1. A subset A of a set B is a set all of whose elements are included in B. **2.** A fuzzy set A is a subset of a fuzzy set B if, for every element x, the value of the membership function of A at x is equal to or less than the value of the membership function of B at x. { 'səb,set }

subsine function A function that is dominated by functions of the form $f(x) = A \sin (x + c)$, where A and c are constants, in the same way that convex functions are dominated by linear functions. { 'səb,sīn ,faŋk·shən }

subspace A subset of a space which, in the appropriate context, is a space in its own right. { 'səb,spās }

substitution group *See* permutation group. { ,səb·stə'tü·shən ,grüp }

subtangent For a given point on a plane curve, the projection on the x axis of a rectangular coordinate system of the segment of the tangent between the point of tangency and the intersection of the tangent with the x axis. { ¦səb'tan·jənt }

subtend A line segment or an arc of a circle subtends an angle with vertex at a specified point if the end points of the line segment or arc lie on the sides of the angle. { səb'tend }

subtraction The addition of one quantity with the negative of another; in a system with an additive operation this is formally the sum of one element with the additive inverse of another. { səb'trak·shən }

subtraction formula An equation expressing a function of the difference of two quantities in terms of functions of the quantities themselves. { səb'trak·shən ,fòr·myə·lə }

subtraction sign The symbol $-$, used to indicate subtraction. Also known as minus sign. { səb'trak·shən ,sīn }

subtrahend A quantity which is to be subtracted from another given quantity. { 'səb·trə,hend }

subtree A subgraph of a tree which is itself a tree. { 'səb,trē }

successive approximations Any method of solving a problem in which an approximate solution is first calculated, this solution is then used in computing an improved approximation, and the process is repeated as many times as desired. { sək'ses·iv ə,präk·sə'mā·shənz }

successor 1. For a vertex a in a directed graph, any vertex b for which there is an arc between a and b directed from a to b. **2.** For a positive integer, n, the next integer, $n + 1$. Also known as consequent. { sək'ses·ər }

sufficiency Condition of an estimator that uses all the information about the population parameter contained in the sample observations. { sə'fish·ən·sē }

sufficient condition A mathematical statement whose truth suffices to assure the truth of a given statement. { sə¦fish·ənt kən'dish·ən }

sufficient statistic A statistic that contains all the information that can possibly be obtained from a sample to estimate a specified parameter of the sampled population. { sə¦fish·ənt stə'tis·tik }

sum 1. The addition of numbers or mathematical objects in context. **2.** The sum of an infinite series is the limit of the sequence consisting of all partial sums of the series. **3.** The sum $A + B$ of two matrices A and B, with the same number of rows and columns, is the matrix whose element c_{ij} in row i and column j is the sum of corresponding elements a_{ij} in A and b_{ij} in B. { səm }

summability method A method, such as Hölder summation or Cesaro summation, of attributing a sum to a divergent series by using some process to average the terms in the series. { ,səm·ə¦bil·əd·ē ,meth·əd }

summable function A function whose Lebesgue integral exists. { ¦səm·ə·bəl 'fəŋk· shən }

summation convention An abbreviated notation used particularly in tensor analysis and relativity theory, in which a product of tensors is to be summed over all possible values of any index which appears twice in the expression. { sə'mā·shən kən¸ven·chən }

summation sign A capital Greek sigma (Σ) that indicates the members of a set are to be added together, and has numbers below and above it indicating the range of values of an index that are to be included in the summation. { sə'mā·shən ¸sīn }

sup *See* least upper bound.

superadditive function A function f such that $f(x + y)$ is greater than or equal to $f(x) + f(y)$ for all x and y in its domain. { ¦sü·pər¸ad·əd·iv 'fəŋk·shən }

superharmonic function A continuous complex function f whose value at a point z_0 exceeds its average values computed by the integral of f around a circle centered at z_0. { ¦sü·pər·här'män·ik 'fəŋk·shən }

superposability *See* congruence. { ¸sü·pər¸pōz·ə'bil·əd·ē }

superposition The principle of superposition states that any given geometric figure in a euclidean space can be so moved about as not to change its size or shape. { ¸sü·pər·pə'zish·ən }

superreflexive Banach space A Banach space, *B*, such that any Banach space that is finitely representable in *B* is a reflexive Banach space. { ¸sü·pər·ri¦flek·siv 'bä¸näk ¸spās }

superset A set whose elements include all the elements of a given set. { 'sü·pər¸set }

supplemental chords Two chords joining a point on the circumference of a circle to the ends of a diameter of the circle.

supplementary angle One angle is supplementary to another angle if their sum is 180°. { ¦səp·lə¦men·trē 'aŋ·gəl }

support The support of a real-valued function f on a topological space is the closure of the set of points where f is not zero. { sə'pȯrt }

support function Relative to a convex body in a real inner product space, a function whose value at a point *P* is the maximum of the inner product of *P* and *Q* for *Q* in the convex body. { sə'pȯrt ¸fəŋk·shən }

supremum *See* least upper bound. { su'prē·məm }

surd A sum of one or more roots of rational numbers, some or all of which are themselves irrational numbers. { sərd }

surface A subset of three-space consisting of those points whose Cartesian coordinates x, y, and z satisfy equations of the form $x = f(u,v)$, $y = g(u,v)$, $z = h(u,v)$, where f, g, and h are differentiable real-valued functions of two parameters u and v which take real values and vary freely in some domain. { 'sər·fəs }

surface harmonics *See* spherical surface harmonics. { 'sər·fəs här¸män·iks }

surface integral The integral of a function of several variables with respect to surface area over a surface in the domain of the function. { 'sər·fəs 'int·ə·grəl }

surface of center The locus of points that are one of the two centers of principal curvature at some point on a given surface. { ¦sər·fəs əv 'sen·tər }

surface of Joachimsthal A surface such that all the members of one of its two families of lines of curvature are plane curves and their planes all pass through a common axis. { ¦sər·fəs əv yō'äk·əmz¸täl }

surface of Liouville A surface that can be assigned parameters u and v such that a linear element ds on the surface is given by $ds^2 = [f(u) + g(v)][du^2 + dv^2]$, where f and g are functions of u and v. { ¦sər·fəs əv 'lyü¸vēl }

surface of Monge A surface generated by a plane curve as its plane rolls without slipping over a developable surface. { ¦sər·fəs əv 'mȯnzh }

surface of negative curvature A surface whose Gaussian curvature is negative at every point. { ¦sər·fəs əv ¦neg·əd·iv 'kər·və·chər }

surface of positive curvature A surface whose Gaussian curvature is positive at every point. { ¦sər·fəs əv 'päs·əd·iv ¸kər·və·chər }

surface of revolution A surface realized by rotating a planar curve about some axis in its plane. { ¦sər·fəs əv ‚rev·ə'lü·shən }

surface of translation A surface that can be generated from two curves by translating either one of them parallel to itself in such a way that each of its points describes a curve that is a translation of the other curve. Also known as translation surface. { ¦sər·fəs əv tranz'lā·shən }

surface of Voss A surface that has a conjugate system of geodesics. { ¦sər·fəs əv 'vȯs }

surface of zero curvature A surface whose Gaussian curvature is zero at every point. { ¦sər·fəs əv 'zir·ō ‚kər·və·chər }

surface patch A surface or a portion of a surface that is bounded by a closed curve. { 'sər·fəs ‚pach }

surjection A mapping f from a set A to a set B such that for every element b of B there is an element a of A such that $f(a) = b$. Also known as surjective mapping. { sər'jek·shən }

surjective mapping See surjection. { sər'jek·tiv 'map·iŋ }

survey design See sample design. { 'sər‚vā di‚zīn }

swastika A plane curve whose equation in Cartesian coordinates x and y is $y^4 - x^4 = xy$. { 'swäs·tə·kə }

switching function A switching function of n variables is a function that assigns to each binary sequence of length n the number 0 or the number 1. { 'swich·iŋ ‚fəŋk·shən }

syllogism A statement together with a conclusion; this usually has the form "if p then q." { 'sil·ə‚jiz·əm }

Sylow subgroup A subgroup H of a given group G such that the order of H is p^n, where p is a prime and n is an integer, and p^n is the highest power of p dividing the order of G. { ¦sī‚lō 'səb‚grüp }

Sylvester's theorem If A is a matrix with distinct eigenvalues $\lambda_1, \ldots, \lambda_n$, then any analytic function $f(A)$ can be realized from the λ_i, $f(\lambda_i)$, and the matrices $A - \lambda_i I$, where I is the identity matrix. { sil'ves·tərz ‚thir·əm }

symbolic logic The formal study of symbolism and its use in the foundations of mathematical logic. { sim'bäl·ik 'läj·ik }

symmetrical distribution A distribution in which observations equidistant from the central maximum have the same frequency. Also known as symmetric distribution. { si¦met·rə·kəl ‚dis·trə'byü·shən }

symmetric chain A sequence of subsets of a set of n elements such that each member of the sequence is a subset of the next one, each member of the sequence has a cardinality one greater than that of the previous member, and the sum of the cardinalities of the first and last members of the sequence equals n. { sə¦me·trik 'chān }

symmetric chain decomposition A partition of the set of all subsets of a finite set, X, into symmetric chains in X. { sə¦me·trik ¦chān dē‚käm·pə'zish·ən }

symmetric design A balanced incomplete block design in which the number b of blocks equals the number v of elements arranged among the blocks. { si¦me·trik di'zīn }

symmetric difference The symmetric difference of two sets consists of all points in one or the other of the sets but not in both. { sə'me·trik 'dif·rəns }

symmetric distribution See symmetrical distribution. { sə¦me·trik ‚di·strə'byü·shən }

symmetric form A bilinear form f which is unchanged under interchange of its independent variables; that is, $f(x,y) = f(y,x)$ for all values of the independent variables x and y. { si¦me·trik 'fȯrm }

symmetric function A function whose value is unchanged for any permutation of its variables. { sə'me·trik 'fəŋk·shən }

symmetric group The group consisting of all permutations of a finite set of symbols. { sə'me·trik 'grüp }

symmetric matrix A matrix which equals its transpose. { sə'me·trik 'mā·triks }

symmetric relation The property of a relation on a set that requires y to be related to x whenever x is related to y. { sə'me·trik ri'lā·shən }

symmetric space A differentiable manifold which has a differentiable multiplication operation that behaves similarly to the multiplication of a complex number and its conjugate. { sə¦me·trik 'spās }

symmetric spherical triangles Spherical triangles whose corresponding angles and corresponding sides are equal but appear in opposite order as viewed from the center of the sphere. { sə¦me·trik ¦sfir·ə·kəl 'trī,aŋ·gəlz }

symmetric tensor A tensor that is left unchanged by the interchange of two contravariant (or covariant) indices. { sə¦me·trik 'ten·sər }

symmetric transformation A transformation T defined on a Hilbert space such that the inner products (x,Ty) and (Tx,y) are equal for any vectors x and y in the domain of T. { sə¦me·trik ,tranz·fər'mā·shən }

symmetry 1. A geometric object G has this property relative to some configuration S of its points if S determines two pieces of G which can be reflected onto each other through S. **2.** A rigid motion of a geometric figure that maps the figure onto itself. { 'sim·ə,trē }

symmetry function *See* symmetry transformation. { 'sim·ə,trē ,fəŋk·shən }

symmetry group A group composed of all rigid motions or similarity transformations of some geometric object onto itself. { 'sim·ə,trē ,grüp }

symmetry plane *See* plane of mirror symmetry. { 'sim·ə,trē ,plān }

symmetry principle The centroid of a geometrical figure (line, area, or volume) is at a point on a line or plane of symmetry of the figure. { 'sim·ə,trē ,prin·sə·pəl }

symmetry transformation A rigid motion sending a geometric object onto itself; examples are rotations and, for the case of a polygon, permutations of the vertices. Also known as symmetry function. { 'sim·ə,trē ,tranz·fər'mā·shən }

symplectic group of dimension n The Lie group of symplectic transformations on an n-dimensional vector space over the quaternions. Symbolized $Sp(n)$. { sim¦plek·tik ,grüp əv di¦men·chən 'en }

symplectic transformation A linear transformation of a vector space over the quaternions that leaves the lengths of vectors unchanged. { sim¦plek·tik ,tranz·fər'mā·shən }

synclastic Property of a surface or portion of a surface for which the centers of curvature of the principal sections at each point lie on the same side of the surface. { sin'klas·tik }

synthetic division A long division process for dividing a polynomial $p(x)$ by a polynomial $(x - a)$ where only the coefficients of these polynomials are used. { sin'thed·ik də'vizh·ən }

systematic error An error which results from some bias in the measurement process and is not due to chance, in contrast to random error. { ,sis·tə'mad·ik 'er·ər }

system of distinct representatives A family of subsets S_i of a given finite set S such that the family has as many members as there are elements in S, and such that it is possible to assign each element x_i of S to a distinct subset S_i with x_i in S_i. { ¦sis·təm əv di¦stinkt ,rep·rə'zen·tə'tivz }

system of stages A collection of nonempty sets that includes the intersection of any two sets that belong to the collection. { ¦sis·təm əv 'stāj·əz }

243

T

T *See* tera-.

tabular interpolation Method of finding from a table the values of the dependent variable for intermediate values of the independent variable. { 'tab·yə·lər in₁tər·pə'lā·shən }

tacnode *See* double cusp. { 'tak₁nōd }

Talbot's curve The negative pedal of an ellipse, with eccentricity greater than $\sqrt{2}/2$, with respect to its center. { 'tal·bəts ₁kərv }

tan *See* tangent. { tan }

tangent **1.** A line is tangent to a curve at a fixed point P if it is the limiting position of a line passing through P and a variable point on the curve Q, as Q approaches P. **2.** The function which is the quotient of the sine function by the cosine function. Abbreviated tan. **3.** The tangent of an angle is the ratio of its sine and cosine. Abbreviated tan. { 'tan·jənt }

tangent bundle The fiber bundle $T(M)$ associated to a differentiable manifold M which is composed of the points of M together with all their tangent vectors. Also known as tangent space. { 'tan·jənt ₁bənd·əl }

tangent circles Two circles that have a single point in common. { ₁tan·jənt 'sər·kəlz }

tangent cone A cone each of whose elements is tangent to a given quadric surface. { 'tan·jənt ₁kōn }

tangential component A component of a given vector acting at right angles to a given radius of a given circle. { tan'jen·chəl kəm'pō·nənt }

tangential coordinates For a surface, a set of four coordinates, three of which are the direction cosines of the normal to the surface, while the fourth is the algebraic distance from the origin to the plane tangent to the surface. { tan'jen·chəl kō'ȯrd·ən·əts }

tangential curvature *See* geodesic curvature. { tan'jen·chəl 'kər·və·chər }

tangential developable *See* tangent surface. { tan'jen·chəl di'vel·əp·ə·bəl }

tangential polar equation An equation of a curve exprepssed in terms of the distance of a point P on the curve from a reference point O and the perpendicular distance from O to the tangent to the curve at P. { tan'jen·chəl ¦pō·lər i'kwā·zhən }

tangent indicatrix *See* spherical indicatrix. { ¦tan·jənt in'dik·ə₁triks }

tangent plane The tangent plane to a surface at a point is the plane having every line in it tangent to some curve on the surface at that point. { 'tan·jənt 'plān }

tangent space **1.** The vector space of all tangent vectors at a given point of a differentiable manifold. **2.** *See* tangent bundle. { 'tan·jənt ₁spās }

tangent surface The ruled surface generated by the tangents to a specified space curve. Also known as tangential developable. { 'tan·jənt ₁sər·fəs }

tangent vector A tangent vector at a point of a differentiable manifold is any vector tangent to a differentiable curve in the manifold at this point; alternatively, a member of the tangent plane to the manifold at the point. { 'tan·jənt ₁vek·tər }

tanh *See* hyperbolic tangent. { ¦tan'āch }

Taylor series The Taylor series corresponding to a function $f(x)$ at a point x_0 is the infinite series whose nth term is $(1/n!) \cdot f^{(n)}(x_0)(x - x_0)^n$, where $f^{(n)}(x)$ denotes the nth derivative of $f(x)$. { 'tā·lər ₁sir·ēz }

Taylor's theorem The theorem that under certain conditions a real or complex function can be represented, in a neighborhood of a point where it is infinitely differentiable, as a power series whose coefficients involve the various order derivatives evaluated at that point. { 'tā·lərz ˌthir·əm }

Tchuprow-Neymann allocation A technique of stratified sampling in which the size of each strata sample is proportional to the size of the strata population and the variance of the strata. { 'chü,prəf 'nā·mən ˌal·ə,kā·shən }

t distribution A distribution used to test a hypothesis about a population mean when the population standard deviation is not known, the sample size is small, and the normal distribution is assumed for the sample mean. { 'tē ˌdis·trə,byü·shən }

telegrapher's equation The partial differential equation $(\partial^2 f/\partial x^2) = a^2(\partial^2 f/\partial y^2) + b(\partial f/\partial y) + cf$, where a, b, and c are constants; appears in the study of atomic phenomena. { tə'leg·rə·fərz i,kwā·zhən }

telescopic series The series whose nth term is $1/[(k + n - 1) (k + n)] = [1/(k + n - 1)] - [1/(k + n)]$, where k is not zero or a negative integer, and whose sum is $1/k$. { ˌtel·ə,skäp·ik 'sir,ēz }

ten's complement In decimal arithmetic, the unique numeral that can be added to a given N-digit numeral to form a sum equal to 10^N (that is, a one followed by N zeros). { 'tenz 'käm·plə·mənt }

tensor 1. An object relative to a locally Euclidean space which possesses a specified system of components for every coordinate system and which changes under a transformation of coordinates. **2.** A multilinear function on the Cartesian product of several copies of a vector space and the dual of the vector space to the field of scalars on the vector space. { 'ten·sər }

tensor analysis The abstract study of mathematical objects having components which express properties similar to those of a geometric tensor; this study is fundamental to Riemannian geometry and the structure of Euclidean spaces.   Also known as tensor calculus. { 'ten·sər ə,nal·ə·səs }

tensor calculus See tensor analysis. { 'ten·sər ˌkal·kyə·ləs }

tensor contraction For a tensor having an upper and a lower index, summation over the components in which these indexes have the same value, in order to obtain a new tensor two lower in rank. { 'ten·sər kən'trak·shən }

tensor differentiation An operation on a tensor in which a term involving a Christoffel symbol is subtracted from the ordinary derivative, to obtain another tensor of one higher rank. { 'ten·sər ˌdif·ə,ren·chē'ā·shən }

tensor field A tensor or collection of tensors defined in some open subset of a Riemann space. { 'ten·sər ˌfēld }

tensorial set Any collection of quantities that are associated with a system of spatial coordinates and which undergo a linear transformation when this system rotates; examples are the components of a tensor and the eigenfunctions of a quantum mechanical operator. { ten'sȯr·ē·əl 'set }

tensor product 1. The product of two tensors is the tensor whose components are obtained by multiplying those of the given tensors. **2.** In algebra, a multiplicative operation performed between modules. { 'ten·sər ˌpräd·əkt }

tensor quantity A quantity mathematically represented by a tensor or possessing properties analogous to a tensor. { 'ten·sər ˌkwän·əd·ē }

tensor space A fiber bundle composed of the points of a Riemannian manifold and tensor fields. { 'ten·sər ˌspās }

tera- A prefix representing 10^{12}, which is equivalent to 1,000,000,000,000 or a million million.   Abbreviated T. { 'ter·ə }

term 1. For an expression, any one of several quantities whose sum is the expression. **2.** For a fraction, either the numerator or the denominator. { tərm }

terminal line One of the two rays that form an angle and may be regarded as having been rotated about a fixed point on another line (the initial line) to form the angle. { 'tərm·ən·əl ˌlīn }

terminal vertex A vertex in a rooted tree that has no successor.   Also known as leaf. { 'tər·mən·əl 'vər,teks }

terminating continued fraction A continued fraction that has a finite number of terms. { ¦tər·mə‚nād·iŋ kən¦tin·yüd 'frak·shən }

terminating decimal A decimal that has only a finite number of nonzero digits to the right of the decimal point. Also known as finite decimal. { ‚tər·mə¦nād·iŋ 'des·məl }

term rank For a matrix in which each entry is either 0 or 1, the largest number of 1's that can be chosen from the matrix so that no two of them lie in the same row or in the same column. { 'tərm ‚raŋk }

ternary expansion The numerical representation of a real number relative to the base 3, the digits determined by how the given number can be written in terms of powers of 3. { 'tər·nə·rē ik'span·chən }

ternary notation A system of notation using the base of 3 and the characters 0, 1, and 2. { 'tər·nə·rē nō'tā·shən }

ternary quantic A quantic that contains three variables. { 'tər·nə·rē ¦kwän·tik }

tesselation A covering of a plane without gaps or overlappings by polygons, all of which have the same size and shape. { ‚tes·ə'lā·shən }

tesseral harmonic A spherical harmonic which is 0 on both a set of equally spaced meridians and a set of parallels of latitude of a sphere with center at the origin of spherical coordinates, dividing the sphere into rectangular and triangular regions. { ¦tes·ə·rəl här'män·ik }

test function An infinitely differentiable function of several real variables used in studying solutions of partial differential equations. { 'test ‚fəŋk·shən }

test of hypothesis A rule for rejecting or accepting a hypothesis concerning a population which is based upon a given sample of data. { 'test əv hī'päth·ə·səs }

test of significance A test of a hypothetical population property against a sample property where an acceptance interval is used as the rule for rejection. { 'test əv sig'nif·ə·kəns }

test statistic A numerical value which summarizes the information contained in the sample data and which is a basis for testing a given hypothesis. { 'test stə‚tis·tik }

tetradic An operator that transforms one dyadic into another. { tə'trad·ik }

tetrahedral angle A polyhedral angle with four faces. { ‚te·trə¦hē·drəl 'aŋ·gəl }

tetrahedral group The group of motions of three-dimensional space that transform a regular tetrahedron into itself. { ‚te·trə'hē·drəl ‚grüp }

tetromino One of the five plane figures that can be formed by joining four unit squares along their sides. { te'trä·mə·nō }

theorem A proven mathematical statement. { 'thir·əm }

theoretical frequency A distributional frequency that would result if the data followed a theoretical distribution law rather than the actual observed frequencies. { ‚thē·ə¦red·ə·kəl 'frē·kwən·sē }

theory The collection of theorems and principles associated with some mathematical object or concept. { 'thē·ə·rē }

theory of equations The study of polynomial equations from the viewpoint of solution methods, relations among roots, and connections between coefficients and roots. { 'thē·ə·rē əv i'kwā·zhənz }

theory of games *See* game theory. { 'thē·ə·rē əv 'gāmz }

theta functions Complex functions used in the study of Riemann surfaces and of elliptic functions and elliptic integrals; they are:

$$\theta_1(z) = 2 \sum_{n=0}^{\infty} (-1)^n q^{(n+1/2)^2} \sin(2n+1)z$$

$$\theta_2(z) = 2 \sum_{n=0}^{\infty} q^{(n+1/2)^2} \cos(2n+1)z$$

$$\theta_3(z) = 1 + 2 \sum_{n=1}^{\infty} q^{n^2} \cos 2nz$$

$$\theta_4(z) = 1 + 2 \sum_{n=1}^{\infty} (-1)^n q^{n^2} \cos 2nz$$

where $q = \exp \pi i \tau$, and τ is a constant complex number with positive imaginary part. { 'thād·ə ˌfəŋk·shənz }

third proportional For numbers a and b, a number x such that $a/b = b/x$. { 'thərd prə'pór·shən·əl }

third quadrant 1. The range of angles from 180 to 270°. **2.** In a plane with a system of Cartesian coordinates, the region in which the x and y coordinates are both negative. { 'thərd 'kwä·drənt }

three-eighths rule 1. An approximation formula for definite integrals which states that the integral of a real-valued function f on an interval $[a,b]$ is approximated by $(3/8)h[f(a) + 3f(a + h) + 3f(a + 2h) + f(b)]$, where $h = (b - a)/3$; this is the integral of a third-degree polynomial whose value equals that of f at a, $a + h$, $a + 2h$, and b. **2.** A method of approximating a definite integral over an interval which is equivalent to dividing the interval into equal subintervals and applying the formula in the first definition to each subinterval. { ˌthrē 'āths ˌrül }

three-decision problem A problem in which a choice must be made among three possible courses of action. { 'thrē di'sizh·ən ˌpräb·ləm }

three-index symbols *See* Christoffel symbols. { 'thrē ˌin·deks 'sim·bəlz }

three-space A vector space over the real numbers whose basis has three vectors. { 'thrē ˌspās }

tied rank If two distinct observations have the same value, thus being given the same rank, they are said to be tied; this presents difficulties in the Wilcoxon two-sample test, the sign test, and the Fisher-Irwin test. { 'tīd 'raŋk }

Tietze extension theorem A topological space X is normal if and only if every continuous function of a closed subset to [0,1] has a continuous extension to all of X. { 'tēt· sə ik'sten·chən ˌthir·əm }

time-reversal test A test used with index numbers that is satisfied when the new index is the reciprocal of the original index if the functions of the base period and given period are interchanged; the advantage of index numbers meeting the criteria of the test is that a symmetric comparison of the two periods is obtained and the results are consistent whether one or the other period is used as a base. { 'tīm ri,vər·səl ˌtest }

time series A statistical process analogous to the taking of data at intervals of time. { 'tīm ˌsir·ēz }

time-series analysis The general study of mathematical systems or processes analogous to that of data taken at time intervals. { 'tīm ˌsir·ēz ə,nal·ə·səs }

times sign *See* multiplication sign. { 'tīmz ˌsīn }

Titchmarsh's theorem The proposition that, if $f(x)$ and $g(x)$ are continuous functions on the positive real numbers and are not identically equal to 0, then their convolution is not identically 0. { 'tich,märsh·əz ˌthir·əm }

topological dynamics The study and application of transformations, or groups of such transformations (particularly topological transformation groups), defined on a topological space (usually compact), with particular regard to properties of interest in the qualitative theory of differential equations. { ˌtäp·ə'läj·ə·kəl dī'nam·iks }

topological groups Groups which also have a topology with the property that the group operation and the inverse operation determine continuous functions. { ˌtäp· ə'läj·ə·kəl 'grüps }

topological K theory *See* K theory. { ˌtäp·ə'läj·ə·kəl 'kā ˌthē·ə·rē }

topological linear space *See* topological vector space. { ˌtäp·ə'läj·ə·kəl 'lin·ē·ər ˌspās }

topologically closed set *See* closed set. { ˌtäp·ə'läj·ə·klē 'klōzd 'set }

topologically complete space A topological space that is homeomorphic to a complete metric space. { ˌtäp·ə'läj·ə·klē kəm'plēt 'spās }

topological mapping *See* homeomorphism. { ˌtäp·ə'läj·ə·kəl 'map·iŋ }

topological product The topological space obtained from taking the Cartesian product of topological spaces. { ¦täp·ə¦läj·ə·kəl 'präd·əkt }

topological property A property that holds true for any topological space homeomorphic to one possessing the property. { ¦täp·ə¦läj·ə·kəl 'präp·ərd·ē }

topological simplex A topological space that is homeomorphic to a simplex. { ¦täp·ə¦läj·ə·kəl 'sim,pleks }

topological simplicial complex See triangulable space. { ¦täp·ə¦läj·ə·kəl sim¦plish·əl 'käm,pleks }

topological space A set endowed with a topology. { ¦täp·ə¦läj·ə·kəl 'spās }

topological vector space A vector space which has a topology with the property that vector addition and scalar multiplication are continuous functions. Also known as linear topological space; topological linear space. { ,täp·ə¦läj·ə·kəl 'vek·tər ,spās }

toric surface A surface generated by rotating an arc of a circle about a line that lies in the plane of the circle but does not pass through its center. Also known as toroidal surface. { 'tȯr·ik ,sər·fəs }

toroidal coordinate system A three-dimensional coordinate system whose coordinate surfaces are the toruses and spheres generated by rotating the families of circles defining a two-dimensional bipolar coordinate system about the perpendicular bisector of the line joining the common points of intersection of one of the families, together with the planes passing through the axis of rotation. { tə¦rȯid·əl kō'ȯrd·ən·ət ,sis·təm }

toroidal surface See toric surface. { tə'rȯid·əl ,sər·fəs }

torsion The rate of change of the positive direction of the binormal of a space curve with respect to arc length along the curve; its sign is defined as positive if it is in the same direction as the principal normal, and negative if it is in the opposite direction. Also known as second curvature. { 'tȯr·shən }

torsion coefficients For a finitely generated abelian group G, the orders of the finite cyclic groups such that G is the direct sum of these groups and infinite cyclic groups. { 'tȯr·shən ,kō·ə,fish·əns }

torsion element 1. A torsion element of an Abelian group G is an element of G with finite period. 2. A torsion element of a module M over an entire, principal ring R is an element x in M for which there exists an element a in R such that $a \neq 0$ and $ax = 0$. { 'tȯr·shən ,el·ə·mənt }

torsion-free group A group whose only torsion element is the unit element. { ¦tȯr·shən ,frē ,grüp }

torsion group 1. A group whose elements all have finite period. 2. For a topological space X, one of a sequence of finite groups $G_n(X)$ such that the homology group $H_n(X)$ is the direct sum of $G_n(X)$ and a number of infinite cyclic groups. { 'tȯr·shən ,grüp }

torsion module A module M over an entire principal ring R is said to be a torsion module if for any element x in M there exists an element a in R such that $a \neq 0$ and $ax = 0$. { 'tȯr·shən ,mä·jül }

torsion subgroup The torsion subgroup of an Abelian group G is the subset of all torsion elements of G. { ¦tȯr·shən 'səb,grüp }

torsion submodule The torsion submodule of a module E over an entire principal ring is the submodule consisting of all torsion elements of E. { ¦tȯr·shən 'səb,mä·jül }

total curvature See Gaussian curvature. { 'tōd·əl 'kər·və·chər }

total differential The total differential of a function of several variables, $f(x_1, x_2, \ldots, x_n)$, is the function given by the sum of terms $(\partial f / \partial x_i) dx_i$ as i runs from 1 to n. Also known as differential. { 'tōd·əl ,dif·ə'ren·chəl }

totally bounded set See precompact set. { 'tōd·əl·ē 'baün·dəd 'set }

totally disconnected A topological space has this property if the largest connected subset containing any given point is only the point itself. { 'tōd·əl·ē ,dis·kə'nek·təd }

totally imaginary field An extension field F of the field of rational numbers such that no embedding of F in the complex numbers is contained in the real numbers. { ¦tōd·əl·ē i,maj·ə,ner·ē 'fēld }

249

total order

total order 1. The total order of an analytic function in a domain D is the algebraic sum of its orders at all poles and zeros in D. **2.** *See* linear order. { 'tōd·əl 'ȯr·dər }

total space The topological space E in the bundle (E,p,B). { ¦tōd·əl 'spās }

total variation For a real function defined on an interval, the least upper bound of the function's variation relative to all possible partitions of the interval. { 'tōd·əl ˌver·ē'ā·shən }

totient *See* Euler's phi function. { 'tō·shənt }

totitive An integer that is less than a given integer and relatively prime to it. { 'tōd·ə‚tiv }

tournament A graph in which there is one line between every pair of points and no loops, and in which a unique direction is assigned to every line. { 'tu̇r·nə·mənt }

tractrix A curve in the plane where every tangent to it has the same length. Also known as equitangential curve. { 'trak‚triks }

trailing zero Any zero following the last nonzero integer of a number. { 'trāl·iŋ 'zir·ō }

transcendence base A transcendence base of a field E over a subfield F is a subset S of E which is algebraically independent over F and is not a proper subset of any other subset S' which is algebraically independent over F. { tran'sen·dəns ‚bās }

transcendence degree The transcendence degree of a field E of a subfield F is the number of elements in a transcendence base of E over F. Also known as transcendence dimension. { tran'sen·dəns di‚grē }

transcendence dimension *See* transcendence degree. { tran'sen·dəns di‚men·chən }

transcendental curve The graph of a transcendental function. { ‚tran·sən¦den·təl 'kərv }

transcendental element An element of a field K is transcendental relative to a subfield F if it satisfies no polynomial whose coefficients come from F. { ¦tran‚sen¦dent·əl 'el·ə·mənt }

transcendental field extension A field extension K of F where the elements of K not in F are all transcendental relative to F. { ‚tran‚sen¦dent·əl 'fēld ik‚sten·chən }

transcendental functions Functions which cannot be given by any algebraic expression involving only their variables and constants. { ¦tran‚sen¦dent·əl 'fəŋk·shənz }

transcendental number An irrational number that is the root of no polynomial with rational-number coefficients. { ¦tran‚sen¦dent·əl 'nəm·bər }

transcendental term In an expression, a term that cannot be expressed solely by numbers and algebraic symbols. { ‚tran·sən¦den·təl 'tərm }

transfinite induction A reasoning process by which if a theorem holds true for the first element of a well-ordered set N and is true for an element n whenever it holds for all predecessors of n, then the theorem is true for all members of N. { tranz'fī‚nīt in'dək·shən }

transfinite number Any ordinal or cardinal number equal to or exceeding aleph null. { tranz'fī‚nīt 'nəm·bər }

transformation group 1. A collection of transformations which forms a group with composition as the operation. **2.** A dynamical system or, more generally, a topological group G together with a topological space X where each g in G gives rise to a homeomorphism of X in a continuous manner with respect to the algebraic structure of G. { ‚tranz·fər'mā·shən ‚grüp }

transformation methods A category of numerical methods for finding the eigenvalues of a matrix, in which a series of orthogonal transformations are used to reduce the matrix to some simpler matrix, usually a triple-diagonal one, before an attempt is made to find the eigenvalues. { ‚tranz·fər'mā·shən ‚meth·ədz }

transformation of similitude *See* homothetic transformation. { ‚tranz·fər¦mā·shən əv si'mil·ə‚tüd }

transition probability Conditional probability concerning a discrete Markov chain giving the probabilities of change from one state to another. { tran'zish·ən ‚präb·ə'bil·əd·ē }

transitive group A group of permutations of a finite set such that for any two elements in the set there exists an element of the group which takes one into the other. { 'tran·səd·iv ‚grüp }

trichotomy property

transitive relation A relation $<$ on a set such that if $a < b$ and $b < c$, then $a < c$. { 'tran·səd·iv ri'lā·shən }

translation 1. A function changing the coordinates of a point in a euclidean space into new coordinates relative to axes parallel to the original. **2.** A function on a group to itself given by operating on each element by some one fixed element. **3.** Let E be a finitely generated extension of a field k, F be an extension of k, and both E and F be contained in a common field; the translation of E to F is the extension EF of F, where EF is the compositum of E and F. Also known as lifting. { tran'slā·shən }

translation surface *See* surface of translation. { tran'slā·shən ‚sər·fəs }

transpose The matrix obtained from a given matrix by interchanging its rows and columns. { 'tranz‚pōz }

transversal 1. A line intersecting a given family of lines. Also known as semisecant. **2.** A curve orthogonal to a hypersurface. **3.** If π is a given map of a set X onto a set Y, a transversal for π is a subset T of X with the property that T contains exactly one point of $\pi^{-1}(y)$ for each $y \in Y$. { trans'vər·səl }

transverse axis The portion of a line passing through the foci of a hyperbola that lies between the two branches of the hyperbola. { trans‚vərs 'ak·səs }

trapezium A quadrilateral where no sides are parallel. { trə'pē·zē·əm }

trapezoid A quadrilateral having two parallel sides. { 'trap·ə‚zȯid }

trapezoidal integration A numerical approximation of an integral by means of the trapezoidal rule. { ‚trap·ə‚zȯid·əl ‚int·ə'grā·shən }

trapezoidal rule The rule that the integral from a to b of a real function $f(x)$ is approximated by

$$\frac{b-a}{2n}\left[f(a) + \sum_{j=1}^{n-1} 2f(x_j) + f(b)\right]$$

where $x_0 = a$, $x_j = x_{j-1} + (b-a)/n$ for $j = 1, 2, \ldots, n-1$. { ‚trap·ə‚zȯid·əl 'rül }

traveling salesman problem The problem of performing successively a number of tasks, represented by vertices of a graph, with the least expenditure on transitions from one task to another, represented by edges of the graph with journey costs attached. { ‚trav·əl·iŋ 'sālz·mən ‚präb·ləm }

trefoil A multifoil consisting of three congruent arcs of a circle arranged around an equilateral triangle. { 'trē‚fȯil }

trial One of a series of duplicate experiments. { trīl }

triangle The figure realized by connecting three noncollinear points by line segments. { 'trī‚aŋ·gəl }

triangle inequality For real or complex numbers or vectors in a normed space x and y, the absolute value or norm of $x + y$ is less than or equal to the sum of the absolute values or norms of x and y. { 'trī‚aŋ·gəl ‚in·i'kwäl·ə·dē }

triangle of vectors A triangle, two of whose sides represent vectors to be added, while the third represents the sum of these two vectors. { 'trī‚aŋ·gəl əv 'vek·tərz }

triangulable space A topological space that is homeomorphic to the set of points that belong to the simplexes of a simplicial complex. Also known as polyhedron; topological simplicial complex. { trī‚aŋ·gyə·lə·bəl 'spās }

triangular matrix A matrix where either all entries above or all entries below the principal diagonal are zero. { trī'aŋ·gyə·lər 'mā·triks }

triangular numbers The numbers 1, 3, 6, 10, . . ., which are the numbers of dots in successive triangular arrays, and are given by the expression $(n+1)(n/2)$, where $n = 1, 2, 3, \ldots$. { trī‚aŋ·gyə·lər 'nəm·bərz }

triangular prism A prism whose bases are triangles. { trī‚aŋ·gyə·lər 'priz·əm }

triangular pyramid A pyramid whose base is a triangle. { trī‚aŋ·gyə·lər 'pir·ə‚mid }

triangulation problem The problem of whether each topological n manifold admits a piecewise linear structure. { trī‚aŋ·gyə'lā·shən ‚präb·ləm }

trichotomy property The property of a linear order $<$ on a set S that for any two elements a and b in S exactly one of the statements $a < b$, $a = b$, or $b < a$ is true. Also known as comparison property. { trī'käd·ə·mē ‚präp·ərd·ē }

trident of Newton

trident of Newton The curve in the plane given by the equation $xy = ax^3 + bx^2 + cx + d$, where $a \neq 0$; this cuts the x axis in one or three points and is asymptotic to the y axis if $d \neq 0$. { 'trīd·ənt əv 'nüt·ən }

tridiagonal matrix A square matrix in which all entries other than those on the principal diagonal and the two adjacent diagonals are zero. { ¦trī·dī'ag·ən·əl 'mā·triks }

trigamma function The derivative of the digamma function. { 'trī,gam·ə ,fəŋk·shən }

trigonometric cofunctions Trigonometric functions that are equal when their arguments are complementary angles, such as sine and cosine, tangent and cotangent, and secant and cosecant. { ¦trig·ə·nə¦me·trik ,kō¦fəŋk·shənz }

trigonometric functions The real-valued functions such as $\sin(x)$, $\tan(x)$, and $\cos(x)$ obtained from studying certain ratios of the sides of a right triangle. Also known as circular functions. { ¦trig·ə·nə¦me·trik 'fəŋk·shənz }

trigonometric polynomial A finite series of functions of the form $a_n \cos nx + b_n \sin nx$; occasionally used synonymously with trigonometric series. { ¦trig·ə·nə¦me·trik ,päl·i'nō·mē·əl }

trigonometric series An infinite series of functions with nth term of the form $a_n \cos nx + b_n \sin nx$. { ¦trig·ə·nə¦me·trik 'sir·ēz }

trigonometric substitutions The substitutions $x = a \sin u$, $x = a \tan u$, and $x = a \sec u$, which are used to rationalize expressions of the form $\sqrt{a^2 - x^2}$, $\sqrt{x^2 + a^2}$, and $\sqrt{x^2 - a^2}$, respectively, when they appear in integrals. { ¦trig·ə·nə¦me·trik ,səb·stə'tü·shənz }

trigonometry The study of triangles and the trigonometric functions. { ,trig·ə'näm·ə·trē }

trihedral Any figure obtained from three noncoplanar lines intersecting in a common point. { trī'hē·drəl }

trihedral angle A polyhedral angle with three faces. { trī¦hē·drəl 'aŋ·gəl }

trillion 1. The number 10^{12}. **2.** In British and German usage, the number 10^{18}. { 'tril·yən }

trinomial A polynomial comprising three terms. { trī'nō·mē·əl }

trinomial distribution A multinomial distribution in which there are three distinct outcomes. { trī'nō·mē·əl ,di·strə·byü·shən }

trinomial surd A sum of three roots of rational numbers, at least two of which are irrational numbers that cannot be combined without evaluating them. { trī'nō·mē·əl 'sərd }

triple-diagonal matrix *See* continuant matrix. { ¦trip·əl di¦ag·ən·əl 'mā·triks }

triple scalar product *See* scalar triple product. { 'trip·əl 'skā·lər ,präd·əkt }

triple vector product The triple vector product of vectors **a**, **b**, and **c** is the cross product of **a** with the cross product of **b** and **c**; written **a** × (**b** × **c**). { ¦trip·əl ¦vek·tər ,präd·əkt }

trirectangular spherical triangle A spherical triangle with three right angles. { ,trī·rek¦taŋ·gyə·lər ¦sfir·ə·kəl 'trī,aŋ·gəl }

trisection The problem of dividing an angle into three equal parts, which is impossible to do with straight edge and compass alone. { trī'sek·shən }

trisectrix The planar curve given by $x^3 + xy^2 + ay^2 - 3ax^2 = 0$ which is symmetric about the x axis and asymptotic to the line $x = -a$; this is useful in studying the trisection of an angle problem. Also known as trisectrix of Maclaurin. { trī'sek·triks }

trisectrix of Catalan *See* Tschirnhausen's cubic. { trī'sek·triks əv 'kad·ə,lan }

trisectrix of Maclaurin *See* trisectrix. { trī'sek·triks əv mə'klòr·ən }

trit A digit in a balanced ternary system, that is, a balanced digit system with base 3. { trit }

trivial graph A graph with one vertex and no edges. { 'triv·ē·əl ¦graf }

trivial solution A solution of a set of homogeneous linear equations in which all the variables have the value zero. { ¦triv·ē·əl sə'lü·shən }

trivial topology For a set S, a topology whose only members are the set itself and the empty set. Also known as indiscreet topology. { ¦triv·ē·əl tə'päl·ə·jē }

tromino One of the two plane figures that can be formed by joining three unit squares along their sides. { 'träm·ə·nō }

true complement *See* radix complement. { 'trü 'käm·plə·mənt }

truncated cone The portion of a cone between two nonparallel planes whose line of intersection lies outside the cone. { ¦trəŋ·kād·əd 'kōn }

truncated distribution A distribution fashioned from another distribution by deleting that part of the distribution to the right or left of a random variable value. { ¦trəŋ‚kād·əd ‚dis·trə'byü·shən }

truncated icosahedron An Archimedean solid with 32 faces (20 regular hexagons and 12 regular pentagons) and 60 vertices, a shape used in the construction of soccer balls. { ¦trəŋ‚kād·əd ‚ī¦käs·ə'hē·drən }

truncated prism The part of a prism that lies between two nonparallel planes that cut the prism and intersect outside the prism. { ¦trəŋ·kād·əd 'priz·əm }

truncated pyramid The part of a pyramid between the base and a plane that is not parallel to the base. { ¦trəŋ·kād·əd 'pir·ə‚mid }

truncation 1. Approximating the sum of an infinite series by the sum of a finite number of its terms. **2.** *See* rounding. { trəŋ'kā·shən }

truth set A set containing all the elements that make a given statement of relationships true when they are substituted in this statement. { 'trüth ‚set }

truth table A table listing statements concerning an event and their respective truth values. { 'trüth ‚tā·bəl }

truth value The result of a logical proposition; either "true" or "false" in classical logic. { 'trüth ‚val·yü }

Tschirnhausen's cubic A plane curve consisting of the envelope of the line through a variable point P on a parabola which is perpendicular to the line from the focus of the parabola to P. Also known as l'Hôpital's cubic; trisectrix of Catalan. { 'chərn‚haůz·ənz 'kyü·bik }

T score A score utilized in setting up norms for standardized tests; obtained by linearly transforming normalized standard scores. { 'tē ‚skór }

T_0 space A topological space where, for each pair of points, at least one has a neighborhood not containing the other. Also known as Kolmogorov space. { ‚tē 'zir·ō ¦spās }

T_1 space A topological space where, for each pair of distinct points, each one has a neighborhood not containing the other. Also known as Fréchet space. { 'tē ‚wən ¦spās }

T_2 space *See* Hausdorff space. { 'tē ‚tü ¦spās }

T_3 space A regular topological space that is also a T_1 space. { 'tē ‚thrē ¦spās }

$T_{3-1/2}$ *See* Tychonoff space. { ¦tē ‚thrē an ə 'haf ¦spās }

T_4 space A normal space that is also a T_1 space. { 'tē ‚fór ¦spās }

t-test A statistical test involving means of normal populations with unknown standard deviations; small samples are used, based on a variable t equal to the difference between the mean of the sample and the mean of the population divided by a result obtained by dividing the standard deviation of the sample by the square root of the number of individuals in the sample. { 'tē‚test }

Tukey lemma The proposition that any nonempty family of finite character has a maximal member. { 'tü·kē‚lem·ə }

Turing computable function A function that can be computed on a Turing machine. { 'tůr·iŋ kəm'pad·ə·bəl 'fəŋk·shən }

Turing's thesis *See* Church's thesis. { 'tůr·iŋz ‚thē·səs }

turning value A relative maximum or relative minimum of a function. { 'tərn·iŋ ‚val·yü }

twelve-color problem The problem of showing that 12 colors are sufficient to color a map on which each country has at most one colony, and that there is such a map which requires 12 colors. { 'twelv ¦kəl·ər ‚präb·ləm }

twin primes A pair of prime numbers that differ by 2. { ¦twin 'prīmz }

twisted curve A curve that does not lie wholly in any one plane. { ¦twis·təd 'kərv }

two-cycle The repetition of numbers generated by a mapping on every second iteration of the mapping. { ˈtü ˈsī·kəl }

two-decision problem The problem of deciding, using statistical information, between two actions or decisions. { 'tü ˌdiˌsizh·ən 'präb·ləm }

two-part experiment An experiment in which two operations or actions are performed; for example, throwing two dice, drawing two marbles from a box, throwing a die and then drawing a marble from a box. { 'tü 'pärt ik'sper·ə·mənt }

two-person game A game consisting of exactly two players with competing interests. { 'tü ˌpər·sən 'gām }

two's complement A number derived from a given n-bit number by requiring the two numbers to sum to a value of 2^n. { 'tüz 'käm·plə·mənt }

two-sided ideal A two-sided ideal I is a sub-ring of a ring R where the products xy and yx are always in I for every x in R and y in I. { 'tü ˌsīd·əd ī'dēl }

two-sided test A test which rejects the null hypothesis when the test statistic T is either less than or equal to c or greater than or equal to d, where c and d are critical values. { 'tü ˌsīd·əd 'test }

two-sphere The surface of a ball; the two-dimensional sphere in Euclidean three-dimensional space obtained from all points whose distance from the origin is one. { 'tü ˌsfir }

two-stage design The design of an experiment which employs a pilot study in order to decide how to design the main experiment. { 'tü ˌstāj di'zīn }

two-stage experiment An experiment in two parts, the outcome of the first part deciding the procedure for the second. { 'tü ˌstāj ik'sper·ə·mənt }

two-stage sampling Sampling from a population whose members are themselves sets of objects and then sampling from the sets selected in the first sampling; for example, to first draw a sample of states and then to draw a sample of representatives to Congress from each state selected. { 'tü ˌstāj 'sam·pliŋ }

two-tailed test A statistical test in which the critical region consists of those values of a test statistic less than a given value as well as those values greater than another given value. Also known as two-tail test. { 'tü ˌtāld ˌtest }

two-tail test See two-tailed test. { 'tü ˌtāl ˌtest }

two-valued logic A system of logic where each statement has two possible values or states, truth or falsehood. { 'tü ˌval·yüd 'läj·ik }

two-valued variable A variable which assumes values in a set containing exactly two elements, often symbolized as 0 and 1. { 'tü ˌval·yüd 'ver·ē·ə·bəl }

two-way series An expression of the form $\cdots + x_{-2} + x_{-1} + x_0 + x_1 + x_2 + \cdots$, where the x_i are real or complex numbers. { ˌtü ˌwā 'sir·ēz }

Tychonoff space A completely regular space that is also a T_1 space. Also known as $T_{3-1/2}$ space. { tī'kä,nȯf ˌspās }

Tychonoff theorem A product of topological spaces is compact if and only if each individual space is compact. { tī'kä,nȯf ˌthir·əm }

type I error One of two types of errors in testing hypotheses: incorrectly rejecting the hypothesis tested when it is true. Also known as error of the first kind. { ˌtīp 'wən ˌer·ər }

type II error One of two types of error in testing hypotheses: incorrectly accepting the hypothesis tested when an alternate hypothesis is true. Also known as error of the second kind. { ˌtīp 'tü ˌer·ər }

U

ultrafilter A filter base which has no properly subordinated filter base. { ¦əl·trə'fil·tər }

ultraspherical polynomials *See* Gegenbauer polynomials. { ¦əl·trə'sfer·ə·kəl ‚päl·i'nō·mē·əlz }

umbilic *See* umbilical point. { úm'bil·ik }

umbilical point A point on a surface at which the normal curvature is the same in all directions. Also known as navel point; umbilic. { əm'bil·ə·kəl ‚póint }

unary operation An operation in which only a single operand is required to produce a unique result; some examples are negation, complementation, square root, transpose, inverse, and conjugate. { 'yü·nə·rē ‚äp·ə‚rā·shən }

unavoidable set of configurations A set of graphs such that any planar graph has at least one member of the set as a subgraph. { ¦ən·ə'vóid·ə·bəl 'set əv kən‚fig·yə'rā·shənz }

unbiased estimate An estimate for a parameter θ whose expected value is θ. { ¦ən'bī·əst 'es·tə·mət }

unbounded manifold A manifold with no boundary. { ¦ən'baún·dəd 'man·ə‚fōld }

unbounded set of real numbers A set with the property that if, *R* is any positive real number, there is a number in the set which is smaller than −*R* or a number larger than *R*. { ¦ən'baún·dəd 'set əv 'rēl 'nəm·bərz }

unconditional convergence A convergent series converges unconditionally if every series obtained by rearranging its terms also converges; equivalent to absolute convergence. { ¦ən·kən'dish·ən·əl kən'vər·jəns }

unconditional inequality An inequality which holds true for all values of the variables involved, or which contains no variables; for example, $y + 2 > y$, or $4 > 3$. Also known as absolute inequality. { ¦ən·kən'dish·ən·əl ‚in·i'kwäl·əd·ē }

unconstrained optimization problem A nonlinear programming problem in which there are no constraint functions. { ¦ən·kən'strānd ‚äp·tə·mə'zā·shən ‚präb·ləm }

uncorrelated random variables Two random variables whose correlation coefficient is zero. { ¦ən'kär·ə‚lād·əd 'ran·dəm 'ver·ē·ə·bəlz }

uncountable set An infinite set which cannot be put in one-to-one correspondence with the set of integers; for example, the set of real numbers. { ¦ən'kaúnt·ə·bəl 'set }

undecagon An 11-sided polygon. { ən'dek·ə‚gän }

undecahedron A polyhedron with 11 faces. { ən‚dek·ə'hē·drən }

undecomino One of the 17,073 plane figures that can be formed by joining 11 unit squares along their sides. { ən·də'käm·ə·nō }

underlying graph A directed graph, that results from replacing each directed arc with an undirected edge. { ¦ən·dər‚lī·iŋ 'graf }

undetermined multipliers *See* Lagrangian multipliers. { ¦ən·di'tər·mənd 'məl·tə‚plī·ərz }

undirected graph A graph whose edges are not assigned directions. { ¦ən·də‚rek·təd 'graf }

unduloid A surface formed by the rotation of a wavy line around a straight line parallel to its axis of symmetry. { 'ən·dyə‚lóid }

ungula A solid bounded by a portion of a circular cylindrical surface and portions of two planes, one of which is perpendicular to the generators of the cylindrical surface. { 'əŋ·gyə·lə }

uniform bound A number M such that $|f_n(x)| < M$ for every x and for every function in a given sequence of functions $\{f_n(x)\}$. { 'yü·nə,fôrm 'baúnd }

uniform boundedness principle A family of pointwise bounded, real-valued continuous functions on a complete metric space X is uniformly bounded on some open subset of X. { 'yü·nə,fôrm 'baún·dəd·nəs ,prin·sə·pəl }

uniform continuity A property of a function f on a set, namely: given any $\epsilon > 0$ there is a $\delta > 0$ such that $|f(x_1) - f(x_2)| < \epsilon$ provided $|x_1 - x_2| < \delta$ for any pair x_1, x_2 in the set. { 'yü·nə,fôrm känt·ən'ü·əd·ē }

uniform convergence A sequence of functions $\{f_n(x)\}$ converges uniformly on E to $f(x)$ if given $\epsilon > 0$ there is an N such that $|f_n(x) - f(x)| < \epsilon$ for all x in E provided $n > N$. { 'yü·nə,fôrm kən'vər·jəns }

uniform distribution The distribution of a random variable in which each value has the same probability of occurrence. Also known as rectangular distribution. { 'yü·nə,fôrm ,di·strə'byü·shən }

uniformity A family of subsets of the direct product of a topological space with itself that is used to derive a uniform topology for the space. Also known as uniform structure. { ,yü·nə'fôr·məd·ē }

uniformity trial The repetition of an experiment under exactly the same controlled conditions as in the original trial. { ,yü·nə'fôr·məd·ē ,trīl }

uniformly convex space A normed vector space such that for any number $\epsilon > 0$ there is a number $\delta > 0$ such that, for any two vectors x and y, if $\| x \| \leq 1 + \delta$, $\| y \| \leq 1 + \delta$, and $\| x + y \| > 2$, then $\| x - y \| < \epsilon$. Also known as uniformly rotund space. { 'yü·nə,fôrm·lē 'kän,veks 'spās }

uniformly equicontinuous family of functions See equicontinuous family of functions. { ,yü·nə'fôrm·lē 'ek·wē·kə'tin·yə·wəs 'fam·lē əv 'faŋk·shənz }

uniformly most powerful test A test which is simultaneously most powerful for all alternatives of interest in an experiment. { ,yü·nə'fôrm·lē ,mōst 'paú·ər·fəl ,test }

uniformly rotund space See uniformly convex space. { 'yü·nə,fôrm·lē rō'tənd 'spās }

uniformly nonsquare Banach space A Banach space for which there is a positive number ϵ such that there are no nonzero elements, x and y, that have the same norm and satisfy the inequalities $\| x + y \| > (2 - \epsilon) \| x \|$ and $\| x - y \| > (2 - \epsilon) \| x \|$. { 'yü·nə,fôrm·lē ,nän,skwer 'bä,näk 'spās }

uniformly summable series For a given summability method and for a given interval, a series for which the sequence that defines the sum converges uniformly on the interval. { 'yü·nə,fôrm·lē 'səm·ə·bəl 'sir,ēz }

uniform scale A scale in which equal distances correspond to equal numerical values. Also known as linear scale. { 'yü·nə,fôrm 'skāl }

uniform space A topological space X whose topology is derived from a family of subsets of $X \times X$, called a uniformity; intuitively, this gives a notion of "nearness" which is uniform throughout the space. { 'yü·nə,fôrm 'spās }

uniform structure See uniformity. { 'yü·nə,fôrm 'strək·chər }

uniform topology The topology of a uniform space. { 'yü·nə,fôrm tə'päl·ə·jē }

unilateral shift A bounded linear operator on the Hilbert space of infinite sequences of complex numbers whose squared moduli form convergent series; it takes the sequence $(x_1, x_2, \ldots)$ to the sequence $(0, x_1, x_2, \ldots)$. { ,yü·nə,lad·ə·rə·l 'shift }

unilateral surface A one-sided surface; equivalently, any nonorientable two-dimensional manifold such as the Möbius strip and the Klein bottle. { 'yü·nə'lad·ə·rəl 'sər·fəs }

unimodal Referring to a distribution with only one mode. { ,yü·nə'mōd·əl }

unimodal sequence A finite sequence of n real numbers, $a_1, a_2, \ldots, a_n$, for which there is a positive integer, j, greater than 1 and less than n, such that a_i is greater than a_{i-1} for i greater than 1 and less than j, a_j is greater than or equal to a_{j-1}, and a_i is less than a_{i-1} for i greater than j and equal to or less than n. { ,yü·nə,mōd·əl 'sē·kwəns }

unimodular matrix A unimodulus matrix with integer entries. { 'yü·nə'mäj·ə·lər 'mā·triks }

unimodulus matrix A square matrix whose determinant is 1. { ¦yü·nə'mäj·ə·ləs 'mā·triks }

union 1. A union of a given family of sets is a set consisting of those elements that are members of at least one set in the family. Also known as join. **2.** For two fuzzy sets A and B, the fuzzy set whose membership function has a value at any element x that is the maximum of the values of the membership functions of A and B at x. **3.** The union of two Boolean matrices A and B, with the same number of rows and columns, is the Boolean matrix whose element c_{ij} in row i and column j is the union of corresponding elements a_{ij} in A and b_{ij} in B. **4.** The union of two graphs is the graph whose set of vertices is the union of the sets of vertices of the two graphs, and whose set of edges is the union of the sets of edges of the two graphs. { 'yün·yən }

union rule of probability The probability that the union of two events E_i and E_j equals the sum total of the probability of the sample points in either E_i or E_j minus the probability of being in both E_i and E_j. { ¦yün·yən ¦rül əv ¸präb·ə'bil·əd·ē }

unit complex number A complex number whose absolute value is 1. { ¦yü·nət ¦käm ¸pleks 'nəm·bər }

unique-factorization domain An integral domain in which every element that is neither a unit nor a prime has an expression as the product of a finite number of primes, and this expression is unique except for unit factors and the order of factors. Also known as factorial ring; unique-factorization ring. { yü¦nēk ¸fak·tə·rə'zā· shə dō¸mān }

unique-factorization ring See unique-factorization domain. { yü¦nēk ¸fak·tə·rə'zā· shən ¸riŋ }

unique factorization theorem A positive integer may be expressed in precisely one way as a product of prime numbers. { yü'nēk ¸fak·tə·rə'zā·shən ¸thir·əm }

unit An element of a ring with identity that has both a left inverse and a right inverse. { 'yü·nət }

unital left module A left module over a ring with a unit element, 1, such that, for any element x of the module, $1 \cdot x = x$.

unital module A module over a ring with a unit element, 1, such that $1 \cdot x = x$ for any element x of the module. { ¦yü·nə·təl 'mäj·əl }

unital left module A left module over a ring with a unit element,1, such that, for any element x of the module, $1 \cdot x = x$. { ¦yü·nət·əl ¦left 'mäj·əl }

unitary group The group of unitary transformations on a k-dimensional complex vector space. Usually denoted U(k). { 'yü·nə¸ter·ē 'grüp }

unitary matrix A matrix whose inverse is equal to the complex conjugate of its transpose. { 'yü·nə¸ter·ē 'mā·triks }

unitary space A finite-dimensional inner-product space over the field of complex numbers. { 'yü·nə¸ter·ē 'spās }

unitary transformation A linear transformation on a vector space which preserves inner products and norms; alternatively, a linear operator whose adjoint is equal to its inverse. { 'yü·nə¸ter·ē ¸tranz·fər'mā·shən }

unit ball The set of all points in Euclidean n-space whose distance from the origin is at most 1. { 'yü·nət 'bȯl }

unit binormal A unit vector in the same direction as the binormal to a point on a surface or space curve. { 'yü·nət bī'nȯr·məl }

unit circle The locus of points in the plane which are precisely one unit from the origin. { 'yü·nət 'sər·kəl }

unit conversion factor See conversion factor. { ¦yü·nət kən'vər·zhən ¸fak·tər }

unit element An element in a ring which acts as a multiplicative identity. { 'yü·nət 'el·ə·mənt }

unit fraction A common fraction whose numerator is unity. { 'yü·nət ¸frak·shən }

unit impulse See delta function. { 'yü·nət 'im¸pəls }

unit normal A unit vector in the direction of the principal normal to a surface or space curve. { 'yü·nət 'nȯr·məl }

unit operator The identity operator. { 'yü·nət 'äp·ə¸rād·ər }

unit sphere The set of points in three-space (more generally n-space) which are precisely one unit distance from the origin. { 'yü·nət 'sfir }

unit tangent A unit vector in the tangent plane at a point of a surface. { 'yü·nət 'tan·jənt }

unit vector A vector whose length is one unit. { 'yü·nət 'vek·tər }

univariate distribution A frequency distribution of only one variate. { ˌyün·ə¦ver·ē·ət ˌdis·trə'byü·shən }

universal algebra The study of algebraic systems such as groups, rings, modules, and fields and the examination of what families of theorems are analogous in each system. { ¦yü·nə¦vər·səl 'al·jə·brə }

universal element An element of a Boolean algebra that includes every element of the algebra. { ¦yü·nə¦vər·səl 'el·ə·mənt }

universally attracting object An object O in a category C such that there exists a unique morphism of each object of C into O. { ¦yü·nə¦vər·sə·lē ə'trak·tiŋ ˌäb¸jekt }

universally repelling object An object O of a category C such that there exists a unique morphism of O into each object of C. { ¦yü·nə¦vər·sə·lē ri'pel·iŋ ˌäb¸jekt }

universal object An object which is universally attracting or universally repelling. { ¦yü·nə'vər·səl ˌäb¸jekt }

universal quantifier A logical relation, often symbolized λ, that may be expressed by the phrase "for all" or "for every"; if P is a predicate, the statement $(\lambda x)P(x)$ is true if $P(x)$ is true for all values of x in the domain of P, and is false otherwise. { ¦yü·nə¦vər·səl 'kwän·tə¸fī·ər }

universal set A set that contains all the elements of concern in the study of a particular problem. { ¦yü·nə¦vər·səl 'set }

unrelated frequencies The long run frequency of any result in one part of an experiment is approximately equal to the long run conditional frequency of that result, given that any specified result has occurred in the other part of the experiment. { ¦ən·ri'lād·əd 'frē·kwən¸sēz }

unsigned integer A whole number that is equal to or greater than zero and does not carry a positive or negative sign. { ən'sīnd 'int·ə·jər }

unsigned real number A number that does not carry a sign indicating whether it is positive or negative, and that is therefore assumed to be positive. { ən'sīnd 'rēl 'nəm·bər }

unstable graph A graph from which it is not possible to delete an edge to produce a subgraph whose group of automorphisms is a subgroup of the group of automorphisms of the original graph. { ¦ən¸stā·bəl 'graf }

unstructured grid In the discretization of partial differential equations, a collection of triangular elements or a random distribution of points at which the equations are expressed in discrete form. { ¦ən·strək·chərd 'grid }

up-and-down method A technique which uses a unit sequential method of testing; the level of variable at each experiment is raised or lowered depending on the outcome of the previous test. { ¦əp ən 'daůn ˌmeth·əd }

upper bound 1. If S is a subset of an ordered set A, an upper bound b for S in A is an element b of A such that $x \leq b$ for all x belonging to A. **2.** An upper bound on a function f with values in a partially ordered set A is an element of A which is larger than every element in the range of f. { ¦əp·ər 'baůnd }

upper integral The upper Riemann integral for a real-valued function $f(x)$ on an interval is computed to be the infimum of all finite sums over all partitions of the interval, the sums having terms given by $(x_i - x_{i-1})y_i$, where the x_i are from a partition, and y_i is the largest value of $f(x)$ over the interval from x_{i-1} to x_i. { 'əp·ər 'int·ə·grəl }

upper limit *See* limit superior. { ¦əp·ər 'lim·ət }

upper semicontinuous decomposition A partition of a topological space with the property that for every member D of the partition and for every open set U containing D there is an open set V containing D which is contained in U and is the union of members of the partition. { 'əp·ər ¦sem·i·kən'tin·yə·wəs dē¸käm·pə'zish·ən }

upper semicontinuous function A real-valued function $f(x)$ is upper semicontinuous at a point x_0 if, for any small positive ϵ, $f(x)$ always is less than $f(x_0) + \epsilon$ for all x in some neighborhood of x_0. { 'əp·ər ¦sem·i·kən'tin·yə·wəs 'faŋk·shən }

upward bias The overestimation or overstatement by a statistical measure of the event it is attempting to describe. { ¦əp·wərd 'bī·əs }

Urysohn lemma If A and B are disjoint, closed sets in a normal space X, there is a real-valued function f such that $0 \le f(x) \le 1$ for all $x \in X$, and $f(A) = 0$ and $f(B) = 1$. { 'ùr·ē,zōn ,lem·ə }

Urysohn theorem The theorem that a regular T_1 space whose topology has a countable base is metrizable. { 'ùr·ē,zōn ,thir·əm }

U-shaped distribution A frequency distribution whose shape approximates that of the letter U. { ¦yü ,shāpt ,dis·trə'byü·shən }

V

valence The number of lines incident on a specified point of a graph. { 'vā·ləns }

validity Correctness; especially the degree of closeness by which iterated results approach the correct result. { və'lid·əd·ē }

valuation A scalar function of a field which has properties similar to those of absolute value. { ˌval·yə'wā·shən }

value 1. The value of a function f at an element x is the element y which f associates with x; that is, $y = f(x)$. **2.** The expected payoff of a matrix game when each player follows an optimal strategy. { 'val·yü }

value group For a discrete valuation v on a field K, this is the group formed by the elements $v(x)$ corresponding to nonzero elements x in K. { 'val·yü ˌgrüp }

value index An index member which is the ratio of the value of all items in a given period to the value of all items in the base period. { 'val·yü ˌin,deks }

Vandermonde determinant The determinant of the $n \times n$ matrix whose ith row appears as 1, x_i, x_i^2, ..., x_i^{n-1} where the x_i^k appear as variables in a given polynomial equation; this provides information about the roots. { 'van·dərˌmōnd di,tər·mə·nənt }

Vandermonde matrix A matrix in which each entry in the first row is 1, and each entry in the ith row is the corresponding entry in the second row to the $(i - 1)$ power. { 'van·dərˌmōnd ˌmā·triks }

Vandermonde's theorem A theorem stating that a binomial $(x + y)^a$, where a is an exponent involving the variables x and y, can be stated in terms of a sum of expressions $x^c y^d$, where the exponents c and d involve the variables x and y also. { 'van·dərˌmōndz ˌthir·əm }

van der Waerden number For two positive integers, k and r, the smallest positive integer, $n(k,r)$, that satisfies van der Waerden's theorem. { ¦van·dər 'werd·ən ˌnəm·bər }

van der Waerden's theorem The theorem that for any positive integers k and r there is a positive integer n such that, if the first n integers are divided into k classes, then there exists an arithmetic progression of r terms that all belong to the same class. { ¦van ·dər 'werd·ənz ˌthir·əm }

variable A symbol which is used to represent some undetermined element from a given set, usually the domain of a function. { 'ver·ē·ə·bəl }

variance The square of the standard deviation. { 'ver·ē·əns }

variance ratio test A technique for comparing the spreads or variabilities of two sets of figures to determine whether the two sets of figures were drawn from the same population. Also known as F test. { 'ver·ē·əns 'rā·shō ˌtest }

variate See random variable. The numerical value of a measurement to be used for statistical handling. { 'ver·ē·ət }

variate difference method A technique for estimating the correlation between the random parts of two given time series. { 'ver·ē,āt 'dif·rəns ˌmeth·əd }

variational principle A technique for solving boundary value problems that is applicable when the given problem can be rephrased as a minimization problem. { ˌver·ē'ā·shən·əl ˌprin·sə·pəl }

vector 1. An element of a vector space. **2.** A matrix consisting of a single row or a single column of entries. { 'vek·tər }

vector analysis The formal study of vectors. { 'vek·tər ə‚nal·ə·səs }

vector bundle A locally trivial bundle whose fibers are isomorphic vector spaces. { 'vek·tər ‚bən·dəl }

vector equation An equation involving vectors. { 'vek·tər i‚kwā·zhən }

vector field 1. The field of vectors arising from considering a system of differential equations on a differentiable manifold. **2.** A function whose range is in a vector space. { 'vek·tər ‚fēld }

vector multiplication The operation which associates to each ordered pair of vectors the cross product of these two vectors. { ‚vek·tər ‚məl·tə·plə'kā·shən }

vector product See cross product. { 'vek·tər ‚präd·əkt }

vector random variable A vector whose entries are random variables that are defined on the same sample space of an experiment. { ‚vek·tər ‚ran·dəm 'ver·ē·ə·bəl }

vector space A system of mathematical objects which have an additive operation producing a group structure and which can be multiplied by elements from a field in a manner similar to contraction or magnification of directed line segments in euclidean space. Also known as linear space. { 'vek·tər ‚spās }

vector sum For a set of located vectors in Euclidean space, $\mathbf{v}_1, \mathbf{v}_2, \ldots, \mathbf{v}_n$, this is the vector whose initial point is the initial point of $\mathbf{v}_1$ and whose terminal point is the terminal point of $\mathbf{v}_n$, when the vectors are laid end to end so that the terminal point of one vector $\mathbf{v}_i$ is the initial point of the next vector $\mathbf{v}_{i+1}$. Also known as resultant. { 'vek·tər ‚səm }

Venn diagram A pictorial representation of set theoretic operations such as union, intersection, and complementation of sets. { 'ven ‚dī·ə‚gram }

vers See versed sine.

versed cosine See coversed sine. { 'vərst 'kō‚sīn }

versed sine The versed sine of A is $1 - \text{cosine } A$. Denoted vers. Also known as versine. { 'vərst 'sīn }

versiera See witch of Agnesi. { ‚vər·sē'er·ə }

versine See versed sine. { 'vər‚sīn }

vertex 1. For a polygon or polyhedron, any of those finitely many points which together with line segments or plane pieces determine the figure or solid. **2.** The common point at which the two sides of an angle intersect. **3.** The fixed point through which pass all the elements of a cone or conical surface. **4.** An intersection of a conic with one of its axes of symmetry. **5.** A member of the set of points that are connected by the edges. Also known as node. **6.** For a simplex, one of the finite set of points on which a simplex is based. { 'vər‚teks }

vertex angle In a triangle, the angle opposite the base. { 'vər‚teks ‚aŋ·gəl }

vertex cover A set of vertices in a graph such that every edge in the graph is incident to at least one vertex in this set. { 'vər‚teks ¦kəv·ər }

vertex-covering number For a graph, the smallest possible number of vertices in a vertex cover. { 'vər‚teks ¦kəv·ər·iŋ ‚nəm·bər }

vertex-disjoint paths In a graph, two paths with common end points that have no other points in common. { ¦vər‚teks 'dis‚jóint ‚paths }

vertex domination number For a graph, the smallest possible number of vertices in a dominating vertex set. Also known as external stability number. { ¦vər‚teks ‚däm·ə'nā·shən ‚nəm·bər }

vertex-induced graph A subgraph whose edges consist of all the edges in the original graph that join pairs of vertices in the subgraph. Also known as induced subgraph. { ¦vər‚teks in‚düst 'graf }

vertical angles The two angles produced by a pair of intersecting lines and lying on opposite sides of the point of intersection. { 'vərd·ə·kəl 'aŋ·gəlz }

Vitali set A set of real numbers such that the difference of any two members of the set is an irrational number and any real number is the sum of a rational number and a member of the set. { vē'täl·ē ‚set }

vol *See* volume.

Volterra equations Given functions $f(x)$ and $K(x,y)$, these are two types of equations with unknown function y:

$$f(x) = \int_a^x K(x,t)y(t)dt$$

$$y(x) = f(x) + \lambda \int_a^x K(x,t)y(t)dt$$

{ vol'ter·ə i͵kwā·shənz }

volume A measure of the size of a body or definite region in three-dimensional space; it is equal to the least upper bound of the sum of the volumes of nonoverlapping cubes that can be fitted inside the body or region, where the volume of a cube is the cube of the length of one of its sides. Abbreviated vol. { 'väl·yəm }

volume by slicing A method of computing the volume of a solid by integrating over the volumes of infinitesimal slices of the solid bounded by parallel planes. { ¦väl·yəm bī 'slīs·iŋ }

volume integral An integral of a function of several variables with respect to volume measure taken over a three-dimensional subset of the domain of the function. { 'väl·yəm 'int·ə·grəl }

von Kármán *See* Kármán. { fȯn 'kär·män }

von Neumann algebra A subalgebra A of the algebra $B(H)$ of bounded linear operators on a complex Hilbert space, such that the adjoint operator of any operator in A is also in A, and A is closed in the strong operator topology in $B(H)$. Also known as ring of operators; W* algebra. { fȯn ¦nȯi·män 'al·jə·brə }

vulgar fraction *See* common fraction. { ¦vəl·gər 'frak·shən }

W

Wald-Wolfowitz run test A procedure used in nonparametric statistics to determine whether the means of two independently drawn samples were taken from the same population. { ¦wȯld ¦wȯlf·ə₁wits 'rən ₁test }

W* algebra *See* von Neumann algebra. { ¦dəb·əl₁yü stär 'al·jə·brə }

walk In graph theory, a set of vertices $(v_0, v_1, \ldots, v_n)$ in a graph, such that v_i and v_{i+1} are joined by a common edge for $i = 0, 1, \ldots, n - 1$. Also known as path. { wȯk }

Wallis formulas Formulas that determine the values of the definite integrals from 0 to $\pi/2$ of the functions $\sin^n (x)$, $\cos^n (x)$, and $\cos^m (x) \sin^n (x)$ for positive integers m and n. Also known as Wallis theorem. { 'wäl·əs ₁fȯr·myə·ləz }

Wallis product An infinite product representation of $\pi/2$, namely,

$$\frac{\pi}{2} = \frac{2}{1}\frac{2}{3}\frac{4}{3}\frac{4}{5} \cdots \frac{2n}{2n - 1}\frac{2n}{2n + 1}$$

{ 'wäl·əs ₁präd·əkt }

Wallis theorem *See* Wallis formulas. { 'wäl·əs ₁thir·əm }

washer method A method of computing the volume of a solid of revolution that is hollow about its axis, by integrating over the volumes of infinitesimal washer-shaped slices bounded by planes perpendicular to the axis of revolution. { 'wash·ər ₁meth·əd }

Watson-Sommerfeld transformation A procedure for transforming a series whose lth term is the product of the lth Legendre polynomial and a coefficient, a_l, having certain properties, into the sum of a contour integral of $a(l)$ and terms involving poles of $a(l)$, where $a(l)$ is a meromorphic function such that $a(l)$ equals a_l at integral values of l; used in studying rainbows, propagation of radio waves around the earth, scattering from various potentials, and scattering of elementary particles. Also known as Sommerfeld-Watson transformation. { 'wät·sən 'zȯm·ər₁felt i₁kwä·zhən }

Watt's curve The curve traced out by the midpoint of a line segment whose end points move along two circles of equal radius. { 'wäts ₁kərv }

wavelet One of a collection of mathematical functions that serve as the elementary building blocks of a mathematical tool for analyzing and synthesizing functions, and for forming representations of signals in both time and frequency. { 'wāv·lət }

weak convergence A sequence of elements $x_1, x_2, \ldots$ from a topological vector space X converges weakly if the sequence $f(x_1), f(x_2), \ldots$ converges for every continuous linear functional f on X. { 'wēk kən'vər·jəns }

weakly complete space A topological vector space in which an element x is associated with any weakly convergent sequence of elements x_n such that the limit of $f(x_n)$ equals $f(x)$ for any continuous linear functional f. { ¦wēk·lē kəm¦plēt 'spās }

weakly connected digraph A directed graph whose underlying graph is a connected graph. { ¦wēk·lē kə₁nek·təd 'dī₁graf }

weak topology A topology on a topological vector space X whose open neighborhoods around a point x are obtained from those points y of X for which every $f_i(x)$ is close to $f_i(y)$, f_i appearing in a finite list of linear functionals. { 'wēk tə'päl·ə·jē }

Weber differential equation A special case of the confluent hypergeometric equation that has as solution a confluent hypergeometric series. Also known as Weber-Hermite equation. { 'vā·bər ˌdif·ə'ren·chəl i'kwā·zhən }

Weber-Hermite equation *See* Weber differential equation. { 'vā·bər er'mēt iˌkwā·zhən }

wedge A polyhedron whose base is a rectangle and whose lateral faces consist of two equilateral triangles and two trapezoids. { wej }

wedge product A product defined on forms such that a wedge product of a p-form and a q-form results in a $p + q$ form. { 'wej ˌpräd·əkt }

Weibull distribution A distribution that describes life-time characteristics of parts and components. { 'wīˌbúl ˌdis·trəˌbyü·shən }

Weierstrass' approximation theorem A continuous real-valued function on a closed interval can be uniformly approximated by polynomials. { 'vī·ərˌshträs ə,präk·sə'mā·shən ˌthir·əm }

Weierstrass functions Used in the calculus of variations, these determine functions satisfying the Euler-Lagrange equation and Jacobi's condition while maximizing a given definite integral. { 'vī·ərˌshträs ˌfəŋk·shənz }

Weierstrassian elliptic function A function that plays a central role in the theory of elliptic functions; for z, g_2 and g_3 real or complex numbers, let y be that number such that

$$z = \int_y^\infty \frac{dt}{\sqrt{4t^3 - g_2 t - g_3}} \, ;$$

the Weierstrassian elliptic function of z with parameters g_2 and g_3 is $p\,(z;\,g_2,\,g_3) = y$. { ˈvī·ərˈshträs·ē·ən i'lip·tik 'fəŋk·shən }

Weierstrass M test An infinite series of numbers will converge or functions will converge uniformly if each term is dominated in absolute value by a nonnegative constant M_n, where these M_n form a convergent series. Also known as Weierstrass' test for convergence. { 'vī·ərˌshträs 'em ˌtest }

Weierstrass' test for convergence *See* Weierstrass M test. { 'vī·ərˌshträs 'test fər kən'vər·jəns }

Weierstrass transform This transform of a real function $f(y)$ is the function given by the integral from $-\infty$ to ∞ of $(4\pi t)^{-1/2} \exp[-(x - y)^2/4]\, f(y)dy$; this is used in studying the heat equation. { 'vī·ərˌshträs 'tranzˌfórm }

weight 1. The unique nonnegative integer assigned to an edge or arc in a network or directed network.   **2.** The sum of the weights (first definition) of all the arcs in an s-t cut.   **3.** The nonnegative integer assigned to a vertex in a generalized s-t network.   **4.** The sum of the weights of all the arcs and vertices in a generalized s-t cut. { wāt }

weighted aggregative index A statistic for a collection of items weighted so as to reflect the relative importance of the items with regard to the overall phenomenon which the index is designed to describe; a price index is an example. { ˈwād·əd ˌa·grəˌgād·iv 'inˌdeks }

weighted average The number obtained by adding the product of α_i times the ith number in a set of N numbers for $i = 1,2, \ldots, N$, where α_i are numbers (weights) such that $\alpha_1 + \alpha_2 + \cdots + \alpha_N = 1$. Also known as weighted mean. { 'wād·əd 'av·rij }

weighted mean *See* weighted average. { ˈwād·əd 'mēn }

weighted moving average A method used for smoothing data in a time series in which each observation being averaged is given a weight which reflects its relative importance in calculating the average. { ˈwād·əd ˈmüv·iŋ 'av·rij }

weight function 1. Two real valued functions f and g are orthogonal relative to a weight function σ on an interval if the integral over the interval of $f \cdot g \cdot \sigma$ vanishes.   **2.** A function defined on the edges of a network or the arcs of a directed network, whose value at each edge or arc is the unique nonnegative integer assigned to that edge or arc.   **3.** A function defined on the vertices of a generalized s-t network, whose value at each vertex is a nonnegative integer. { 'wāt ˌfəŋk·shən }

Weingarten formulas Equations concerning the normals to a surface at a point.
{ 'wīn,gart·ən ,for·myə·ləs }
Weingarten surface A surface such that either of the principal radii is uniquely determined by the other. { 'wīn,gärt·ən ,sər·fəs }
weird number An abundant number that is not a semiperfect number. { ¦wird 'nəm·bər }
well-formed formula A finite sequence or string of symbols that is grammatically or syntactically correct for a given set of grammatical or syntactical rules. { 'wel ¦förmd ,for·myə·lə }
well-ordered set A linearly ordered set where every subset has a least element. { 'wel ¦ör·dərd 'set }
well-ordering principle The proposition that every set can be endowed with an order so that it becomes a well-ordered set; this is equivalent to the axiom of choice. { 'wel ¦ör·dər·iŋ 'prin·sə·pəl }
well-posed problem A problem that has a unique solution which depends continuously on the initial data. { 'wel ¦pōzd 'präb·ləm }
Wheweel equation An equation which relates the arc length along a plane curve to the angle of inclination of the tangent to the curve. { 'wā,wēl i,kwā·zhən }
white stochastic process A stochastic process such that there is no correlation between any of its components at different times, including autocorrelations. { 'wīt stō'kas·tik 'prä,səs }
Whitney number The kth Whitney number of a ranked poset is the number of elements of rank k. { 'wit·nē ,nəm·bər }
Whitney sum A tangent bundle TX over a differentiable manifold X is a Whitney sum of continuous bundles A and B over X if for each x the fibers of A and B at x are complementary subspaces of the tangent space at x. { 'wit·nē ,səm }
Whittaker differential equation A special form of Gauss' hypergeometric equation with solutions as special cases of the confluent hypergeometric series. { 'wid·ə·kər ,dif·ə¦ren·chəl i'kwā·zhən }
whole number An integer equal to or greater than zero; one of the numbers 0, 1, 2, 3, ... { 'hōl ¦nəm·bər }
width For a plane convex, the greatest lower bound on the distance separating two parallel lines such that the set lies between them. { width }
Wiener-Hopf equations Integral equations arising in the study of random walks and harmonic analysis; they are

$$g(x) = \int_0^\infty K(|x - t|)f(t)\, dt$$

$$f(x) = \int_0^\infty K(|x + t|)f(t)\, dt + g(x)$$

where g and K are known functions on the positive real numbers and f is the unknown function. { 'vē·nər 'höpf i,kwā·zhənz }
Wiener-Hopf technique A method used in solving certain integral equations, boundary-value problems, and other problems, which involves writing a function that is holomorphic in a vertical strip of the complex z plane as the product of two functions, one of which is holomorphic both in the strip and everywhere to the right of the strip, while the other is holomorphic in the strip and everywhere to the left of the strip. { ¦vēn·ər 'höpf ,tek,nēk }
Wiener-Khintchine theorem The theorem that determines the form of the correlation function of a given stationary stochastic process. { 'vē·nər kin'chēn ,thir·əm }
Wiener process A stochastic process with normal density at each stage, arising from the study of Brownian motion, which represents the limit of a sequence of experiments. Also known as Gaussian noise. { 'vē·nər ,prä·səs }

Wilcoxon one-sample test A rank test for testing the hypothesis $\mu = \mu_H$ against the alternative $\mu > \mu_H$ under the assumption that observations are symmetrically distributed about μ_H; here μ_H is a given number and μ is the (unknown) mean of a random variable. { 'wil,käk·sən ¦wən ¦sam·pəl ,test }

Wilcoxon paired comparison distribution The distribution of the rank sum V_- (or V_+) of the negative differences (or positive differences) of observations in paired comparisons. { 'wil,käk·sən ¦perd kəm¦par·ə·sən ,di·strə,byü·shən }

Wilcoxon paired comparison test The test based upon the rank sum V_- (or V_+) of the negative differences (or positive differences) of observations in paired comparisons. { 'wil,käk·sən ¦perd kəm¦par·ə·sən ,test }

Wilcoxon two-sample distribution The distribution of the Wilcoxon two-sample test statistic; it consists of the rank sums of treated subjects. { 'wil,käk·sən ¦tü ¦sam·pəl ,di·strə,byü·shən }

Wilcoxon two-sample test The test based upon the rank sum of treated (or untreated) subjects. { 'wil,käk·sən ¦tü ¦sam·pəl ,test }

Wilson's theorem The number $(n - 1)! + 1$ is divisible by n if and only if n is a prime. { 'wil·sənz ,thir·əm }

winding number The number of times a given closed curve winds in the counterclockwise direction about a designated point in the plane. Also known as index. { 'wīnd·iŋ ,nəm·bər }

witch of Agnesi The curve, symmetric about the y axis and asymptotic in both directions to the x axis, given by $x^2y = 4a^2(2a - y)$. Also known as versiera. { 'wich əv än'nyä·zē }

Witt-Grothendieck group The Grothendieck group of the monoid consisting of isometry classes of nondegenerate symmetric forms on vector spaces over a given field, where the product of two such forms is given by their orthogonal sum. { ¦wit 'grōt·ən,dēk ,grüp }

Witt group The group of isometry classes of symmetric forms on vector spaces over a given field, where the product of two such forms is given by their orthogonal sum. { 'wit ,grüp }

Witt's theorem If F and F' are subspaces of a vector space E with a nondegenerate, symmetric form g, then an isometry of g from F onto F' can be extended to an isometry of g from E onto itself. { 'wits ,thir·əm }

Wronskian An $n \times n$ matrix whose ith row is a list of the $(i - 1)$st derivatives of a set of functions $f_1, \ldots, f_n$; ordinarily used to determine linear independence of solutions of linear homogeneous differential equations. { 'vrän·skē·ən }

x axis 1. A horizontal axis in a system of rectangular coordinates. **2.** That line on which distances to the right or left (east or west) of the reference line are marked, especially on a map, chart, or graph. { 'eks ,ak·səs }

x component The projection of a vector quantity on the x axis of a coordinate system. { 'eks kəm,pō·nənt }

x coordinate One of the coordinates of a point in a two- or three-dimensional Cartesian coordinate system, equal to the directed distance of a point from the y axis in a two-dimensional system, or from the plane of the y and z axes in a three-dimensional system, measured along a line parallel to the x axis. { 'eks kō'órd·ən·ət }

X test A one-sample test which rejects the hypothesis $\mu = \mu_H$ in favor of the alternative $\mu > \mu_H$ if $X - \mu_H \geq c$ where c is an appropriate critical value, X is the arithmetic mean of observations, μ_H is a given number, and μ is the (unknown) expected value of the random variable X. { 'eks ,test }

y axis 1. A vertical axis in a system of rectangular coordinates. 2. That line on which distances above or below (north or south) the reference line are marked, especially on a map, chart, or graph. { 'wī ˌak·səs }

y component The projection of a vector quantity on the y axis of a coordinate system. { 'wī kəmˌpō·nənt }

y coordinate One of the coordinates of a point in a two- or three-dimensional coordinate system, equal to the directed distance of a point from the x axis in a two-dimensional system, or from the plane of the x and z axes in a three-dimensional coordinate system, measured along a line parallel to the y axis. { 'wī kō,órd·ən·ət }

Yonden square An experimental design that is an incomplete block design derived from a Latin square by dropping one or more rows and by treating columns as blocks. Also known as incomplete Latin square. { 'yän·dən ˌskwer }

Young's inequality An inequality that applies to a function $y = f(x)$ that is continuous and strictly increasing for $x \geqq 0$ and satisfies $f(0) = 0$, with inverse function $x = g(y)$; it states that, for any positive numbers a and b in the ranges of x and y, respectively, the product ab is equal to or less than the sum of the integral from 0 to a of $f(x)dx$ and the integral from 0 to b of $g(y)dy$. { 'yəŋz ˌin·ə'kwäl·əd·ē }

Z

z axis One of the three axes in a three-dimensional Cartesian coordinate system; in a rectangular coordinate system it is perpendicular to the x and y axes. { 'zē ˌak·səs }

z component The projection of a vector quantity on the z axis of a coordinate system. { 'zē kəmˌpō·nənt }

z coordinate One of the coordinates of a point in a three-dimensional coordinate system, equal to the directed distance of a point from the plane of the x and y axes, measured along a line parallel to the z axis. { 'zē kōˌȯrd·ən·ət }

Zeckendorf's theorem The theorem that any positive integer can be expressed as a sum of distinct Fibonacci numbers, no two of which are consecutive. { 'zek·ənˌdȯrfs ˌthir·əm }

Zeno's paradox An erroneous group of paradoxes dealing with motion; the most famous one concerns two objects, one chasing the other which has a given head start, where the chasing one moves faster yet seemingly never catches the other. { 'zē·nōz 'par·əˌdäks }

zero 1. The additive identity element of an algebraic system. **2.** Any point where a given function assumes the value zero. { 'zir·ō }

zero divisor *See* divisor of zero. { ¦zir·ō di'vīz·ər }

zero geodesic *See* null geodesic. { 'zir·ō ˌjē·ə'des·ik }

zero point A complex number at which an analytic function assumes the value zero. { 'zir·ō ˌpȯint }

zero-sum game A two-person game where the sum of the payoffs to the two players is zero for each move. { 'zir·ō ¦səm ˌgām }

zero vector The element 0 of a vector space such that, for any vector v in the space, the vector sum of 0 and v is v. { 'zir·ō ˌvek·tər }

zeta function *See* Riemann zeta function. { 'zād·ə ˌfəŋk·shən }

zonal harmonics Spherical harmonics which do not depend on the azimuthal angle; they are proportional to Legendre polynomials of $\cos \theta$, where θ is the colatitude. { 'zōn·əl här'män·iks }

zone The portion of a sphere lying between two parallel planes that intersect the sphere. { zōn }

Zorn's lemma If every linearly ordered subset of a partially ordered set has a maximal element in the set, then the set has a maximal element. { 'zȯrnz 'lem·ə }

Z score A measure of how many standard deviations a raw score is from the mean. { 'zē ˌskȯr }

z-transform The z-transform of a sequence whose general term is f_n is the sum of a series whose general term is $f_n z^{-n}$, where z is a complex variable; n runs over the positive integers for a one-sided transform, over all the integers for a two-sided transform. { 'zē 'tranzˌfȯrm }

Appendix

Equivalents of commonly used units for the U.S. Customary System and the metric system

1 inch = 2.5 centimeters (25 millimeters)
1 foot = 0.3 meter (30 centimeters)
1 yard = 0.9 meter
1 mile = 1.6 kilometers

1 centimeter = 0.4 inch
1 meter = 3.3 feet
1 meter = 1.1 yards
1 kilometer = 0.62 mile

1 inch = 0.083 foot
1 foot = 0.33 yard (12 inches)
1 yard = 3 feet (36 inches)
1 mile = 5280 feet (1760 yards)

1 acre = 0.4 hectare
1 acre = 4047 square meters

1 hectare = 2.47 acres
1 square meter = 0.00025 acre

1 gallon = 3.8 liters
1 fluid ounce = 29.6 milliliters
32 fluid ounces = 946.4 milliliters

1 liter = 1.06 quarts = 0.26 gallon
1 milliliter = 0.034 fluid ounce

1 quart = 0.25 gallon (32 ounces; 2 pints)
1 pint = 0.125 gallon (16 ounces)
1 gallon = 4 quarts (8 pints)

1 quart = 0.95 liter
1 ounce = 28.35 grams
1 pound = 0.45 kilogram
1 ton = 907.18 kilograms

1 gram = 0.035 ounce
1 kilogram = 2.2 pounds
1 kilogram = 1.1×10^{-3} ton

1 ounce = 0.0625 pound
1 pound = 16 ounces
1 ton = 2000 pounds

$$°C = (°F - 32) \div 1.8$$

$$°F = (1.8 \times °C) + 32$$

Appendix

Conversion factors for the U.S. Customary System, metric system, and International System

A. Units of length

Units		cm	m	$in.$	ft	yd	mi
1 cm	=	1	0.01	0.3937008	0.03280840	0.01093613	6.213712×10^{-6}
1 m	=	100.	1	39.37008	3.280840	1.093613	6.213712×10^{-4}
1 in.	=	2.54	0.0254	1	0.08333333...	0.02777777...	1.578283×10^{-5}
1 ft	=	30.48	0.3048	12.	1	0.3333333...	$1.893939... \times 10^{-4}$
1 yd	=	91.44	0.9144	36.	3.	1	$5.681818... \times 10^{-4}$
1 mi	=	1.609344×10^{5}	1.609344×10^{3}	6.336×10^{4}	5280.	1760.	1

B. Units of area

Units		cm^2	m^2	$in.^2$	ft^2	yd^2	mi^2
1 cm²	=	1	10^{-4}	0.1550003	1.076391×10^{-3}	1.195990×10^{-4}	3.861022×10^{-11}
1 m²	=	10^{4}	1	1550.003	10.76391	1.195990	3.861022×10^{-7}
1 in.²	=	6.4516	6.4516×10^{-4}	1	$6.944444... \times 10^{-3}$	7.716049×10^{-4}	2.490977×10^{-10}
1 ft²	=	929.0304	0.09290304	144.	1	0.1111111...	3.587007×10^{-8}
1 yd²	=	8361.273	0.8361273	1296.	9.	1	3.228306×10^{-7}
1 mi²	=	2.589988×10^{10}	2.589988×10^{6}	4.014490×10^{9}	2.78784×10^{7}	3.0976×10^{6}	1

C. Units of volume

Units	m^3	cm^3	liter	$in.^3$	ft^3	qt	gal
1 m^3 =	1	10^6	10^3	6.102374×10^4	35.31467×10^{-3}	1.056688	264.1721
1 cm^3 =	10^{-6}	1	10^{-3}	0.06102374	3.531467×10^{-5}	1.056688×10^{-3}	2.641721×10^{-4}
1 liter =	10^{-3}	1000.	1	61.02374	0.03531467	1.056688	0.2641721
1 $in.^3$ =	1.638706×10^{-5}	16.38706	0.01638706	1	5.787037×10^{-4}	0.01731602	4.329004×10^{-3}
1 ft^3 =	2.831685×10^{-2}	28316.85	28.31685	1728.	1	2.992208	7.480520
1 qt =	9.463529×10^{-4}	946.3529	0.9463529	57.75	0.03342014	1	0.25
1 gal (U.S.) =	3.785412×10^{-3}	3785.412	3.785412	231.	0.1336806	4.	1

D. Units of mass

Units	g	kg	oz	lb	metric ton	ton
1 g =	1	10^{-3}	0.03527396	2.204623×10^{-3}	10^{-6}	1.102311×10^{-6}
1 kg =	1000.	1	35.27396	2.204623	10^{-3}	1.102311×10^{-3}
1 oz (avdp) =	28.34952	0.02834952	1	0.0625	2.834952×10^{-5}	3.125×10^{-5}
1 lb (avdp) =	453.5924	0.4535924	16.	1	4.535924×10^{-4}	$5. \times 10^{-4}$
1 metric ton =	10^8	1000.	35273.96	2204.623	1	1.102311
1 ton =	907184.7	907.1847	32000.	2000.	0.9071847	1

Appendix

Conversion factors for the U.S. Customary System, metric system, and International System (cont.)

E. Units of density

Units	$g \cdot cm^{-3}$	$g \cdot L^{-1}, kg \cdot m^{-3}$	$oz \cdot in.^{-3}$	$lb \cdot in.^{-3}$	$lb \cdot ft^{-3}$	$lb \cdot gal^{-1}$
$1 \ g \cdot cm^{-3}$	$= 1$	1000.	0.5780365	0.03612728	62.42795	8.345403
$1 \ g \cdot L^{-1}, kg \cdot m^{-3}$	$= 10^{-3}$	1	5.780365×10^{-4}	3.612728×10^{-5}	0.06242795	8.345403×10^{-3}
$1 \ oz \cdot in.^{-3}$	$= 1.729994$	1729.994	1	0.0625	108.	14.4375
$1 \ lb \cdot in.^{-3}$	$= 27.67991$	27679.91	16.	1	1728.	231.
$1 \ lb \cdot ft^{-3}$	$= 0.01601847$	16.01847	9.259259×10^{-3}	5.787037×10^{-4}	1	0.1336806
$1 \ lb \cdot gal^{-1}$	$= 0.1198264$	119.8264	4.749536×10^{-3}	4.329004×10^{-3}	7.480519	1

F. Units of pressure

Units	$Pa, N \cdot m^{-2}$	$dyn \cdot cm^{-2}$	bar	atm	$kgf \cdot cm^{-2}$	$mmHg \ (torr)$	$in. \ Hg$	$lbf \cdot in.^{-2}$
$1 \ Pa, 1 \ N \cdot m^{-2}$	$= 1$	10	10^{-5}	9.869233×10^{-6}	1.019716×10^{-5}	7.500617×10^{-3}	2.952999×10^{-4}	1.450377×10^{-4}
$1 \ dyn \cdot cm^{-2}$	$= 0.1$	1	10^{-6}	9.869233×10^{-7}	1.019716×10^{-6}	7.500617×10^{-4}	2.952999×10^{-5}	1.450377×10^{-5}
$1 \ bar$	$= 10^{5}$	10^{6}	1	0.9869233	1.019716	750.0617	29.52999	14.50377
$1 \ atm$	$= 101325$	1013250.0	1.01325	1	1.033227	760.	29.92126	14.69595
$1 \ kgf \cdot cm^{-2}$	$= 98066.5$	980665	0.980665	0.9678411	1	735.5592	28.95903	14.22334
$1 \ mmHg \ (torr)$	$= 133.3224$	1333.224	1.333224×10^{-3}	1.315789×10^{-3}	1.359510×10^{-3}	1	0.03937008	0.01933678
$1 \ in. \ Hg$	$= 3386.388$	33863.88	0.03386388	0.03342105	0.03453155	25.4	1	0.4911541
$1 \ lbf \cdot in.^{-2}$	$= 6894.757$	68947.57	0.06894757	0.06804596	0.07030696	51.71493	2.036021	1

G. Units of energy

Units	g mass (energy equiv)	J	eV	cal	cal$_\text{IT}$	Btu$_\text{IT}$	kWh	hp-h	ft-lbf	ft³·lbf·in^{-2}	liter-atm
1 g mass = 1 (energy equiv)	1	8.987552×10^{13}	5.609589×10^{32}	2.148076×10^{3}	2.146640×10^{13}	8.518555×10^{10}	2.496542×10^{7}	3.347918×10^{7}	6.628878×10^{13}	4.603388×10^{11}	8.870024×10^{11}
1 J =	1.112650×10^{-14}	1	6.241510×10^{18}	0.2390057	0.2388459	9.478172×10^{-4}	$2.777777\ldots \times 10^{-7}$	3.725062×10^{-7}	0.7375622	5.121960×10^{-3}	9.869233×10^{-3}
1 eV =	1.782662×10^{-33}	1.602176×10^{-19}	1	3.829293×10^{-20}	3.826733×10^{-20}	1.518570×10^{-22}	4.450490×10^{-26}	5.968206×10^{-26}	1.181705×10^{-19}	8.206283×10^{-22}	1.581225×10^{-21}
1 cal =	4.655328×10^{-14}	4.184	2.611448×10^{19}	1	0.9993312	3.965667×10^{-3}	1.1622222×10^{-6}	1.558562×10^{-6}	3.085960	2.143028×10^{-2}	0.04129287
1 cal$_\text{IT}$ =	4.658443×10^{-14}	4.1868	2.613195×10^{19}	1.000669	1	3.968321×10^{-3}	1.163×10^{-6}	1.559609×10^{-6}	3.088025	2.144462×10^{-2}	0.04132050
1 Btu$_\text{IT}$ =	1.173908×10^{-11}	1055.056	6.585141×10^{21}	252.1644	251.9958	1	2.930711×10^{-4}	3.930148×10^{-4}	778.1693	5.403953	10.41259
1 kWh =	4.005540×10^{-8}	3600000.	2.246944×10^{25}	860420.7	859845.2	3412.142	1	1.341022	2655224	18349.06	35529.24
1 hp-h =	2.986931×10^{-8}	2384519.	1.675545×10^{25}	641615.6	641186.5	2544.33	0.7456998	1	1980000.	13750.	26494.15
1 ft-lbf =	1.508551×10^{-14}	1.355818	8.462351×10^{18}	0.3240483	0.3238315	1.285067×10^{-3}	3.766161×10^{-7}	$5.050505\ldots \times 10^{-7}$	1	$6.944444\ldots \times 10^{-3}$	0.01338088
1 ft³·lbf·in^{-2} =	2.172313×10^{-12}	195.2378	1.218579×10^{21}	46.66295	46.63174	0.1850497	5.423272×10^{-5}	$7.272727\ldots \times 10^{-5}$	144.	1	1.926847
1 liter-atm =	1.127393×10^{-12}	101.325	6.324210×10^{20}	24.21726	24.20106	0.09603757	2.814583×10^{-5}	3.774419×10^{-5}	74.73349	0.5189825	1

Appendix

Mathematical notation, with definitions

Signs and symbols

$+$	Plus (sign of addition)
$+$	Positive
$-$	Minus (sign of subtraction)
$-$	Negative
$\pm$ ($\mp$)	Plus or minus (minus or plus)
$\times$	Times, by (multiplication sign)
$\cdot$	Multiplied by
$\div$	Sign of division
$/$	Divided by
$:$	Ratio sign, divided by, is to
$::$	Equals, as (proportion)
$<$	Less than
$>$	Greater than
$\ll$	Much less than
$\gg$	Much greater than
$=$	Equals
$\equiv$	Identical with
$\sim$	Similar to
$\approx$	Approximately equals
$\cong$	Approximately equals, congruent
$\leq$	Equal to or less than
$\geq$	Equal to or greater than
$\neq$ $\neq$	Not equal to
$\rightarrow$ $\doteq$	Approaches
$\propto$	Varies as
∞	Infinity
$\sqrt{}$	Square root of
$\sqrt[3]{}$	Cube root of
$\therefore$	Therefore
$\parallel$	Parallel to
( ) [ ] { }	Parentheses, brackets and braces; quantities enclosed by them to be taken together in multiplying, dividing, etc.
$\overline{AB}$	Length of line from A to B
π	pi $= 3.14159...$
$\circ$	Degrees
$'$	Minutes

Mathematical notation, with definitions (cont.)

Signs and symbols (cont.)

$''$	Seconds
$\angle$	Angle
dx	Differential of x
Δ	(delta) difference
Δx	Increment of x
$\partial u/\partial x$	Partial derivative of u with respect to x
$\int$	Integral of
$\displaystyle\int_b^a$	Integral of, between limits a and b
$\oint$	Line integral around a closed path
Σ	(sigma) summation of
$f(x), F(x)$	Functions of x
∇	Del or nabla, vector differential operator
∇^2	Laplacian operator
$£$	Laplace operational symbol
$4!$	Factorial $4 = 1 \times 2 \times 3 \times 4$
$\lvert x \rvert$	Absolute value of x
$\dot{x}$	First derivative of x with respect to time
$\ddot{x}$	Second derivative of x with respect to time
$\mathbf{A} \times \mathbf{B}$	Vector-product; magnitude of $\mathbf{A}$ times magnitude of $\mathbf{B}$ times sine of the angle from $\mathbf{A}$ to $\mathbf{B}$; $AB \sin \overline{AB}$
$\mathbf{A} \cdot \mathbf{B}$	Scalar product of $\mathbf{A}$ and $\mathbf{B}$; magnitude of $\mathbf{A}$ times magnitude of $\mathbf{B}$ times cosine of the angle from $\mathbf{A}$ to $\mathbf{B}$; $AB \cos \overline{AB}$

Mathematical logic

$p, q, P(x)$	Sentences, propositional functions, propositions
$-p, \sim p$, non p, Np	Negation red "not p" ($\neq$: read "not equal")
$p \vee q, p + q, Apq$	Disjunction, read "p or q," "p, q," or both
$p \wedge q, p \cdot q, p\&q, Kpq$	Conjunction, read "p and q"
$p \to q, p \supset q, p \Rightarrow q, Cpq$	Implication, read "p implies q" or "if p then q"
$p \leftrightarrow q, p \equiv q, p \Leftrightarrow q, Epq,$ p iff q	Equivalence, read "p is equivalent to q" or "p if and only if q"
n.a.s.c.	Read "necessary and sufficient condition"
$( ), [ ], \{ \}, \ldots, \cdots$	Parentheses
$\forall, \Sigma$	Universal quantifier, read "for all" or "for every"
$\exists, \Pi$	Existential quantifier, read "there is a" or "there exists"
$\vdash$	Assertion sign ($p \vdash q$: read "q follows from p"; $\vdash p$: read "p is or follows from an axiom," or "p is a tautology"

Appendix

<table>
<tr><td colspan="2">Mathematical notation, with definitions (cont.)</td></tr>
</table>

Mathematical logic (cont.)

0, 1	Truth, falsity (values)
$=$	Identity
$\overset{Df}{=}$, $\overset{df}{=}$, $\underset{df}{=}$, $\equiv$	Definitional identity
■	"End of proof"; "QED"

Set theory, relations, functions

X, Y	Sets
$x \in X$	x is a member of the set X
$x \notin X$	x is not a member of X
$A \subset X$, $A \subseteq X$	Set A is contained in set X
$A \not\subset X$, $A \not\subseteq X$	A is not contained in X
$X \cup Y$, $X + Y$	Union of sets X and Y
$X \cap Y$, $X \cdot Y$	Intersection of sets X and Y
$+$, $\dot{+}$, $\circ$	Symmetric difference of sets
$\cup X_i$, ΣX_i	Union of all the sets X_i
$\cap X_i$, ΠX_i	Intersection of all the sets X_i
$\varnothing$, 0, Λ	Null set, empty set
X', $\mathbf{C}X$, CX	Complement of the set X
$X - Y$, $X \backslash Y$	Difference of sets X and Y
$\hat{x}(P(x))$, $\{x | P(x)\}$, $\{x{:}P(x)\}$	The set of all x with the property P
(x,y,z), $\langle x,y,z \rangle$	Ordered set of elements x, y, and z; to be distinguished from (x,z,y) for example
$\{x,y,z\}$	Unordered set, the set whose elements are x, y, z, and no others
$\{a_1, a_2, \ldots, a_n\}$, $\{a_i\}_{i=1,2\ldots n}$, $\{a_i\}_{i=1}^{n}$	The set whose members are a_i, where i is any number whole from 1 to n
$\{a_1, a_2, \ldots\}$, $\{a_i\}_{i=1,2,\ldots}$, $\{a_i\}_{i=1}^{\infty}$	The set whose members are a_i, where i is any whole positive number
$X \times Y$	Cartesian product, set of all (x,y) such that $x \in X$, $y \in Y$
$\{a_i\}_{i \in I}$	The set whose elements are a_i, where $i \in I$
xRy, $R(x,y)$	Relation
$\equiv$, $\cong$, $\sim$, $\simeq$	Equivalence relations, for example, congruence
$\geqq$, $\geq$, &, $\gg$, $\leqq$, $\leq$, $<$	Transitive relations, for example, numerical order
$f{:}X \to Y$, $X \overset{f}{\to} Y$, $X \to Y$, $f \in Y^X$	Function, mapping, transformation
f^{-1}, $\overset{-1}{f}$, $X \overset{f^{-1}}{\leftarrow} Y$	Inverse mapping
$g \circ f$	Composite functions: $(g \circ f)(x) = g(f(x))$

Mathematical notation, with definitions (cont.)

Set theory, relations, functions (cont.)

$f(X)$	Image of X by f
$f^{-1}(X)$	Inverse-image set, counter image
1-1, one-one	Read "one-to-one correspondence"
$\begin{array}{ccc} X & \xrightarrow{f} & Y \\ \phi\downarrow & & \downarrow\psi \\ W & \xrightarrow{g} & Z \end{array}$	Diagram: the diagram is commutative in case $\psi \circ f = g \circ \phi$
$f|A$	Partial mapping, restriction of function f to set A
$\overline{\overline{X}}$, card X, $|X|$	Cardinal of the set A
$\aleph_0$, d	Denumerable infinity
c, $\mathfrak{c}$, $2^{\aleph_0}$	Power of continuum
ω	Order type of the set of positive integers
σ^-	Read "countably"

Number, numerical functions

1.4; 1,4; 1 · 4	Read "one and four-tenths"
1(1)20(10)100	Read "from 1 to 20 in intervals of 1, and from 20 to 100 in intervals of 10"
const	Constant
$A \geqq 0$	The number A is nonnegative, or, the matrix A is positive definite, or, the matrix A has nonnegative entries
$x|y$	Read "x divides y"
$x \equiv y \bmod p$	Read "x congruent to y modulo p"
$a_0 + \dfrac{1}{a_1} + \dfrac{1}{a_2} + \cdots,$	Continued fractions
$a_0 + \dfrac{1|}{|a_1} + \dots$	
$[a,b]$	Closed interval
$[a,b)$, $[a,b[$	Half-open interval (open at the right)
(a,b), $]a,b[$	Open interval
$[a,\infty)$, $[a,\rightarrow[$	Interval closed at the left, infinite to the right
$(-\infty,\infty)$, $]\leftarrow,\rightarrow[$	Set of all real numbers
$\max_{x \in X} f(x),$ $\max \{f(x)|x \in X\}$	Maximum of $f(x)$ when x is in the set X
min	Minimum
sup, l.u.b.	Supremum, least upper bound
inf, g.l.b.	Infimum, greatest lower bound
$\lim_{x\to a} f(x) = b,$ $\lim_{x=a} f(x) = b,$ $f(x) \to b$ as $x \to a$	b is the limit of $f(x)$ as x approaches a

Appendix

Mathematical notation, with definitions (cont.)	

Number, numerical functions (cont.)

$\lim_{x \to a-} f(x)$, $\lim_{x=a-0} f(x)$, $f(a-)$	Limit of $f(x)$ as x approaches a from the left
$\lim \sup$, $\overline{\lim}$	Limit superior
$\lim \inf$, $\underline{\lim}$	Limit inferior
l.i.m.	Limit in the mean
$z = x + iy = re^{i\theta}$, $\zeta = \xi + i\eta$, $w = u + iv = \rho e^{i\phi}$	Complex variables
$\bar{z}$, z^*	Complex conjugate
Re, $\Re$	Real part
Im, $\Im$	Imaginary part
arg	Argument
$\dfrac{\partial(u,v)}{\partial(x,y)}$, $\dfrac{D(u,v)}{D(x,y)}$	Jacobian, functional determinant
$\displaystyle\int_E f(x)d\mu(x)$	Integral (for example, Lebesgue integral) of function f over set E with respect to measure μ
$f(n) \sim \log n$ as $n \to \infty$	$f(n)/\log n$ approaches 1 as $n \to \infty$
$f(n) = O(\log n)$ as $n \to \infty$	$f(n)/\log n$ is bounded as $n \to \infty$.
$f(n) = o(\log n)$	$f(n)/\log n$ approaches zero
$f(x) \nearrow b$, $f(x) \uparrow b$	$f(x)$ increases, approaching the limit b
$f(x) \downarrow b$, $f(x) \searrow b$	$f(x)$ decreases, approaching the limit b
a.e., p.p.	Almost everywhere
ess sup	Essential supremum
C^0, $C^0(X)$, $C(X)$	Space of continuous functions
C^k, $C^k[a,b]$	The class of functions having continuous kth derivative (on $[a,b]$)
C'	Same as C^1
Lip_α, $\text{Lip } \alpha$	Lipschitz class of functions
L^p, L_p, $L^p[a,b]$	Space of functions having integrable absolute pth power (on $[a,b]$)
L'	Same as L^1
(C,α), (C,p)	Cesàro summability

Special functions

$[x]$	The integral part of x
$\dbinom{n}{k}$, ^nC_k, $_nC_k$	Binomial coefficient $n!/k!(n - k)!$
$\left(\dfrac{n}{p}\right)$	Legendre symbol
e^x, $\exp x$	Exponential function

Mathematical notation, with definitions (cont.)

Special functions (cont.)

$\sinh x$, $\cosh x$, $\tanh x$	Hyperbolic functions
sn x, cn x, dn x	Jacobi elliptic functions
$\wp(x)$	Weierstrass elliptic function
$\Gamma(x)$	Gamma function
$J_\nu(x)$	Bessel function
$\chi_X(x)$	Characteristic function of the set X; $\chi_X(x) = 1$ in case $x \in X$, otherwise $\chi_X(x) = 0$
sgn x	Signum: sgn $= 0$, while sgn $x = x/|x|$ for $x \neq 0$
$\delta(x)$	Dirac delta function

Algebra, tensors, operators

$+$, $\cdot$, $\times$, $\circ$, $\top$, τ	Laws of composition in algebraic systems
e, 0	Identity, unit, neutral element (of an additive system)
e, 1, I	Identity, unit, neutral element (of a general algebraic system)
e, ϵ, E, P	Idempotent
a^{-1}	Inverse of a
$\text{Hom}(M,N)$	Group of all homomorphisms of M into N
G/H	Factor group, group of cosets
$[K:k]$	Dimension of K over k
$\oplus$, $\dotplus$	Direct sum
$\otimes$	Tensor product, Kronecker product
$\wedge$	Exterior product, Grassmann product
$\vec{x}$, $\mathbf{x}$, $\mathfrak{x}$, x	Vector
$\vec{x} \cdot \vec{y}$, $\mathbf{x} \cdot \mathbf{y}$, $(\mathfrak{x},\mathfrak{y})$	Inner product, scalar product, dot product
$\mathbf{x} \times \mathbf{y}$, $[\mathfrak{x},\mathfrak{y}]$, $\mathbf{x} \wedge \mathbf{y}$	Outer product, vector product, cross product
$\|x\|$, $|x|$, $\|\mathbf{x}\|$, $\|\mathbf{x}\|_p$	Norm of the vector x
Ax, xA	The image of x under the transformation A
δ_{ij}	Kronecker delta: $\delta_{ii} = 1$, while $\delta_{ij} = 0$ for $i \neq j$
A', ^tA, A^t, ^tA	Transpose of the matrix A
A^*, $\tilde{A}$	Adjoint, Hermitian conjugate of A
tr A, Sp A	Trace of the matrix A
det A, $|A|$	Determinant of the matrix A
$\Delta^n f(x)$, $\Delta_h^n f$, $\underset{h}{\Delta^n} f(x)$	Finite differences
$[x_0,x_1]$, $[x_0,x_1,x_2]$, $\underset{x_1}{\Delta} u_{x_0}$, $[x_0,x_1]_f$	Divided differences
∇f, grad f	Read "gradient of f"
$\nabla \cdot \mathbf{v}$, div $\mathbf{v}$	Read "divergence of $\mathbf{v}$"
$\nabla \times \mathbf{v}$, curl $\mathbf{v}$, rot $\mathbf{v}$	Read "curl of $\mathbf{v}$"

Appendix

Mathematical notation, with definitions (cont.)

Algebra, tensors, operators (cont.)

∇^2, Δ, div grad	Laplacian
[X,Y]	Poisson bracket, or commutator, or Lie product
$GL(n,R)$	Full linear group of degree n over field R
$O(n,R)$	Full orthogonal group
$SO(n,R)$ $O^+(n,R)$	Special orthogonal group

Topology

E^n	Euclidean n space
S^n	n sphere
$\rho(p,q)$, $d(p,q)$	Metric, distance (between points p and q)
$\overline{X}$, X^-, cl X, X^c	Closure of the set X
FrX, fr$X$, ∂X, bdry X	Frontier, boundary of X
int X, $\mathring{X}$	Interior of X
T_2 space	Hausdorff space
F_σ	Union of countably many closed sets
G_δ	Intersection of countably many open sets
dim X	Dimensionality, dimension of X
$\pi_1(X)$	Fundamental group of the space X
$\pi_n(X)$, $\pi_n(X,A)$	Homotopy groups
$H_n(X)$, $H_n(X,A;G)$, $H_*(X)$	Homology groups
$H^n(X)$, $H^n(X,A;G)$, $H^*(X)$	Cohomology groups

Probability and statistics

X, Y	Random variables
$P(X \leq 2)$, $\Pr(X \leq 2)$	Probability that $X \leq 2$
$P(X \leq 2\|Y \geq 1)$	Conditional probability
$E(X)$, $\mathscr{E}(X)$	Expectation of X
$E(X\|Y \geq 1)$	Conditional expectation
c.d.f.	Cumulative distribution function
p.d.f.	Probability density function
c.f.	Characteristic function
$\bar{x}$	Mean (especially, sample mean)
σ, s.d.	Standard deviation
σ^2, Var, var	Variance
μ_1, μ_2, μ_3, μ_i, μ_{ij}	Moments of a distribution
ρ	Coefficient of correlation
$\rho_{12 \cdot 34}$	Partial correlation coefficient

Appendix

Symbols commonly used in geometry

Example	Meaning	Comments
$\overline{AB}$	Line segment having end points A and B	Could be named $\overline{BA}$
AB	Length of $\overline{AB}$	$AB = 5$ means $\overline{AB}$ is 5 units long
$\overleftrightarrow{XY}$	Line containing points X and Y	Could be named $\overleftrightarrow{YX}$
$\overrightarrow{PQ}$	Ray with end point P and containing Q	
$\angle V$	Angle with vertex V	Use only if no more than two rays have end point V
$\angle RST$	Angle formed by $\overrightarrow{SR}$ and $\overrightarrow{ST}$	Could be named $\angle TSR$
$\angle x$	Angle named x	
$\angle x = 30°$	Angle named x has measure 30°	
$\overset{\frown}{JK}$	Minor arc with end points J and K	
$\overset{\frown}{JLK}$	Major arc that contains point L	
$\odot C$	Circle with center C	
$\triangle XYZ$	Triangle with vertices X, Y, and Z	Could be named $\triangle YZX$, $\triangle XZY$, and so forth

Example	Meaning	Comments
$\square ABCD$	Rectangle with vertices A, B, C, and D	Adjacent letters must be adjacent vertices P and R are not adjacent vertices
$\square PQRS$	Square with vertices P, Q, R, and S	
$\square ABCD$	Parallelogram with vertices A, B, C, and D	
$ABCD \dots X$	Polygon with vertices A, B, $\dots$, X	Adjacent letters must be adjacent vertices
$\overline{AB} \perp j$	A segment ($\overline{AB}$) is perpendicular to a line (j)	
$\overrightarrow{ZP} \parallel \overline{TV}$	A ray ($\overrightarrow{ZP}$) is parallel to a segment ($\overline{TV}$)	
$\overline{AB} \cong \overline{XY}$	Two segments ($\overline{AB}$ and $\overline{XY}$) are congruent	Congruent segments have equal lengths
$\triangle ABC \sim \triangle XYZ$	Two triangles ($\triangle ABC$ and $\triangle XYZ$) are similar	

Appendix

Formulas for trigonometric (circular) functions*

Definitions

$$\sin z = \frac{e^{iz} - e^{-iz}}{2i} \quad (z = x + iy)$$

$$\cos z = \frac{e^{iz} + e^{-iz}}{2}$$

$$\tan z = \frac{\sin z}{\cos z}$$

$$\csc z = \frac{1}{\sin z}$$

$$\sec z = \frac{1}{\cos z}$$

$$\cot z = \frac{1}{\tan z}$$

Periodic properties

$$\sin (z + 2k\pi) = \sin z \quad (k \text{ any integer})$$
$$\cos (z + 2k\pi) = \cos z$$
$$\tan (z + k\pi) = \tan z$$

Relations between circular functions

$$\sin^2 z + \cos^2 z = 1$$
$$\sec^2 z - \tan^2 z = 1$$
$$\csc^2 z - \cot^2 z = 1$$

Negative angle formulas

$$\sin (-z) = -\sin z$$
$$\cos (-z) = \cos z$$
$$\tan (-z) = -\tan z$$

Addition formulas

$$\sin (z_1 + z_2) = \sin z_1 \cos z_2 + \cos z_1 \sin z_2$$
$$\cos (z_1 + z_2) = \cos z_1 \cos z_2 - \sin z_1 \sin z_2$$
$$\tan (z_1 + z_2) = \frac{\tan z_1 + \tan z_2}{1 - \tan z_1 \tan z_2}$$
$$\cot (z_1 + z_2) = \frac{\cot z_1 \cot z_2 - 1}{\cot z_2 + \cot z_1}$$

Half-angle formulas

$$\sin \frac{z}{2} = \pm\left(\frac{1 - \cos z}{2}\right)^{\frac{1}{2}}$$

$$\cos \frac{z}{2} = \pm\left(\frac{1 + \cos z}{2}\right)^{\frac{1}{2}}$$

Half-angle formulas (cont.)

$$\tan \frac{z}{2} = \pm\left(\frac{1 - \cos z}{1 + \cos z}\right)^{\frac{1}{2}}$$

$$= \frac{1 - \cos z}{\sin z} = \frac{\sin z}{1 + \cos z}$$

The ambiguity in sign may be resolved with the aid of a diagram.

Transformation of trigonometric integrals

If $\tan \frac{u}{2} = z$ then

$$\sin u = \frac{2z}{1 + z^2}, \qquad \cos u = \frac{1 - z^2}{1 + z^2},$$

$$du = \frac{2}{1 + z^2} dz$$

Multiple-angle formulas

$$\sin 2z = 2 \sin z \cos z = \frac{2 \tan z}{1 + \tan^2 z}$$
$$\cos 2z = 2 \cos^2 z - 1 = 1 - 2 \sin^2 z$$
$$= \cos^2 z - \sin^2 z = \frac{1 - \tan^2 z}{1 + \tan^2 z}$$
$$\tan 2z = \frac{2 \tan z}{1 - \tan^2 z} = \frac{2 \cot z}{\cot^2 z - 1}$$
$$= \frac{2}{\cot z - \tan z}$$
$$\sin 3z = 3 \sin z - 4 \sin^3 z$$
$$\cos 3z = -3 \cos z + 4 \cos^3 z$$
$$\sin 4z = 8 \cos^3 z \sin z - 4 \cos z \sin z$$
$$\cos 4z = 8 \cos^4 z - 8 \cos^2 z + 1$$

Products of sines and cosines

$$2 \sin z_1 \sin z_2 = \cos (z_1 - z_2) - \cos (z_1 + z_2)$$
$$2 \cos z_1 \cos z_2 = \cos (z_1 - z_2) + \cos (z_1 + z_2)$$
$$2 \sin z_1 \cos z_2 = \sin (z_1 - z_2) + \sin (z_1 + z_2)$$

Addition and subtraction of two functions

$$\sin z_1 + \sin z_2$$
$$= 2 \sin \left(\frac{z_1 + z_2}{2}\right) \cos \left(\frac{z_1 - z_2}{2}\right)$$
$$\sin z_1 - \sin z_2$$
$$= 2 \cos \left(\frac{z_1 + z_2}{2}\right) \sin \left(\frac{z_1 - z_2}{2}\right)$$

Formulas for trigonometric (circular) functions* (cont.)

Addition and subtraction of two functions (cont.)

$\cos z_1 + \cos z_2$

$$= 2 \cos \left(\frac{z_1 + z_2}{2} \right) \cos \left(\frac{z_1 - z_2}{2} \right)$$

$\cos z_1 - \cos z_2$

$$= -2 \sin \left(\frac{z_1 + z_2}{2} \right) \sin \left(\frac{z_1 - z_2}{2} \right)$$

$$\tan z_1 \pm \tan z_2 = \frac{\sin (z_1 \pm z_2)}{\cos z_1 \cos z_2}$$

$$\cot z_1 \pm \cot z_2 = \frac{\sin (z_2 \pm z_1)}{\sin z_1 \sin z_2}$$

Relations between squares of sines and cosines

$\sin^2 z_1 - \sin^2 z_2 = \sin (z_1 + z_2) \sin (z_1 - z_2)$

$\cos^2 z_1 - \cos^2 z_2 = -\sin (z_1 + z_2) \sin (z_1 - z_2)$

$\cos^2 z_1 - \sin^2 z_2 = \cos (z_1 + z_2) \cos (z_1 - z_2)$

Formulas for solution of plane triangles

In a triangle with angles A, B, and C and sides opposite a, b, and c respectively,

$$\frac{a}{\sin A} = \frac{b}{\sin B} = \frac{c}{\sin C}$$

Formulas for solution of plane triangles

$$\cos A = \frac{c^2 + b^2 - a^2}{2bc}$$

$a = b \cos C + c \cos B$

$$\frac{a + b}{a - b} = \frac{\tan \frac{1}{2} (A + B)}{\tan \frac{1}{2} (A - B)}$$

$$\text{area} = \frac{bc \sin A}{2} = [s(s - a)(s - b)(s - c)]^{1/2}$$

$s = \frac{1}{2} (a + b + c)$

Formulas for solution of spherical triangles

If A, B, and C are the three angles and a, b, and c the opposite sides,

$$\frac{\sin A}{\sin a} = \frac{\sin B}{\sin b} = \frac{\sin C}{\sin c}$$

$\cos a = \cos b \cos c + \sin b \sin c \cos A$

$$= \frac{\cos b \cos (c \pm \theta)}{\cos \theta}$$

where $\tan \theta = \tan b \cos A$

$\cos A = -\cos B \cos C + \sin B \sin C \cos a$

---

*From M. Abramowitz and I. A. Stegun (eds.), *Handbook of Mathematical Functions (with Formulas, Graphs, and Mathematical Tables)*, 10th printing, National Bureau of Standards, 1972.

Appendix

Values of trigonometric functions*

Degrees	Radians	Sin	Csc	Tan	Cot	Sec	Cos		Degrees
0° 0'	.0000	.0000	—	.0000	—	1.000	1.0000	1.5708	90° 0'
10'	.0029	.0029	343.8	.0029	343.8	1.000	1.0000	1.5679	50'
20'	.0058	.0058	171.9	.0058	171.9	1.000	1.0000	1.5650	40'
30'	.0087	.0087	114.6	.0087	114.5	1.000	1.0000	1.5621	30'
40'	.0116	.0116	85.95	.0116	85.94	1.000	.9999	1.5592	20'
50'	.0145	.0145	68.76	.0145	68.75	1.000	.9999	1.5563	10'
1° 0'	.0175	.0175	57.30	.0175	57.29	1.000	.9998	1.5533	89° 0'
10'	.0204	.0204	49.11	.0204	49.10	1.000	.9998	1.5504	50'
20'	.0233	.0233	42.98	.0233	42.96	1.000	.9997	1.5475	40'
30'	.0262	.0262	38.20	.0262	38.19	1.000	.9997	1.5446	30'
40'	.0291	.0291	34.38	.0291	34.37	1.000	.9996	1.5417	20'
50'	.0320	.0320	31.26	.0320	31.24	1.001	.9995	1.5388	10'
2° 0'	.0349	.0349	28.65	.0349	28.64	1.001	.9994	1.5359	88° 0'
10'	.0378	.0378	26.45	.0378	26.43	1.001	.9993	1.5330	50'
20'	.0407	.0407	24.56	.0407	24.54	1.001	.9992	1.5301	40'
30'	.0436	.0436	22.93	.0437	22.90	1.001	.9990	1.5272	30'
40'	.0465	.0465	21.49	.0466	21.47	1.001	.9989	1.5243	20'
50'	.0495	.0495	20.23	.0495	20.21	1.001	.9988	1.5213	10'
3° 0'	.0524	.0523	19.11	.0524	19.08	1.001	.9986	1.5184	87° 0'
10'	.0553	.0552	18.10	.0553	18.07	1.002	.9985	1.5155	50'
20'	.0582	.0581	17.20	.0582	17.17	1.002	.9983	1.5126	40'
30'	.0611	.0610	16.38	.0612	16.35	1.002	.9981	1.5097	30'
40'	.0640	.0640	15.64	.0641	15.60	1.002	.9980	1.5068	20'
50'	.0669	.0669	14.96	.0670	14.92	1.002	.9978	1.5039	10'
4° 0'	.0698	.0698	14.34	.0699	14.30	1.002	.9976	1.5010	86° 0'
10'	.0727	.0727	13.76	.0729	13.73	1.003	.9974	1.4981	50'
20'	.0756	.0756	13.23	.0758	13.20	1.003	.9971	1.4952	40'
30'	.0785	.0785	12.75	.0787	12.71	1.003	.9969	1.4923	30'
40'	.0814	.0814	12.29	.0816	12.25	1.003	.9967	1.4893	20'
50'	.0844	.0843	11.87	.0846	11.83	1.004	.9964	1.4864	10'

(top →)	Radians	Sin	Csc	Tan	Cot	Sec	Cos	Radians	(← bottom)
85° 0'	1.4835	.9962	1.004	11.43	.0875	11.47	.0872	.0873	5° 0'
50'	806	959	004	11.06	904	11.10	901	902	10'
40'	777	957	004	10.71	934	10.76	929	931	20'
30'	1.4748	9954	1.005	10.39	.0963	10.43	.0958	.0960	**30'**
20'	719	951	005	10.08	992	10.13	987	989	40'
10'	690	948	005	9.788	1022	9.839	1016	1018	50'
84° 0'	1.4661	.9945	1.006	9.514	.1051	9.567	.1045	.1047	6° 0'
50'	632	942	006	9.255	080	9.309	074	076	10'
40'	603	939	006	9.010	110	9.065	103	105	20'
30'	1.4573	9936	1.006	8.777	.1139	8.834	.1132	.1134	**30'**
20'	544	932	007	8.556	169	8.614	161	164	40'
10'	515	929	007	8.345	198	8.405	190	193	50'
83° 0'	1.4486	.9925	1.008	8.144	.1228	8.206	.1219	.1222	7° 0'
50'	457	922	008	7.953	257	8.016	248	251	10'
40'	428	918	008	7.770	287	7.834	276	280	20'
30'	1.4399	9914	1.009	7.596	.1317	7.661	.1305	.1309	**30'**
20'	370	911	009	7.429	346	7.496	334	338	40'
10'	341	907	009	7.269	376	7.337	363	367	50'
82° 0'	1.4312	.9903	1.010	7.115	.1405	7.185	.1392	.1396	8° 0'
50'	283	899	010	6.968	435	7.040	421	425	10'
40'	254	894	011	6.827	465	6.900	449	454	20'
30'	1.4224	9890	1.011	6.691	.1495	6.765	.1478	.1484	**30'**
20'	195	886	012	6.561	524	6.636	507	513	40'
10'	166	881	012	6.435	554	6.512	536	542	50'
81° 0'	1.4137	.9877	1.012	6.314	.1584	6.392	.1564	.1571	9° 0'
Degrees	Radians	Cos	Sec	Cot	Tan	Csc	Sin		Degrees

Values of trigonometric functions* (cont.)

Degrees	Radians	Sin	Csc	Tan	Cot	Sec	Cos		Degrees
9° 0'	.1571	.1564	6.392	.1584	6.314	1.012	.9877	1.4137	81° 0'
10'	600	593	277	614	197	013	872	108	50'
20'	629	622	166	644	084	013	868	079	40'
30'	1658	1650	6.059	1673	5.976	014	9863	1.4050	30'
40'	687	679	5.955	703	871	014	858	4021	20'
50'	716	708	855	733	769	015	853	992	10'
10° 0'	.1745	.1736	5.759	.1763	5.671	1.015	.9848	1.3963	80° 0'
10'	774	765	665	793	576	016	843	934	50'
20'	804	794	575	823	485	016	838	904	40'
30'	1833	1822	5.487	1853	5.396	017	9833	1.3875	30'
40'	862	851	403	883	309	018	827	846	20'
50'	891	880	320	914	226	018	822	817	10'
11° 0'	.1920	.1908	5.241	.1944	5.145	1.019	.9816	1.3788	79° 0'
10'	949	937	164	974	066	019	811	759	50'
20'	978	965	089	2004	4.989	020	805	730	40'
30'	2007	1994	5.016	2035	4.915	020	9799	1.3701	30'
40'	036	2022	4.945	065	843	021	793	672	20'
50'	065	051	876	095	773	022	787	643	10'
12° 0'	.2094	.2079	4.810	.2126	4.705	1.022	.9781	1.3614	78° 0'
10'	123	108	745	156	638	023	775	584	50'
20'	153	136	682	186	574	024	769	555	40'
30'	2182	2164	4.620	2217	4.511	024	9763	1.3526	30'
40'	211	193	560	247	449	025	757	497	20'
50'	240	221	502	278	390	026	750	468	10'
13° 0'	.2269	.2250	4.445	.2309	4.331	1.026	.9744	1.34329	77° 0'
10'	298	278	390	339	275	027	737	410	50'
20'	327	306	336	370	219	028	730	381	40'
30'	2356	2334	4.284	2401	4.165	028	9724	1.3352	30'
40'	385	363	232	432	113	029	717	323	20'
50'	414	391	182	462	061	030	710	294	10'

Degrees	Radians	Sin	Csc	Tan	Cot	Sec	Cos	Radians	Degrees
14° 0'	.2443	.2419	4.134	.2493	4.011	1.031	.9703	1.3265	76° 0'
10'	.2473	.2447	4.086	.2524	3.962	1.031	.9696	1.3235	50'
20'	.2502	.2476	4.039	.2555	3.914	1.032	.9689	1.3206	40'
30'	.2531	.2504	3.994	.2586	3.867	1.033	.9681	1.3177	30'
40'	.2560	.2532	3.950	.2617	3.821	1.034	.9674	1.3148	20'
50'	.2589	.2560	3.906	.2648	3.776	1.034	.9667	1.3119	10'
15° 0'	.2618	.2588	3.864	.2679	3.732	1.035	.9659	1.3090	75° 0'
10'	.2647	.2616	3.822	.2711	3.689	1.036	.9652	1.3061	50'
20'	.2676	.2644	3.782	.2742	3.647	1.037	.9644	1.3032	40'
30'	.2705	.2672	3.742	.2773	3.606	1.038	.9636	1.3003	30'
40'	.2734	.2700	3.703	.2805	3.566	1.039	.9628	1.2974	20'
50'	.2763	.2728	3.665	.2836	3.526	1.039	.9621	1.2945	10'
16° 0'	.2793	.2756	3.628	.2867	3.487	1.040	.9613	1.2915	74° 0'
10'	.2822	.2784	3.592	.2899	3.450	1.041	.9605	1.2886	50'
20'	.2851	.2812	3.556	.2931	3.412	1.042	.9596	1.2857	40'
30'	.2880	.2840	3.521	.2962	3.376	1.043	.9588	1.2828	30'
40'	.2909	.2868	3.487	.2994	3.340	1.044	.9580	1.2799	20'
50'	.2938	.2896	3.453	.3026	3.305	1.045	.9572	1.2770	10'
17° 0'	.2967	.2924	3.420	.3057	3.271	1.046	.9563	1.2741	73° 0'
10'	.2996	.2952	3.388	.3089	3.237	1.047	.9555	1.2712	50'
20'	.3025	.2979	3.357	.3121	3.204	1.048	.9546	1.2683	40'
30'	.3054	.3007	3.326	.3153	3.172	1.048	.9537	1.2654	30'
40'	.3083	.3035	3.295	.3185	3.140	1.049	.9528	1.2625	20'
50'	.3113	.3062	3.265	.3217	3.108	1.050	.9520	1.2595	10'
18° 0'	.3142	.3090	3.236	.3249	3.078	1.051	.9511	1.2566	72° 0'
	Cos	Sec	Cot	Tan	Csc	Sin	Radians		Degrees

Appendix

Values of trigonometric functions* (cont.)

Degrees	Radians	Sin	Csc	Tan	Cot	Sec	Cos		
18° 0'	.3142	.3090	3.236	.3249	3.078	1.051	.9511	1.2566	72° 0'
10'	.171	.118	.207	.281	.047	.052	.502	.537	50'
20'	.200	.145	.179	.314	.018	.053	.492	.508	40'
30'	.3229	.3173	3.152	.3346	2.989	1.054	.9483	1.2479	30'
40'	.258	.201	.124	.378	.960	.056	.474	.450	20'
50'	.287	.228	.098	.411	.932	.057	.465	.421	10'
19° 0'	.3316	.3256	3.072	.3443	2.904	1.058	.9455	1.2392	71° 0'
10'	.345	.283	.046	.476	.877	.059	.446	.363	50'
20'	.374	.311	.021	.508	.850	.060	.436	.334	40'
30'	.3403	.3338	2.996	.3541	2.824	1.061	.9426	1.2305	30'
40'	.432	.365	.971	.574	.798	.062	.417	.275	20'
50'	.462	.393	.947	.607	.773	.063	.407	.246	10'
20° 0'	.3491	.3420	2.924	.3640	2.747	1.064	.9397	1.2217	70° 0'
10'	.520	.448	.901	.673	.723	.065	.387	.188	50'
20'	.549	.475	.878	.706	.699	.066	.377	.159	40'
30'	.3578	.3502	2.855	.3739	2.675	1.068	.9367	1.2130	30'
40'	.607	.529	.833	.772	.651	.069	.356	.101	20'
50'	.636	.557	.812	.805	.628	.070	.346	.072	10'
21° 0'	.3665	.3584	2.790	.3839	2.605	1.071	.9336	1.2043	69° 0'
10'	.694	.611	.769	.872	.583	.072	.325	1.2014	50'
20'	.723	.638	.749	.906	.560	.074	.315	.985	40'
30'	.3752	.3665	2.729	.3939	2.539	1.075	.9304	1.1956	30'
40'	.782	.692	.709	.973	.517	.076	.293	.926	20'
50'	.811	.719	.689	.4006	.496	.077	.283	6897	10'
22° 0'	.3840	.3746	2.669	.4040	2.475	1.079	.9272	1.1868	68° 0'
10'	.869	.773	.650	.074	.455	.080	.261	.839	50'
20'	.898	.800	.632	.108	.434	.081	.250	.810	40'
30'	.3927	.3827	2.613	.4142	2.414	1.082	.9239	1.1781	30'
40'	.956	.854	.595	.176	.394	.084	.228	.752	20'
50'	.985	.881	.577	.210	.375	.085	.216	.723	10'

Degrees	Radians	Sin	Csc	Tan	Cot	Sec	Cos	Radians	Degrees
23° 0'	.4014	.3907	2.559	.4245	2.356	1.086	.9205	1.1694	67° 0'
10'	043	934	542	279	337	088	194	665	50'
20'	072	961	525	314	318	089	182	636	40'
30'	.4102	.3987	2.508	4343	2.300	1.090	9171	1.1606	30'
40'	131	.4014	491	383	282	092	159	577	20'
50'	160	041	475	417	264	093	147	548	10'
24° 0'	.4189	.4067	2.459	4452	2.246	1.095	9135	1.1519	66° 0'
10'	218	094	443	487	229	096	124	490	50'
20'	247	120	427	522	211	097	112	461	40'
30'	4276	4147	2.411	4557	2.194	1.099	9100	1.1432	30'
40'	305	173	396	592	177	100	088	403	20'
50'	334	200	381	628	161	102	075	374	10'
25° 0'	4363	4226	2.366	4663	2.145	1.103	9063	1.1345	65° 0'
10'	392	253	352	699	128	105	051	316	50'
20'	422	279	337	734	112	106	038	286	40'
30'	4451	4305	2.323	4770	2.097	1.108	9026	1.1257	30'
40'	480	331	309	806	081	109	013	228	20'
50'	509	358	295	841	066	111	001	199	10'
26° 0'	4538	4384	2.281	4877	2.050	1.113	8988	1.1170	64° 0'
10'	567	410	268	913	035	114	975	141	50'
20'	596	436	254	950	020	116	962	112	40'
30'	4625	4462	2.241	4986	2.006	1.117	8949	1.1083	30'
40'	654	488	228	5022	1.991	119	936	054	20'
50'	683	514	215	059	977	121	923	1.1025	10'
27° 0'	.4712	.4540	2.203	5095	1.963	1.122	8910	1.0996	63° 0'
Degrees	Radians	Cos	Sec	Cot	Tan	Csc	Sin	Radians	Degrees

Values of trigonometric functions* (cont.)

Degrees	Radians	Sin	Csc	Tan	Cot	Sec	Cos		Degrees
27° 0'	.4712	.4540	2.203	.5095	1.963	1.122	.8910	1.0996	63° 0'
10'	741	566	190	132	949	124	897	966	50'
20'	771	592	178	169	935	126	884	937	40'
30'	.4800	.4617	2.166	.5206	1.921	1.127	.8870	1.0908	30'
40'	829	643	154	243	907	129	857	879	20'
50'	858	669	142	280	894	131	843	850	10'
28° 0'	.4887	.4695	2.130	.5317	1.881	1.133	.8829	1.0821	62° 0'
10'	916	720	118	354	868	134	816	792	50'
20'	945	746	107	392	855	136	802	763	40'
30'	.4974	.4772	2.096	.5430	1.842	1.138	.8788	1.0734	30'
40'	.5003	797	085	467	829	140	774	705	20'
50'	032	823	074	505	816	142	760	676	10'
29° 0'	.5061	.4848	2.063	.5543	1.804	1.143	.8746	1.0647	61° 0'
10'	091	874	052	581	792	145	732	617	50'
20'	120	899	041	619	780	147	718	588	40'
30'	.5149	.4924	2.031	.5658	1.767	1.149	.8704	1.0559	30'
40'	178	950	020	696	756	151	689	530	20'
50'	207	975	010	735	744	153	675	501	10'
30° 0'	.5236	.5000	2.000	.5774	1.732	1.155	.8660	1.0472	60° 0'
10'	265	025	1.990	812	720	157	646	443	50'
20'	294	050	980	851	709	159	631	414	40'
30'	.5323	.5075	1.970	.5890	1.698	1.161	.8616	1.0385	30'
40'	352	100	961	930	686	163	601	356	20'
50'	381	125	951	969	675	165	587	327	10'
31° 0'	.5411	.5150	1.942	.6009	1.664	1.167	.8572	1.0297	59° 0'
10'	440	175	932	048	653	169	557	268	50'
20'	469	200	923	088	643	171	542	239	40'
30'	.5498	.5225	1.914	.6128	1.632	1.173	.8526	1.0210	30'
40'	527	250	905	168	621	175	511	181	20'
50'	556	275	896	208	611	177	496	152	10'

Degrees	Radians	Sin	Csc	Tan	Cot	Sec	Cos	Radians	Degrees
32° 0'	5585	5299	1.887	.6249	1.600	1.179	8480	1.0123	**58° 0'**
10'	614	324	878	289	590	181	465	094	50'
20'	643	348	870	330	580	184	450	065	40'
30'	5672	5373	1.861	6371	570	186	8434	1.0036	**30'**
40'	701	398	853	412	560	188	418	1.0007	20'
50'	730	422	844	453	550	190	403	977	10'
33° 0'	5760	5446	1.836	6494	1.540	1.192	8387	.9948	**57° 0'**
10'	789	471	828	536	530	195	371	919	50'
20'	818	495	820	577	520	197	355	890	40'
30'	5847	5519	1.812	6619	1.511	1.199	8339	.9861	**30'**
40'	876	544	804	661	501	202	323	832	20'
50'	905	568	796	703	492	204	307	803	10'
34° 0'	5934	5592	1.788	6745	1.483	1.206	8290	.9774	**56° 0'**
10'	963	616	781	787	473	209	274	745	50'
20'	992	640	773	830	464	211	258	716	40'
30'	6021	5664	1.766	6873	1.455	1.213	8241	.9687	**30'**
40'	050	688	758	916	446	216	225	657	20'
50'	080	712	751	959	437	218	208	628	10'
35° 0'	6109	5736	1.743	7002	1.428	1.221	8192	.9599	**55° 0'**
10'	138	760	736	046	419	223	175	570	50'
20'	167	783	729	089	411	226	158	541	40'
30'	6196	5807	1.722	7133	1.402	1.228	8141	.9512	**30'**
40'	225	831	715	177	393	231	124	483	20'
50'	254	854	708	221	385	233	107	454	10'
36° 0'	6283	5878	1.701	7265	1.376	1.236	8090	.9425	**54° 0'**
Degrees	Radians	Cos	Sec	Cot	Tan	Csc	Sin	Radians	Degrees

Appendix

Values of trigonometric functions* (cont.)

Degrees	Radians	Sin	Csc	Tan	Cot	Sec	Cos		Degrees
36° 0'	.6283	.5878	1.701	.7265	1.376	1.236	.8090	.9425	54° 0'
10'	312	901	695	310	368	239	073	396	50'
20'	341	925	688	355	360	241	056	367	40'
30'	6370	5948	681	.7400	.351	1.244	8039	.9338	30'
40'	400	972	675	445	343	247	021	308	20'
50'	429	995	668	490	335	249	004	279	10'
37° 0'	.6458	.6018	1.662	.7536	1.327	1.252	.7986	.9250	53° 0'
10'	487	041	655	581	319	255	969	221	50'
20'	516	065	649	627	311	258	951	192	40'
30'	6545	6088	643	7673	303	1.260	7934	.9153	30'
40'	574	111	636	720	295	263	916	134	20'
50'	603	134	630	766	288	266	898	105	10'
38° 0'	.6632	.6157	1.624	.7813	1.280	1.269	.7880	.9076	52° 0'
10'	661	180	618	860	272	272	862	047	50'
20'	690	202	612	907	265	275	844	9018	40'
30'	6720	6225	1.606	7954	257	1.278	7826	.8988	30'
40'	749	248	601	8002	250	281	808	959	20'
50'	778	271	595	050	242	284	790	930	10'
39° 0'	.6807	.6293	1.589	.8098	1.235	1.287	.7771	.8901	51° 0'
10'	836	316	583	146	228	290	753	872	50'
20'	865	338	578	195	220	293	735	843	40'
30'	6894	6361	1.572	8243	213	1.296	7716	.8814	30'
40'	923	383	567	292	206	299	698	785	20'
50'	952	406	561	342	199	302	679	756	10'
40° 0'	.6981	.6428	1.556	.8391	1.192	1.305	.7660	.8727	50° 0'
10'	7010	450	550	441	185	309	642	698	50'
20'	039	472	545	491	178	312	623	668	40'
30'	7069	6494	1.540	8541	171	1.315	7604	.8639	30'
40'	098	517	535	591	164	318	585	610	20'
50'	127	539	529	642	157	322	566	581	10'

Degrees	Radians	Sin	Csc	Tan	Cot	Sec	Cos	Radians	Degrees
41° 0'	.7156	.6561	1.524	.8693	1.150	1.325	.7547	.8552	49° 0'
10'	185	583	519	744	144	328	528	523	50'
20'	214	604	514	796	137	332	509	494	40'
30'	.7243	.6626	1.509	.8847	1.130	1.335	.7490	.8465	30'
40'	272	648	504	899	124	339	470	436	20'
50'	301	670	499	952	117	342	451	407	10'
42° 0'	.7330	.6691	1.494	.9004	1.111	1.346	.7431	.8378	48° 0'
10'	359	713	490	057	104	349	412	348	50'
20'	389	734	485	110	098	353	392	319	40'
30'	.7418	.6756	1.480	.9163	1.091	1.356	.7373	.8290	30'
40'	447	777	476	217	085	360	353	261	20'
50'	476	799	471	271	079	364	333	232	10'
43° 0'	.7505	.6820	1.466	.9325	1.072	1.367	.7314	.8203	47° 0'
10'	534	841	462	380	066	371	294	174	50'
20'	563	862	457	435	060	375	274	145	40'
30'	.7592	.6884	1.453	.9490	1.054	1.379	.7254	.8116	30'
40'	621	905	448	545	048	382	234	087	20'
50'	650	926	444	601	042	386	214	058	10'
44° 0'	.7679	.6947	1.440	.9657	1.036	1.390	.7193	.8029	46° 0'
10'	709	967	435	713	030	394	173	999	50'
20'	738	988	431	770	024	398	153	970	40'
30'	.7767	.7009	1.427	.9827	1.018	1.402	.7133	.7941	30'
40'	796	030	423	884	012	406	112	912	20'
50'	825	050	418	942	006	410	092	883	10'
45° 0'	.7854	.7071	1.414	1.000	1.000	1.414	.7071	.7854	45° 0'
Degrees	Radians	Cos	Sec	Cot	Tan	Csc	Sin		

*From G. E. F. Sherwood and A. E. Taylor, *Calculus*, ed ed. Prentice-Hall, 1954.

Appendix

Special constants

$\pi = 3.14159\ 26535\ 89793\ 23846\ 2643\ldots$

$e = 2.71828\ 18284\ 59045\ 23536\ 0287\ldots = \lim\limits_{n\to\infty}\left(1 + \frac{1}{n}\right)^n$

$\quad$ = natural base of logarithms

$\sqrt{2} = 1.41421\ 35623\ 73095\ 0488\ldots$

$\sqrt{3} = 1.73205\ 08075\ 68877\ 2935\ldots$

$\sqrt{5} = 2.23606\ 79774\ 99789\ 6964\ldots$

$\sqrt[3]{2} = 1.25992\ 1050\ldots$

$\sqrt[3]{3} = 1.44224\ 9570\ldots$

$\sqrt[5]{2} = 1.14869\ 8355\ldots$

$\sqrt[5]{3} = 1.24573\ 0940\ldots$

$e^{\pi} = 23.14069\ 26327\ 79269\ 006\ldots$

$\pi^{e} = 22.45915\ 77183\ 61045\ 47342\ 715\ldots$

$e^{e} = 15.15426\ 22414\ 79264\ 190\ldots$

$\log_{10} 2 = 0.30102\ 99956\ 63981\ 19521\ 37389\ldots$

$\log_{10} 3 = 0.47712\ 12547\ 19662\ 43729\ 50279\ldots$

$\log_{10} e = 0.43429\ 44819\ 03251\ 82765\ldots$

$\log_{10} \pi = 0.49714\ 98726\ 94133\ 85435\ 12683\ldots$

$\log_{e} 10 = \ln 10 = 2.30258\ 50929\ 94045\ 68401\ 7991\ldots$

$\log_{e} 2 = \ln 2 = 0.69314\ 71805\ 59945\ 30941\ 7232\ldots$

$\log_{e} 3 = \ln 3 = 1.09861\ 22886\ 68109\ 69139\ 5245\ldots$

$\gamma = 0.57721\ 56649\ 01532\ 86060\ 6512\ldots$ = Euler's constant

$\quad = \lim\limits_{n\to\infty}\left(1 + \frac{1}{2} + \frac{1}{3} + \cdots + \frac{1}{n} - \ln n\right)$

$e^{\gamma} = 1.78107\ 24179\ 90197\ 9852\ldots$

$\sqrt{e} = 1.64872\ 12707\ 00128\ 1468\ldots$

$\sqrt{\pi} = \Gamma(\tfrac{1}{2}) = 1.77245\ 38509\ 05516\ 02729\ 8167\ldots$

where Γ is the gamma function

$\Gamma(\tfrac{1}{3}) = 2.67893\ 85347\ 07748\ldots$

$\Gamma(\tfrac{1}{4}) = 3.62560\ 99082\ 21908\ldots$

1 radian $= 180°/\pi = 57.29577\ 95130\ 8232\ldots°$

$1° = \pi/180$ radians $= 0.01745\ 32925\ 19943\ 29576\ 92\ldots$ radians

---

SOURCE: Murray R. Spiegel and John Liu, *Mathematical Handbook of Formulas and Tables*, 2d ed., Schaum's Outline Series, McGraw-Hill, 1999.

302

Common logarithm table, giving log (a + b)

a	b: .00	.01	.02	.03	.04	.05	.06	.07	.08	.09
1.0	.0000	.0043	.0086	.0128	.0170	.0212	.0253	.0294	.0334	.0374
1.1	.0414	.0453	.0492	.0531	.0569	.0607	.0646	.0682	.0719	.0755
1.2	.0792	.0828	.0864	.0899	.0934	.0969	.1004	.1038	.1072	.1106
1.3	.1139	.1173	.1206	.1239	.1271	.1303	.1335	.1367	.1399	.1430
1.4	.1461	.1492	.1523	.1553	.1584	.1614	.1644	.1673	.1703	.1732
1.5	.1761	.1790	.1818	.1847	.1875	.1903	.1931	.1959	.1987	.2014
1.6	.2041	.2068	.2095	.2122	.2148	.2175	.2201	.2227	.2253	.2279
1.7	.2304	.2330	.2355	.2380	.2405	.2430	.2455	.2480	.2504	.2529
1.8	.2553	.2577	.2601	.2625	.2648	.2672	.2695	.2718	.2742	.2765
1.9	.2788	.2810	.2833	.2856	.2878	.2900	.2923	.2945	.2967	.2989
2.0	.3010	.3032	.3054	.3075	.3096	.3118	.3139	.3160	.3181	.3201
2.1	.3222	.3243	.3263	.3284	.3304	.3324	.3345	.3365	.3385	.3404
2.2	.3424	.3444	.3464	.3483	.3502	.3522	.3541	.3560	.3579	.3598
2.3	.3617	.3636	.3655	.3674	.3692	.3711	.3729	.3747	.3766	.3784
2.4	.3802	.3820	.3838	.3856	.3874	.3892	.3909	.3927	.3945	.3962
2.5	.3979	.3997	.4014	.4031	.4048	.4065	.4082	.4099	.4116	.4133
2.6	.4150	.4166	.4183	.4200	.4216	.4232	.4249	.4265	.4281	.4298
2.7	.4314	.4330	.4346	.4362	.4378	.4393	.4409	.4425	.4440	.4456
2.8	.4472	.4487	.4502	.4518	.4533	.4548	.4564	.4579	.4594	.4609
2.9	.4624	.4639	.4654	.4669	.4683	.4698	.4713	.4728	.4742	.4757
3.0	.4771	.4786	.4800	.4814	.4829	.4843	.4857	.4871	.4886	.4900
3.1	.4914	.4928	.4942	.4955	.4969	.4983	.4997	.5011	.5024	.5038
3.2	.5052	.5065	.5079	.5092	.5105	.5119	.5132	.5145	.5159	.5172
3.3	.5185	.5198	.5211	.5224	.5237	.5250	.5263	.5276	.5289	.5302
3.4	.5315	.5328	.5340	.5353	.5366	.5378	.5391	.5403	.5416	.5428
3.5	.5441	.5453	.5465	.5478	.5490	.5502	.5515	.5527	.5539	.5551
3.6	.5563	.5575	.5587	.5599	.5611	.5623	.5635	.5647	.5658	.5670
3.7	.5682	.5694	.5705	.5717	.5729	.5740	.5752	.5763	.5775	.5786
3.8	.5798	.5809	.5821	.5832	.5843	.5855	.5866	.5877	.5888	.5899
3.9	.5911	.5922	.5933	.5944	.5955	.5966	.5977	.5988	.5999	.6010
4.0	.6021	.6031	.6042	.6053	.6064	.6075	.6085	.6096	.6107	.6117
4.1	.6128	.6138	.6149	.6160	.6170	.6180	.6191	.6201	.6212	.6222
4.2	.6232	.6243	.6253	.6263	.6274	.6284	.6294	.6304	.6314	.6325
4.3	.6335	.6345	.6355	.6365	.6375	.6385	.6395	.6405	.6415	.6425
4.4	.6435	.6444	.6454	.6464	.6474	.6484	.6493	.6503	.6513	.6522
4.5	.6532	.6542	.6551	.6561	.6571	.6580	.6590	.6599	.6609	.6618
4.6	.6628	.6637	.6646	.6656	.6665	.6675	.6684	.6693	.6702	.6712
4.7	.6721	.6730	.6739	.6749	.6758	.6767	.6776	.6785	.6794	.6803
4.8	.6812	.6821	.6830	.6839	.6848	.6857	.6866	.6875	.6884	.6893
4.9	.6902	.6911	.6920	.6928	.6937	.6946	.6955	.6964	.6972	.6981
5.0	.6990	.6998	.7007	.7016	.7024	.7033	.7042	.7050	.7059	.7067
5.1	.7076	.7084	.7093	.7101	.7110	.7118	.7126	.7135	.7143	.7152
5.2	.7160	.7168	.7177	.7185	.7193	.7202	.7210	.7218	.7226	.7235
5.3	.7243	.7251	.7259	.7267	.7275	.7284	.7292	.7300	.7308	.7316
5.4	.7324	.7332	.7340	.7348	.7356	.7364	.7372	.7380	.7388	.7396

Appendix

Common logarithm table, giving log (a + b) (cont.)

a	b: .00	.01	.02	.03	.04	.05	.06	.07	.08	.09
5.5	.7404	.7412	.7419	.7427	.7435	.7443	.7451	.7459	.7466	.7474
5.6	.7482	.7490	.7497	.7505	.7513	.7520	.7528	.7536	.7543	.7551
5.7	.7559	.7566	.7574	.7582	.7589	.7597	.7604	.7612	.7619	.7627
5.8	.7634	.7642	.7649	.7657	.7664	.7672	.7679	.7686	.7694	.7701
5.9	.7709	.7716	.7723	.7731	.7738	.7745	.7752	.7760	.7767	.7774
6.0	.7782	.7789	.7796	.7803	.7810	.7818	.7825	.7832	.7839	.7846
6.1	.7853	.7860	.7868	.7875	.7882	.7889	.7896	.7903	.7910	.7917
6.2	.7924	.7931	.7938	.7945	.7952	.7959	.7966	.7973	.7980	.7987
6.3	.7993	.8000	.8007	.8014	.8021	.8028	.8035	.8041	.8048	.8055
6.4	.8062	.8069	.8075	.8082	.8089	.8096	.8102	.8109	.8116	.8122
6.5	.8129	.8136	.8142	.8149	.8156	.8162	.8169	.8176	.8182	.8189
6.6	.8195	.8202	.8209	.8215	.8222	.8228	.8235	.8241	.8248	.8254
6.7	.8261	.8267	.8274	.8280	.8287	.8293	.8299	.8306	.8312	.8319
6.8	.8325	.8331	.8338	.8344	.8351	.8357	.8363	.8370	.8376	.8382
6.9	.8388	.8395	.8401	.8407	.8414	.8420	.8426	.8432	.8439	.8445
7.0	.8451	.8457	.8463	.8470	.8476	.8482	.8488	.8494	.8500	.8506
7.1	.8513	.8519	.8525	.8531	.8537	.8543	.8549	.8555	.8561	.8567
7.2	.8573	.8579	.8585	.8591	.8597	.8603	.8609	.8615	.8621	.8627
7.3	.8633	.8639	.8645	.8651	.8657	.8663	.8669	.8675	.8681	.8686
7.4	.8692	.8698	.8704	.8710	.8716	.8722	.8727	.8733	.8739	.8745
7.5	.8751	.8756	.8762	.8768	.8774	.8779	.8785	.8791	.8797	.8802
7.6	.8808	.8814	.8820	.8825	.8831	.8837	.8842	.8848	.8854	.8859
7.7	.8865	.8871	.8876	.8882	.8887	.8893	.8899	.8904	.8910	.8915
7.8	.8921	.8927	.8932	.8938	.8943	.8949	.8954	.8960	.8965	.8971
7.9	.8976	.8982	.8987	.8993	.8998	.9004	.9009	.9015	.9020	.9025
8.0	.9031	.9036	.9042	.9047	.9053	.9058	.9063	.9069	.9074	.9079
8.1	.9085	.9090	.9096	.9101	.9106	.9112	.9117	.9122	.9128	.9133
8.2	.9138	.9143	.9149	.9154	.9159	.9165	.9170	.9175	.9180	.9186
8.3	.9191	.9196	.9201	.9206	.9212	.9217	.9222	.9227	.9232	.9238
8.4	.9243	.9248	.9253	.9258	.9263	.9269	.9274	.9279	.9284	.9289
8.5	.9294	.9299	.9304	.9309	.9315	.9320	.9325	.9330	.9335	.9340
8.6	.9345	.9350	.9355	.9360	.9365	.9370	.9375	.9380	.9385	.9390
8.7	.9395	.9400	.9405	.9410	.9415	.9420	.9425	.9430	.9435	.9440
8.8	.9445	.9450	.9455	.9460	.9465	.9469	.9474	.9479	.9484	.9489
8.9	.9494	.9499	.9504	.9509	.9513	.9518	.9523	.9528	.9533	.9538
9.0	.9542	.9547	.9552	.9557	.9562	.9566	.9571	.9576	.9581	.9586
9.1	.9590	.9595	.9600	.9605	.9609	.9614	.9619	.9624	.9628	.9633
9.2	.9638	.9643	.9647	.9652	.9657	.9661	.9666	.9671	.9675	.9680
9.3	.9685	.9689	.9694	.9699	.9703	.9708	.9713	.9717	.9722	.9727
9.4	.9731	.9736	.9741	.9745	.9750	.9754	.9759	.9764	.9768	.9773
9.5	.9777	.9782	.9786	.9791	.9795	.9800	.9805	.9809	.9814	.9818
9.6	.9823	.9827	.9832	.9836	.9841	.9845	.9850	.9854	.9859	.9863
9.7	.9868	.9872	.9877	.9881	.9886	.9890	.9894	.9899	.9903	.9908
9.8	.9912	.9917	.9921	.9926	.9930	.9934	.9939	.9943	.9948	.9952
9.9	.9956	.9961	.9965	.9969	.9974	.9978	.9983	.9987	.9991	.9996

General rules of integration*

$$\int a\,dx = ax$$

$$\int af(x)\,dx = a\int f(x)\,dx$$

$$\int (u \pm v \pm w \pm \cdots)\,dx = \int u\,dx \pm \int v\,dx \pm \int w\,dx \pm \cdots$$

$$\int u\,dv = uv - \int v\,du \quad \text{[integration by parts]}$$

$$\int f(ax)\,dx = \frac{1}{a}\int f(u)\,du \quad \text{where } u = ax$$

$$\int F(u)\frac{dx}{du}\,du = \int \frac{F(u)}{f'(x)}\,du \quad \text{where } u = f(x)$$

$$\int u^n\,du = \frac{u^{n+1}}{n+1}, \quad n \neq -1$$

$$\int \frac{du}{u} = \ln u \quad \text{if } u > 0 \text{ or } \ln(-u) \text{ if } u < 0$$

$$= \ln |u|$$

$$\int e^u\,du = e^u$$

$$\int a^u\,du = \int e^{u\ln a}\,du = \frac{e^{u\ln a}}{\ln a} = \frac{a^u}{\ln a}, \quad a > 0,\ a \neq 1$$

$$\int \sin u\,du = -\cos u$$

$$\int \cos u\,du = \sin u$$

$$\int \tan u\,du = \ln \sec u = -\ln \cos u$$

$$\int \cot u\,du = \ln \sin u$$

$$\int \sec u\,du = \ln (\sec u + \tan u) = \ln \tan\left(\frac{u}{2} + \frac{\pi}{4}\right)$$

$$\int \csc u\,du = \ln (\csc u - \cot u) = \ln \tan\frac{u}{2}$$

$$\int \sec^2 u\,du = \tan u$$

$$\int \csc^2 u\,du = -\cot u$$

$$\int \tan^2 u\,du = \tan u - u$$

$$\int \cot^2 u\,du = -\cot u - u$$

$$\int \sin^2 u\,du = \frac{u}{2} - \frac{\sin 2u}{4} = \frac{1}{2}(u - \sin u \cos u)$$

$$\int \cos^2 u\,du = \frac{u}{2} + \frac{\sin 2u}{4} = \frac{1}{2}(u + \sin u \cos u)$$

$$\int \sec u \tan u\,du = \sec u$$

$$\int \csc u \cot u\,du = -\csc u$$

$$\int \sinh u\,du = \cosh u$$

$$\int \cosh u\,du = \sinh u$$

$$\int \tanh u\,du = \ln \cosh u$$

$$\int \coth u\,du = \ln \sinh u$$

Appendix

General rules of integration* (cont.)

$$\int \operatorname{sech} u \, du = \sin^{-1}(\tanh u) \quad \text{or} \quad 2 \tan^{-1} e^u$$

$$\int \operatorname{csch} u \, du = \ln \tanh \frac{u}{2} \quad \text{or} \quad -\coth^{-1} e^u$$

$$\int \operatorname{sech}^2 u \, du = \tanh u$$

$$\int \operatorname{csch}^2 u \, du = -\coth u$$

$$\int \tanh^2 u \, du = u - \tanh u$$

$$\int \coth^2 u \, du = u - \coth u$$

$$\int \sinh^2 u \, du = \frac{\sinh 2u}{4} - \frac{u}{2} = \frac{1}{2}(\sinh u \cosh u - u)$$

$$\int \cosh^2 u \, du = \frac{\sinh 2u}{4} + \frac{u}{2} = \frac{1}{2}(\sinh u \cosh u + u)$$

$$\int \operatorname{sech} u \tanh u \, du = -\operatorname{sech} u$$

$$\int \operatorname{csch} u \coth u \, du = -\operatorname{csch} u$$

$$\int \frac{du}{u^2 + a^2} = \frac{1}{a} \tan^{-1} \frac{u}{a}$$

$$\int \frac{du}{u^2 - a^2} = \frac{1}{2a} \ln\left(\frac{u - a}{u + a}\right) = -\frac{1}{a} \coth^{-1} \frac{u}{a} \quad u^2 > a^2$$

$$\int \frac{du}{a^2 - u^2} = \frac{1}{2a} \ln\left(\frac{a + u}{a - u}\right) = \frac{1}{a} \tanh^{-1} \frac{u}{a} \quad u^2 < a^2$$

$$\int \frac{du}{\sqrt{a^2 - u^2}} = \sin^{-1} \frac{u}{a}$$

$$\int \frac{du}{\sqrt{u^2 + a^2}} = \ln(u + \sqrt{u^2 + a^2}) \quad \text{or} \quad \sinh^{-1} \frac{u}{a}$$

$$\int \frac{du}{\sqrt{u^2 - a^2}} = \ln(u + \sqrt{u^2 - a^2})$$

$$\int \frac{du}{u\sqrt{u^2 - a^2}} = \frac{1}{a} \sec^{-1} \left|\frac{u}{a}\right|$$

$$\int \frac{du}{u\sqrt{u^2 + a^2}} = -\frac{1}{a} \ln\left(\frac{a + \sqrt{u^2 + a^2}}{u}\right)$$

$$\int \frac{du}{u\sqrt{a^2 - u^2}} = -\frac{1}{a} \ln\left(\frac{a + \sqrt{a^2 - u^2}}{u}\right)$$

$$\int f^{(n)}g \, dx = f^{(n-1)}g - f^{(n-2)}g' + f^{(n-3)}g'' - \cdots (-1)^n \int fg^{(n)} \, dx$$

This is called generalized integration by parts.

*Here, u, v, w are functions of x; a, b, p, q, n any constants, restricted if indicated; $e = 2.71828\ldots$ is the natural base of logarithms; ln denotes the natural logarithm of u where it is assumed that $u > 0$ [in general, to extend formulas to cases where $u < 0$ as well, replace ln u by ln $|u|$]; all angles are in radians; all constants of integration are omitted but implied.

SOURCE: Murray R. Spiegel and John Liu, *Mathematical Handbook of Formulas and Tables*, 2d ed., Schaum's Outline Series, McGraw-Hill, 1999.

Regular polytopes in n dimensions

Polytope	Schläfli symbol	Vertices	Edges	Faces	Solid cells	Hypersolid cells
$n = 2$						
p-gon	$\{p\}$	p	p			
$n = 3$						
Tetrahedron	$\{3,3\}$	4	6	4		
Cube	$\{4,3\}$	8	12	6		
Octahedron	$\{3,4\}$	6	12	8		
Dodecahedron	$\{5,3\}$	20	30	12		
Icosahedron	$\{3,5\}$	12	30	20		
$n = 4$						
5-cell	$\{3,3,3\}$	5	10	10	5	
8-cell	$\{4,3,3\}$	16	32	24	8	
16-cell	$\{3,3,4\}$	8	24	32	16	
24-cell	$\{3,4,3\}$	24	96	96	24	
120-cell	$\{5,3,3\}$	600	1200	720	120	
600-cell	$\{3,3,5\}$	120	720	1200	600	
$n > 4$						
Simplex	$\{3,3,\ldots,3\}$	$n + 1$	$\frac{1}{2} n(n + 1)$	$\ldots$		$n + 1$
Hypercube	$\{4,3,\ldots,3\}$	2^n	$2^{n-1}n$	$\ldots$		$2n$
Cross polytope	$\{3,\ldots,3,4\}$	$2n$	$2n(n - 1)$	$\ldots$		2^n